Ajith Abraham, Crina Grosan, Vitorino Ramos (Eds.)

Swarm Intelligence in Data Mining

Studies in Computational Intelligence, Volume 34

Editor-in-chief
Prof. Janusz Kacprzyk
Systems Research Institute
Polish Academy of Sciences
ul. Newelska 6
01-447 Warsaw
Poland
E-mail: kacprzyk@ibspan.waw.pl

Further volumes of this series
can be found on our homepage:
springer.com

Vol. 16. Yaochu Jin (Ed.)
Multi-Objective Machine Learning, 2006
ISBN 3-540-30676-5

Vol. 17. Te-Ming Huang, Vojislav Kecman,
Ivica Kopriva
*Kernel Based Algorithms for Mining Huge
Data Sets,* 2006
ISBN 3-540-31681-7

Vol. 18. Chang Wook Ahn
Advances in Evolutionary Algorithms, 2006
ISBN 3-540-31758-9

Vol. 19. Ajita Ichalkaranje, Nikhil
Ichalkaranje, Lakhmi C. Jain (Eds.)
*Intelligent Paradigms for Assistive and
Preventive Healthcare,* 2006
ISBN 3-540-31762-7

Vol. 20. Wojciech Penczek, Agata Półrola
*Advances in Verification of Time Petri Nets
and Timed Automata,* 2006
ISBN 3-540-32869-6

Vol. 21. Cândida Ferreira
*Gene Expression on Programming: Mathematical
Modeling by an Artificial Intelligence,* 2006
ISBN 3-540-32796-7

Vol. 22. N. Nedjah, E. Alba, L. de Macedo
Mourelle (Eds.)
Parallel Evolutionary Computations, 2006
ISBN 3-540-32837-8

Vol. 23. M. Last, Z. Volkovich, A. Kandel (Eds.)
Algorithmic Techniques for Data Mining, 2006
ISBN 3-540-33880-2

Vol. 24. Alakananda Bhattacharya, Amit Konar,
Ajit K. Mandal
Parallel and Distributed Logic Programming,
2006
ISBN 3-540-33458-0

Vol. 25. Zoltán Ésik, Carlos Martín-Vide,
Victor Mitrana (Eds.)
*Recent Advances in Formal Languages
and Applications,* 2006
ISBN 3-540-33460-2

Vol. 26. Nadia Nedjah, Luiza de Macedo Mourelle
(Eds.)
Swarm Intelligent Systems, 2006
ISBN 3-540-33868-3

Vol. 27. Vassilis G. Kaburlasos
*Towards a Unified Modeling and Knowledge-
Representation based on Lattice Theory,* 2006
ISBN 3-540-34169-2

Vol. 28. Brahim Chaib-draa, Jörg P. Müller (Eds.)
Multiagent based Supply Chain Management, 2006
ISBN 3-540-33875-6

Vol. 29. Sai Sumathi, S.N. Sivanandam
*Introduction to Data Mining and its
Applications,* 2006
ISBN 3-540-34689-9

Vol. 30. Yukio Ohsawa, Shusaku Tsumoto (Eds.)
*Chance Discoveries in Real World Decision
Making,* 2006
ISBN 3-540-34352-0

Vol. 31. Ajith Abraham, Crina Grosan, Vitorino
Ramos (Eds.)
Stigmergic Optimization, 2006
ISBN 3-540-34689-9

Vol. 32. Akira Hirose
Complex-Valued Neural Networks, 2006
ISBN 3-540-33456-4

Vol. 33. Martin Pelikan, Kumara Sastry, Erick
Cantú-Paz (Eds.)
*Scalable Optimization via Probabilistic
Modeling,* 2006
ISBN 3-540-34953-7

Vol. 34. Ajith Abraham, Crina Grosan, Vitorino
Ramos (Eds.)
Swarm Intelligence in Data Mining, 2006
ISBN 3-540-34955-3

Ajith Abraham
Crina Grosan
Vitorino Ramos (Eds.)

Swarm Intelligence in Data Mining

With 91 Figures and 73 Tables

Springer

Dr. Ajith Abraham
IITA Professorship Program
School of Computer Science
and Engineering
Chung-Ang University, 221
Heukseok-dong
Dongjak-gu Seoul 156-756
Republic of Korea
E-mail: ajith.abraham@ieee.org;
abraham.ajith@acm.org

Dr. Vitorino Ramos
CVRM-IST, IST
Technical University of Lisbon
Av. Rovisco Pais
1049-001, Lisboa
Portugal
E-mail: vitorino.ramos@alfa.ist.utl.pt

Dr. Crina Grosan
Department of Computer Science
Faculty of Mathematics
and Computer Science
Babes-Bolyai University
Cluj-Napoca, Kogalniceanu 1
400084 Cluj - Napoca
Romania
E-mail: cgrosan@cs.ubbcluj.ro

Library of Congress Control Number: 2006928619

ISSN print edition: 1860-949X
ISSN electronic edition: 1860-9503
ISBN-10   3-540-34955-3 Springer Berlin Heidelberg New York
ISBN-13   978-3-540-34955-6 Springer Berlin Heidelberg New York

This work is subject to copyright. All rights are reserved, whether the whole or part of the material is concerned, specifically the rights of translation, reprinting, reuse of illustrations, recitation, broadcasting, reproduction on microfilm or in any other way, and storage in data banks. Duplication of this publication or parts thereof is permitted only under the provisions of the German Copyright Law of September 9, 1965, in its current version, and permission for use must always be obtained from Springer-Verlag. Violations are liable to prosecution under the German Copyright Law.

Springer is a part of Springer Science+Business Media
springer.com
© Springer-Verlag Berlin Heidelberg 2006

The use of general descriptive names, registered names, trademarks, etc. in this publication does not imply, even in the absence of a specific statement, that such names are exempt from the relevant protective laws and regulations and therefore free for general use.

Cover design: deblik, Berlin
Typesetting by the authors and SPi
Printed on acid-free paper   SPIN: 11613589   89/SPi   5 4 3 2 1 0

Foreword

Science is a swarm.

To the layperson, the stereotypical scientist is logical, clear-thinking, well-informed but perhaps socially awkward, carefully planning his or her experiments and then analyzing the resulting data deliberately, with precision. The scientist works alone, emotion-free, searching only for truth, having been well advised about the pitfalls and temptations that lie along the path to discovery and the expansion of human knowledge.

Those who work in science understand how inaccurate this stereotype is. In reality, researchers' daily routines follow a process better described as collective trial-and-error, nearly random at times. A most salient feature of scientific behavior is its collaborative nature. From applying for grants to seeking tenure, from literature reviews to peer review to conference presentations, every bit of the scientific enterprise is social, every step of the process is designed to make scientists aware of one another's work, to force researchers to compare, to communicate, to study the work that others are doing, in order to push the paradigm forward - not as independent, isolated seekers-of-truth, but more like a swarm.

If we plotted a group of scientists as points on a space of dimensions of theories and methods, and ran the plot so we could see changes over time, we would see individuals colliding and crossing, escaping the group's gravity field and returning, disintegrating but simultaneously cohering in some mysterious way and moving as a deliberate, purposeful bunch, across the space - constantly pushing toward a direction that improves the state of knowledge, sometimes stepping in the wrong direction, but relentlessly insisting toward an epistemological optimum.

The book you hold in your hand is a snapshot of the swarm that is the swarm paradigm, a flash photograph of work by researchers from all over the world, captured in mid-buzz as they search, using collective trial and error, for ways to take advantage of processes that are observed in nature and instantiated in computer programs.

In this volume you will read about a number of different kinds of computer programs that are called "swarms." It really wouldn't be right for something as messy as a swarm to have a crisp, precise definition. In general the word swarm is probably more connotative than denotative; there is more to the way swarms feel than to any actual properties that may characterize them. A swarm is going to have some randomness in it - it will not be perfectly choreographed like a flock or a school. A swarm is going to contain a good number of members. The members of the swarm will interact with one another in some way, that is, they will affect one another's behaviors. As they influence one another, there will be some order and some chaos in the population. This is what a swarm is.

The swarm intelligence literature has mostly arisen around two families of algorithms. One kind develops knowledge about a problem by the accumulation of artifacts, often metaphorically conceptualized as pheromones. Individuals respond to signs of their peers' behaviors, leaving signs themselves; those signs increase or decay depending, in the long run, on how successfully they indicate a good solution for a given problem. The movements of swarm population members are probabilistically chosen as a function of the accumulation of pheromone along a decision path.

In another kind of swarm algorithm each individual is a candidate problem solution; in the beginning the solutions are random and not very good, but they improve over time. Individuals interact directly with their peers, emulating their successes; each individual serves as both teacher and learner, and in the end the researcher can interrogate the most successful member of the population to find, usually, a good problem solution.

It is important that both of these kinds of algorithms, ant colony swarms and the particle swarms, are included together in one volume, along with other kinds of swarms. In the forward push of knowledge it is useful for researchers to look over and see what the others are doing; the swarm of science works through the integration of disparate points of view. Already we are seeing papers describing hybrids of these approaches, as well as other evolutionary and heuristic methods - this is an inevitable and healthy direction for the research to take. Add to this the emergence of new swarm methods, based for instance on honeybee behaviors, and you see in this volume the upward trajectory of a rich, blooming new field of research.

Science is a way of searching, and should not be mistaken for a list of answers - it is a fountain of questions, and the pursuit of answers. No chapter in this book or any other will give you the full, final explanation about how swarms learn, optimize, and solve problems; every chapter will give you insights into how the unpredictable and messy process of swarming can accomplish these things.

As the stereotype of the scientist as a lone intellect has been challenged, revising the stereotype should change the way we think about knowledge, as well. Knowledge is not a package of information stored in a brain, it is a process distributed across many brains. Knowing is something that only living beings can do, and knowing in the scientific sense only takes place when individuals participate in the game. Every paradigm has its leaders and its followers, its innovators and its drones, but no scientific paradigm can exist without communication and all the living behaviors that go with that - collaboration, competition, conflict, collision, coordination, caring.

These chapters are technical and challenging, and rewarding. Here our basic task is data-mining, where we have some information and want to make sense of it, however we have defined that. Swarm methods are generally good in high dimensions, with lots of variables; they tend to be robust in noisy spaces; swarms are unafraid of multimodal landscapes, with lots of good-but-not-best solutions. Researchers in this volume are pushing this new paradigm into highly demanding data sets, reporting here what they are able to get it to do.

May 05, 2006 *James Kennedy, USA*

Preface

Swarm Intelligence (SI) is an innovative distributed intelligent paradigm for solving optimization problems that originally took its inspiration from the biological examples by swarming, flocking and herding phenomena in vertebrates. Particle Swarm Optimization (PSO) incorporates swarming behaviors observed in flocks of birds, schools of fish, or swarms of bees, and even human social behavior, from which the idea is emerged. Ant Colony Optimization (ACO) deals with artificial systems that is inspired from the foraging behavior of real ants, which are used to solve discrete optimization problems.

Historically the notion of finding useful patterns in data has been given a variety of names including data mining, knowledge discovery, information extraction, etc. Data Mining is an analytic process designed to explore large amounts of data in search of consistent patterns and/or systematic relationships between variables, and then to validate the findings by applying the detected patterns to new subsets of data. In order to achieve this, data mining uses computational techniques from statistics, machine learning and pattern recognition.

Data mining and *Swarm intelligence* may seem that they do not have many properties in common. However, recent studies suggests that they can be used together for several real world data mining problems especially when other methods would be too expensive or difficult to implement.

This book deals with the application of swarm intelligence methodologies in data mining. Addressing the various issues of swarm intelligence and data mining using different intelligent approaches is the novelty of this edited volume. This volume comprises of 11 chapters including an introductory chapters giving the fundamental definitions and some important research challenges. Chapters were selected on the basis of fundamental ideas/concepts rather than the thoroughness of techniques deployed. The eleven chapters are organized as follows.

In Chapter 1, *Grosan et al.* present the biological motivation and some of the theoretical concepts of swarm intelligence with an emphasis on particle swarm optimization and ant colony optimization algorithms. The basic data mining terminologies are explained and linked with some of the past and ongoing works using swarm intelligence techniques.

Martens et al. in Chapter 2 introduce a new algorithm for classification, named AntMiner+, based on an artificial ant system with inherent self-organizing capabilities. AntMiner+ differs from the previously proposed AntMiner classification technique in three aspects. Firstly, AntMiner+ uses a MAX-MIN ant system which is an improved version of the originally proposed ant system, yielding better performing classifiers. Secondly, the complexity of the environment in which the ants operate has substantially decreased. Finally, AntMiner+ leads to fewer and better performing rules.

In Chapter 3, *Jensen* presents a feature selection mechanism based on ant colony optimization algorithm to determine a minimal feature subset from a problem domain while retaining a suitably high accuracy in representing the original features. The proposed method is applied to two very different challenging tasks, namely web classification and complex systems monitoring.

Galea and *Shen* in the fourth chapter present an ant colony optimization approach for the induction of fuzzy rules. Several ant colony optimization algorithms are run simultaneously, with each focusing on finding descriptive rules for a specific class. The final outcome is a fuzzy rulebase that has been evolved so that individual rules complement each other during the classification process.

In the fifth chapter *Tsang and Kwong* present an ant colony based clustering model for intrusion detection. The proposed model improves existing ant-based clustering algorithms by incorporating some meta-heuristic principles. To further improve the clustering solution and alleviate the curse of dimensionality in network connection data, four unsupervised feature extraction algorithms are also studied and evaluated.

Omran et al. in the sixth chapter present particle swarm optimization algorithms for pattern recognition and image processing problems. First a clustering method that is based on PSO is discussed. The application of the proposed clustering algorithm to the problem of unsupervised classification and segmentation of images is investigated. Then PSO-based approaches that tackle the color image quantization and spectral unmixing problems are discussed.

In the seventh chapter *Azzag et al.* present a new model for data clustering, which is inspired from the self-assembly behavior of real ants. Real ants can build complex structures by connecting themselves to each others. It is shown is this paper that this behavior can be used to build a hierarchical tree-structured partitioning of the data according to the similarities between those data. Authors have also introduced an incremental version of the artificial ants algorithm.

Kazemian et al. in the eighth chapter presents a new swarm data clustering method based on Flowers Pollination by Artificial Bees (FPAB). FPAB does not require any parameter settings and any initial information such as the number of classes and the number of partitions on input data. Initially, in FPAB, bees move the pollens and pollinate them. Each pollen will grow in proportion to its garden flowers. Better growing will occur in better conditions. After some iterations, natural selection reduces the pollens and flowers and the gardens of the same type of flowers will be formed. The prototypes of each gardens are taken as the initial cluster centers for Fuzzy C Means algorithm which is used to reduce obvious misclassification errors. In the next stage, the prototypes of gardens are assumed as a single flower and FPAB is applied to them again.

Palotai et al. in the ninth chapter propose an Alife architecture for news foraging. News foragers in the Internet were evolved by a simple internal selective algorithm: selection concerned the memory components, being finite in size and containing the list of most promising supplies. Foragers received reward for locating not yet found news and crawled by using value estimation. Foragers were allowed to multiply if they passed a given productivity threshold. A particular property of this community is that there is no direct interaction (here, communication) amongst foragers that allowed us to study compartmentalization, assumed to be important for scalability, in a very clear form.

Veenhuis and Köppen in the tenth chapter introduce a data clustering algorithm based on species clustering. It combines methods of particle swarm optimization and flock algorithms. A given set of data is interpreted as a multi-species swarm which wants to separate into single-species swarms, i.e., clusters. The data to be clustered are assigned to datoids which form a swarm on a two-dimensional plane. A datoid can be imagined as a bird carrying a piece of data on its back. While swarming, this swarm divides into sub-swarms moving over the plane and consisting of datoids carrying similar data. After swarming, these sub swarms of datoids can be grouped together as clusters.

In the last chapter *Yang et al.* present a clustering ensemble model using ant colony algorithm with validity index and ART neural network. Clusterings are visually formed on the plane by ants walking, picking up or dropping down projected data objects with different probabilities. Adaptive Resonance Theory (ART) is employed to combine the clusterings produced by ant colonies with different moving speeds.

We are very much grateful to the authors of this volume and to the reviewers for their tremendous service by critically reviewing the chapters. The editors would like to thank Dr. Thomas Ditzinger (Springer Engineering Inhouse Editor, Studies in Computational Intelligence Series), Professor Janusz Kacprzyk (Editor-in-Chief, Springer Studies in Computational Intelligence Series) and Ms. Heather King (Editorial Assistant, Springer Verlag, Heidelberg) for the editorial assistance and excellent cooperative collaboration to produce this important scientific work. We hope that the reader will share our excitement to present this volume on '*Swarm Intelligence in Data Mining*' and will find it useful.

April, 2006   *Ajith Abraham, Chung-Ang University, Seoul, Korea*
Crina Grosan, Cluj-Napoca, Babeş-Bolyai University, Romania
Vitorino Ramos, Technical University of Lisbon, Portugal

Contents

1 Swarm Intelligence in Data Mining
Crina Grosan, Ajith Abraham and Monica Chis 1
1.1 Biological Collective Behavior 1
1.2 Swarms and Artificial Life 4
    1.2.1 Particle Swarm Optimization (PSO) 4
    1.2.2 Ant Colonies Optimization 9
1.3 Data mining 10
    1.3.1 Steps of Knowledge Discovery 10
1.4 Swarm Intelligence and Knowledge Discovery 11
1.5 Ant Colony Optimization and Data mining 15
1.6 Conclusions 16
References 16

2 Ants Constructing Rule-Based Classifiers
David Martens, Manu De Backer, Raf Haesen, Bart Baesens, Tom Holvoet 21
2.1 Introduction 21
2.2 Ant Systems and Data Mining 22
    2.2.1 Ant Systems 22
    2.2.2 Data Mining 24
    2.2.3 Data Mining with Ant Systems 25
2.3 AntMiner+ 27
    2.3.1 The Construction Graph 28
    2.3.2 Edge Probabilities 29
    2.3.3 Heuristic Value 29
    2.3.4 Pheromone Updating 30
    2.3.5 Early Stopping 32
2.4 Distributed Data Mining With AntMiner+:
    a Credit Scoring Case 33
2.5 Experiments and Results 34
    2.5.1 Experimental Set-Up 34
    2.5.2 Datasets 35

		Credit Scoring	35
		Toy Problems	36
	2.5.3	Software Implementation	38
	2.5.4	Discussion	39
2.6	Conclusion and Future Research		40
References			41

3 Performing Feature Selection with ACO
Richard Jensen 45

3.1	Introduction		45
3.2	Rough Feature Selection		47
	3.2.1	Theoretical Background	47
	3.2.2	Reduction Method	48
3.3	Fuzzy-Rough Feature Selection		50
	3.3.1	Fuzzy Equivalence Classes	50
	3.3.2	Fuzzy Lower and Upper Approximations	51
	3.3.3	Fuzzy-Rough Reduction Method	52
	3.3.4	A Worked Example	53
3.4	Ant-based Feature Selection		56
	3.4.1	ACO Framework	57
	3.4.2	Feature Selection	58
		Selection Process	59
		Complexity Analysis	59
		Pheromone Update	60
3.5	Crisp Ant-based Feature Selection Evaluation		60
	3.5.1	Experimental Setup	61
	3.5.2	Experimental Results	62
3.6	Fuzzy Ant-based Feature Selection Evaluation		63
	3.6.1	Web Classification	63
		System Overview	63
		Experimentation and Results	65
	3.6.2	Systems Monitoring	66
		Comparison of Fuzzy-Rough Methods	68
		Comparison with Entropy-based Feature Selection	69
		Comparison with the use of PCA	70
		Comparison with the use of a Support Vector Classifier	70
3.7	Conclusion		71
References			72

4 Simultaneous Ant Colony Optimization Algorithms for Learning Linguistic Fuzzy Rules
Michelle Galea, Qiang Shen 75

4.1	Introduction		75
4.2	Background		76
	4.2.1	Fuzzy Rules and Rule-Based Systems	76

		Fuzzy Sets and Operators...................................	77
		Linguistic Variables and Fuzzy Rules.......................	78
		Classification using Fuzzy Rules	79
		A Rule-Matching Example..........................	80
	4.2.2	Ant Colony Optimization and Rule Induction	81
4.3	Simultaneous Fuzzy Rule Learning		84
	4.3.1	Why Simultaneous Rule Learning	84
	4.3.2	*FRANTIC-SRL*	86
		Rule Construction.......................................	86
		Heuristic..	87
		Pheromone Updating	88
		Transition Rule	88
		Rule Evaluation......................................	89
4.4	Experiments and Analyses		90
	4.4.1	Experiment Setup	90
		The Datasets	90
		Other Induction Algorithms...........................	91
		FRANTIC-SRL Parameters.............................	92
	4.4.2	Saturday Morning Problem Results	93
	4.4.3	Water Treatment Plant Results	93
4.5	Conclusions and Future Work		95
References			97

5 Ant Colony Clustering and Feature Extraction for Anomaly Intrusion Detection
Chi-Ho Tsang, Sam Kwong 101

5.1	Introduction		101
5.2	Related Works		103
5.3	Ant Colony Clustering Model		104
	5.3.1	Basics and Problems of Ant-based Clustering Approach	104
	5.3.2	Measure of Local Regional Entropy	106
	5.3.3	Pheromone Infrastructure.................................	107
	5.3.4	Modified Short-term Memory and α-adaptation	109
	5.3.5	Selection Scheme, Parameter Settings and Cluster Retrieval	110
5.4	Experiments and Results...		111
	5.4.1	Dataset Description and Preprocessing	111
	5.4.2	Metrics of Cluster Validity and Classification Performance......	112
	5.4.3	Cluster Analysis on Benchmark Datasets.....................	114
	5.4.4	ACCM with Feature Extraction for Intrusion Detection.........	116
5.5	Conclusions		120
5.6	Future Works		121
References			121

6 Particle Swarm Optimization for Pattern Recognition and Image Processing

Mahamed G.H. Omran, Andries P. Engelbrecht, Ayed Salman 125
6.1 Introduction 125
6.2 Background 126
    6.2.1 The clustering problem 126
        The K-means Algorithm 128
        The Fuzzy C-means Algorithm 129
        Swarm Intelligence Approaches 130
    6.2.2 Color Image Quantization 130
    6.2.3 Spectral Unmixing 132
        Linear Pixel Unmixing (or Linear Mixture Modeling) 132
        Selection of the End-Members 133
6.3 Particle Swarm Optimization 134
6.4 A PSO-based Clustering Algorithm with Application to Unsupervised Image Classification 135
    6.4.1 Experimental Results 137
6.5 A PSO-based Color Image Quantization (PSO-CIQ) Algorithm 138
    6.5.1 Experimental Results 140
6.6 The PSO-based End-Member Selection (PSO-EMS) Algorithm 141
    6.6.1 The Generation of Abundance Images 143
    6.6.2 Experimental results 143
6.7 Conclusion 148
References 148

7 Data and Text Mining with Hierarchical Clustering Ants

Hanene Azzag, Christiane Guinot, Gilles Venturini 153
7.1 Introduction 153
7.2 Biological and computer models 154
    7.2.1 Ants based algorithms for clustering 154
    7.2.2 Self-assembly in real ants 155
    7.2.3 A computer model of ants self-assembly for hierarchical clustering 155
    7.2.4 Self-assembly and robotics 157
7.3 Two stochastic and deterministic algorithms 158
    7.3.1 Common principles 158
    7.3.2 Stochastic algorithm: AntTree$_{STOCH}$ 158
    7.3.3 Deterministic algorithm with no thresholds and no parameters : AntTree$_{NO-THRESHOLDS}$ 161
    7.3.4 Properties 162
7.4 Experimental results with numeric, symbolic and textual databases 164
    7.4.1 Testing methodology 164
    7.4.2 Parameters study 166
    7.4.3 Tested algorithms 166
    7.4.4 Results with numeric databases 168
    7.4.5 Results with symbolic databases 168

		7.4.6	Processing times	169
		7.4.7	Comparison with biomimetic methods	170
		7.4.8	Comparative study on textual databases	172
	7.5	Real world applications		175
		7.5.1	Human skin analysis	175
		7.5.2	Web usage mining	177
		7.5.3	Generation and interactive exploration of a portal site	179
	7.6	Incremental clustering of a large data set		182
		7.6.1	Principles of AntTree$_{INC}$	182
		7.6.2	Results with incremental and large data sets	184
	7.7	Conclusions		186
References				186

8 Swarm Clustering Based on Flowers Pollination by Artificial Bees
Majid Kazemian, Yoosef Ramezani, Caro Lucas, Behzad Moshiri 191

	8.1	Introduction		191
	8.2	Clustering		192
		8.2.1	What is clustering?	192
		8.2.2	Why swarm intelligence?	193
		8.2.3	Swarm clustering	193
		8.2.4	Some artificial models	194
	8.3	FPAB		195
		8.3.1	FPAB underlying algorithms	196
			Picking up pollen	197
			Pollinating	198
			Natural selection	198
			Merge algorithm	199
	8.4	Experimental results		199
	8.5	Conclusion and future works		200
References				201

9 Computer study of the evolution of 'news foragers' on the Internet
Zsolt Palotai, Sándor Mandusitz, András Lőrincz 203

	9.1	Introduction		203
	9.2	Related work		204
	9.3	Forager architecture		205
		9.3.1	Algorithms	206
		9.3.2	Reinforcing agent	208
		9.3.3	Foragers	209
	9.4	Experimental results		210
		9.4.1	Environment	210
		9.4.2	Time lag and multiplication	210
		9.4.3	Compartmentalization	211
	9.5	Discussion		213
	9.6	Conclusions		217

References 217

10 Data Swarm Clustering
Christian Veenhuis, Mario Köppen 221
10.1 Introduction 221
10.2 Data Clustering 223
10.3 Flock Algorithms 223
10.4 Particle Swarm Optimization 225
10.5 Data Swarm Clustering 226
    10.5.1 Initialization 227
    10.5.2 Iteration 228
    10.5.3 Cluster Retrieval 234
10.6 Experimental Setup 234
    10.6.1 Synthetical Datasets 234
    10.6.2 Real Life Datasets 236
    10.6.3 Parameters 236
10.7 Results 237
10.8 Conclusion 240
References 241

11 Clustering Ensemble Using ANT and ART
Yan Yang, Mohamed Kamel, Fan Jin 243
11.1 Introduction 243
11.2 Ant Colony Clustering Algorithm with Validity Index (ACC-VI) ... 245
    11.2.1 Ant Colony Clustering Algorithm 245
    11.2.2 Clustering Validity Index 247
    11.2.3 ACC-VI Algorithm 248
11.3 ART Algorithm 249
11.4 Clustering Ensemble Model 253
    11.4.1 Consensus Functions 253
    11.4.2 ART Ensemble Aggregation Model 253
11.5 Experimental Analysis 255
    11.5.1 Artificial Data Set (2D3C) 256
    11.5.2 Real Data Set (Iris) 256
    11.5.3 Reuter-21578 Document Collection 258
11.6 Conclusions 262
            Acknowledgements 262
References 262

Index 265

1

Swarm Intelligence in Data Mining

Crina Grosan[1], Ajith Abraham[2] and Monica Chis[3]

[1] Department of Computer Science
  Babeş-Bolyai University, Cluj-Napoca, 3400, Romania
  cgrosan@cs.ubbcluj.ro
[2] IITA Professorship Program,
  School of Computer Science and Engineering
  Chung-Ang University, Seoul 156-756, Korea
  ajith.abraham@ieee.org
[3] Avram Iancu University, Ilie Macelaru 1, 3400
  Cluj-Napoca, Romania
  mchis@artelecom.net

Summary

This chapter presents the biological motivation and some of the theoretical concepts of swarm intelligence with an emphasis on particle swarm optimization and ant colony optimization algorithms. The basic data mining terminologies are explained and linked with some of the past and ongoing works using swarm intelligence techniques.

1.1 Biological Collective Behavior

Swarm behavior can be seen in bird flocks, fish schools, as well as in insects like mosquitoes and midges. Many animal groups such as fish schools and bird flocks clearly display structural order, with the behavior of the organisms so integrated that even though they may change shape and direction, they appear to move as a single coherent entity [11]. The main principles of the collective behavior as presented in Figure 1.1 are:

- *Homogeneity*: every bird in flock has the same behavior model. The flock moves without a leader, even though temporary leaders seem to appear.
- *Locality*: the motion of each bird is only influenced by its nearest flock mates. Vision is considered to be the most important senses for flock organization.
- *Collision Avoidance*: avoid with nearby flock mates.
- *Velocity Matching* : attempt to match velocity with nearby flock mates.
- *Flock Centering*: attempt to stay close to nearby flock mates

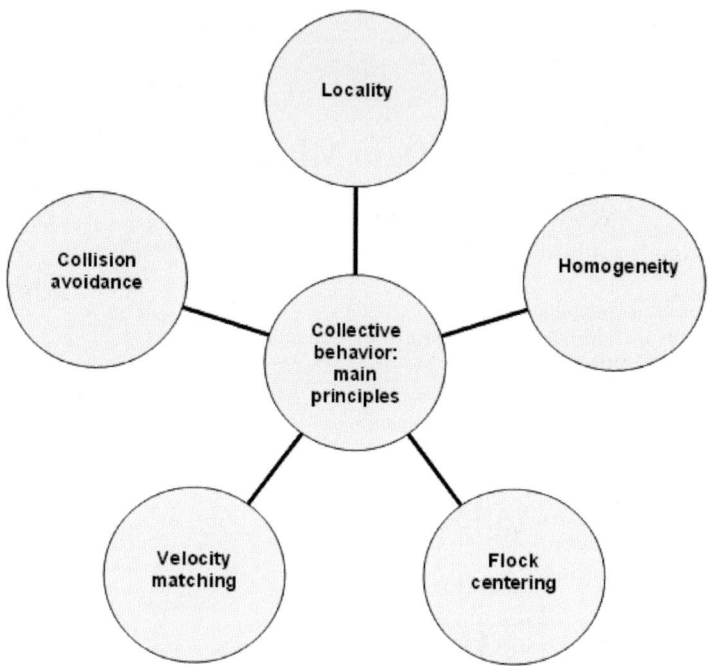

Fig. 1.1. The main principles of collective behavior.

Individuals attempt to maintain a minimum distance between themselves and others at all times. This rule has the highest priority and corresponds to a frequently observed behavior of animals in nature [36]. If individuals are not performing an avoidance manoeuvre, they tend to be attracted towards other individuals (to avoid being isolated) and to align themselves with neighbors [50], [51].

Couzin et al. [11] identified four collective dynamical behaviors as illustrated in Figure 1.1:

- *Swarm* : an aggregate with cohesion, but a low level of polarization (parallel alignment) among members
- *Torus*: individuals perpetually rotate around an empty core (milling). The direction of rotation is random.
- *Dynamic parallel group*: the individuals are polarized and move as a coherent group, but individuals can move throughout the group and density and group form can fluctuate [42], [50].
- *Highly parallel group*: much more static in terms of exchange of spatial positions within the group than the dynamic parallel group and the variation in density and form is minimal.

As mentioned in [22], at a high-level, a swarm can be viewed as a group of agents cooperating to achieve some purposeful behavior and achieve some goal (see Figure 1.3). This collective intelligence seems to emerge from what are often large

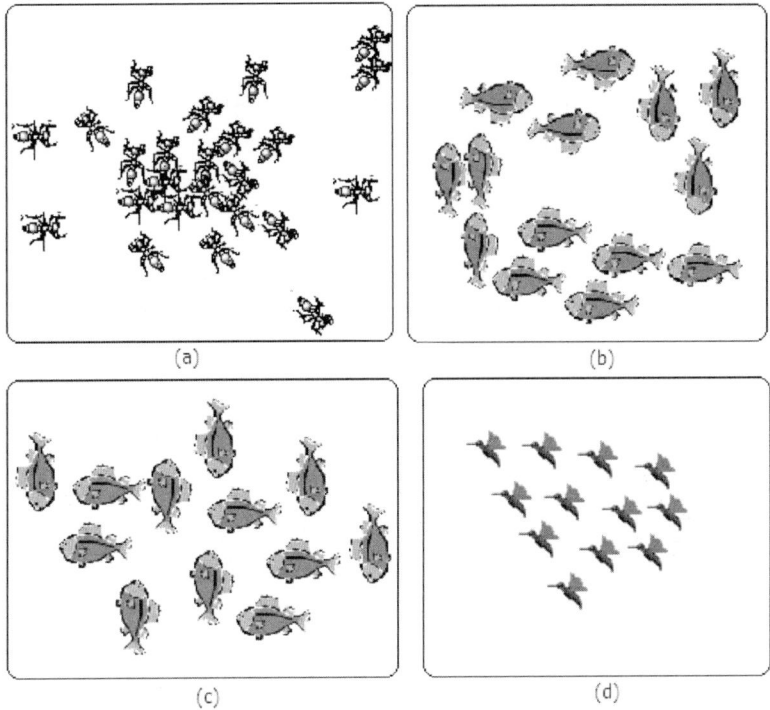

Fig. 1.2. Several models of collective behavior: (a) swarm (b) torus (c) dynamic parallel group and (d) highly parallel group.

groups of relatively simple agents. The agents use simple local rules to govern their actions and via the interactions of the entire group, the swarm achieves its objectives. A type of self-organization emerges from the collection of actions of the group.

An autonomous agent is a subsystem that interacts with its environment, which probably consists of other agents, but acts relatively independently from all other agents [22]. The autonomous agent does not follow commands from a leader, or some global plan [23]. For example, for a bird to participate in a flock, it only adjusts its movements to coordinate with the movements of its flock mates, typically its neighbors that are close to it in the flock. A bird in a flock simply tries to stay close to its neighbors, but avoid collisions with them. Each bird does not take commands from any leader bird since there is no lead bird. Any bird can in the front, center and back of the swarm. Swarm behavior helps birds take advantage of several things including protection from predators (especially for birds in the middle of the flock), and searching for food (essentially each bird is exploiting the eyes of every other bird) [22].

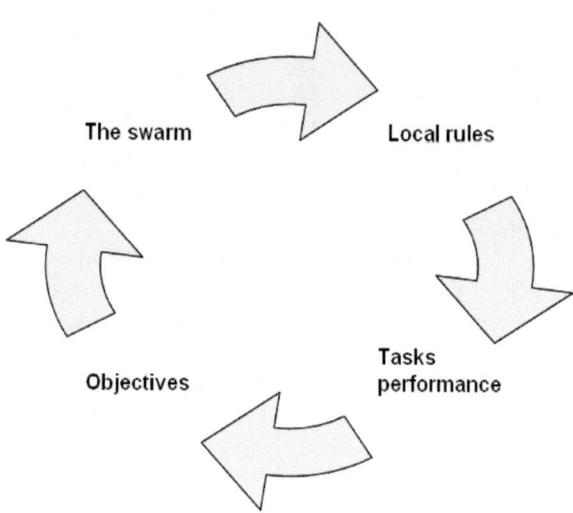

Fig. 1.3. The simple scheme of a swarm.

1.2 Swarms and Artificial Life

Since 1990, several collective behavior (like social insects, bird flocking) inspired algorithms have been proposed. The application areas of these algorithms refer to well studied optimization problems like NP-hard problems (Traveling Salesman Problem, Quadratic Assignment Problem, Graph problems), network routing, clustering, data mining, job scheduling etc.

(PSO) and Ant Colonies Optimization (ACO) are currently the most popular algorithms in the swarm intelligence domain.

1.2.1 Particle Swarm Optimization (PSO)

PSO is a population-based search algorithm and is initialized with a population of random solutions, called particles [26]. Unlike in the other evolutionary computation techniques, each particle in PSO is also associated with a velocity. Particles fly through the search space with velocities which are dynamically adjusted according to their historical behaviors. Therefore, the particles have the tendency to fly towards the better and better search area over the course of search process. The PSO was first designed to simulate birds seeking food which is defined as a 'cornfield vector' [29], [30], [31], [32], [33].

Assume the following scenario: a group of birds are randomly searching food in an area. There is only one piece of food in the area being searched. The birds do not know where the food is. But they know how far the food is and their peers' positions. So what's the best strategy to find the food? An effective strategy is to follow the bird which is nearest to the food.

PSO learns from the scenario and uses it to solve the optimization problems. In PSO, each single solution is like a 'bird' in the search space, which is called 'particle'. All particles have fitness values which are evaluated by the fitness function to be optimized, and have velocities which direct the flying of the particles. (The particles fly through the problem space by following the particles with the best solutions so far). PSO is initialized with a group of random particles (solutions) and then searches for optima by updating each generation.

Each individual is treated as a volume-less particle (a point) in the D-dimensional search space. The ith particle is represented as $X_i = (x_{i1}, x_{i2}, \ldots, x_{iD})$. At each generation, each particle is updated by the following two 'best' values. The first one is the best previous location (the position giving the best fitness value) a particle has achieved so far. This value is called *pBest*. The *pBest* of the ith particle is represented as $P_i = (p_{i1}, p_{i2}, \ldots, p_{iD})$. At each iteration, the P vector of the particle with the best fitness in the neighborhood, designated l or g, and the P vector of the current particle are combined to adjust the velocity along each dimension, and that velocity is then used to compute a new position for the particle. The portion of the adjustment to the velocity influenced by the individual's previous best position (P) is considered as the *cognition* component, and the portion influenced by the best in the neighborhood is the *social* component. With the addition of the inertia factor ω, by Shi and Eberhart [59] (brought in for balancing the global and the local search), these equations are:

$$v_{id} = \omega^* v_{id} + \eta_1^* rand()^* (p_{id} - x_{id}) + \eta_2^* Rand()^* (p_{gd} - x_{id}) \quad (1.1)$$

$$x_{id} = x_{id} + v_{id} \quad (1.2)$$

where *rand()* and *Rand()* are two random numbers independently generated within the range [0,1] and η_1 and η_2 are two learning factors which control the influence of the social and cognitive components. In (1.1), if the sum on the right side exceeds a constant value, then the velocity on that dimension is assigned to be $\pm V_{max}$. Thus, particles' velocities are clamped to the range $[-V_{max}, V_{max}]$ which serves as a constraint to control the global exploration ability of particle swarm. Thus, the likelihood of particles leaving the search space is reduced. Note that this is not to restrict the values of X_i within the range $[-V_{max}, V_{max}]$; it only limits the maximum distance that a particle will move during one iteration ([19], [20], [21]). The main PSO algorithm as described by Pomeroy [52] is given below:

/* set up particles' next location */
for each particle p **do** {
    for $d = 1$ **to** dimensions **do** {
        p.next[d] = random()
        p.velocity[d] = random(deltaMin, deltaMax)

```
            }
        p.bestSoFar = initialFitness
    }
    /* set particles' neighbors */
    for each particle p do {
        for n = 1 to numberOfNeighbors do {
            p.neighbor[n] = getNeighbor(p,n)
        }
    }
    /* run Particle Swarm Optimizer */
    while iterations ≤ maxIterations do {
    /* Make the "next locations" current and then*/
    /* test their fitness. */
        for each particle p do {
            for d = 1 to dimensions do {
                p.current[d] = p.next[d]
            }
            fitness = test(p)
            if fitness > p.bestSoFar then do {
                p.bestSoFar = fitness
                for d = 1 to dimensions do {
                    p.best[d] = p.current[d]
                }
            }
            if fitness = targetFitness then do {
                ...
                /* e.g., write out solution and quit */
            }
        } /* end of: for each particle p */
        for each particle p do {
            n = getNeighborWithBestFitness(p)
            for d = 1 to dimensions do {
                iFactor = iWeight * random(iMin, iMax)
                sFactor = sWeight * random(sMin, sMax)
                pDelta[d] = p.best[d] - p.current[d]
                nDelta[d] = n.best[d] - p.current[d]
                delta = (iFactor * pDelta[d]) + (sFactor * nDelta[d])
                delta = p.velocity[d] + delta
                p.velocity[d] = constrict(delta)
                p.next[d] = p.current[d] + p.velocity[d]
            }
        } /* end of: for each particle p */
    } /* end of: while iterations ≤ maxIterations */
```

end /* end of main program */

/* Return neighbor n of particle p */
function getNeighbor(p, n) {
...
return neighborParticle
}

/* Return particle in p's neighborhood */
/* with the best fitness */
function getNeighborWithBestFitness(p) {
...
return neighborParticle
}

/* Limit the change in a particle's */
/* dimension value */
function constrict(delta) {
if delta < deltaMin **then**
return deltaMin
else
if delta > deltaMax **then**
return deltaMax
else
return delta
}

The basic scheme of PSO algorithm is presented in Figure 1.4. The PSO algorithm can be seen as a set of vectors whose trajectories oscillate around a region defined by each individual previous best position and the best position of some other individuals [34]. There are different neighborhood topologies used to identify which particles from the swarm can influence the individuals. The most common ones are known as the *gbest* and *lbest:*

In the *gbest* swarm, the trajectory of each individual (particle) is influenced by the best individual found in the entire swarm. It is assumed that *gbest* swarms converge fast, as all the particles are attracted simultaneously to the best part of the search space. However, if the global optimum is not close to the best particle, it may be impossible for the swarm to explore other areas and, consequently, the swarm can be trapped in a local optima [35].

In the *lbest* swarm, each individual is influenced by a smaller number of its neighbors (which are seen as adjacent members of the swarm array). Typically, *lbest* neighborhoods comprise of two neighbors: one on the right side and one on the left side (a ring lattice). This type of swarm will converge slower but can locate the global optimum with a greater chance. *lbest* swarm is able to flow around local optima, sub-swarms being able to explore different optima [35]. A graphical representation of a

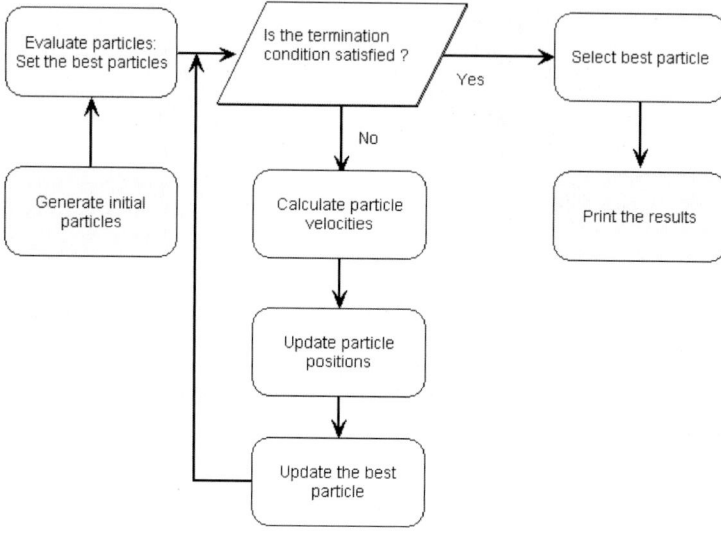

Fig. 1.4. The basic structure of PSO.

gbest swarm and a *lbest* swarm respectively is depicted in Figure 1.5 (taken from [35]). If we consider social and geographical neighborhoods as presented in Figure 1.6, then both *gbest* and *lbest* may be viewed as forms of social neighborhoods.

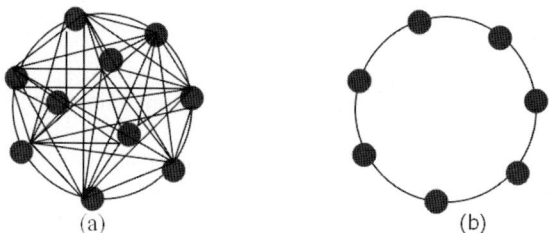

Fig. 1.5. Graphical representation of (a) *gbest* swarm (b) *lbest* swarm.

Watts [70], [71] introduced the small-world network model which allows to interpolate between regular low-dimensional lattices and random networks, by introducing a certain amount of random long-range connections into an initially regular network [14]. Starting from here, several models have been developed: icing model [5], spreading of epidemics [44], [45], evolution of random walks [27] are some of them. Watts identifies two factors influencing the information exchange between the small-world network members:

- the degree of connectivity : the behavior of each individual will be influenced by the behavior of its k neighbors.

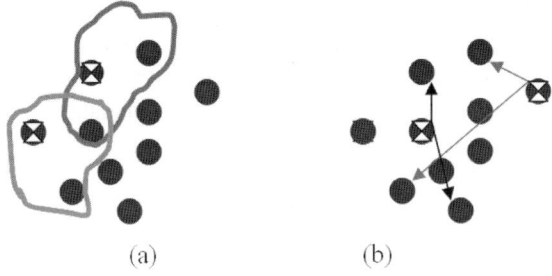

Fig. 1.6. Examples of (a) geographical neighborhood (b) social neighborhood in a swarm.

- the amount of clustering : clustering refers to the neighbors in common with some other individuals.

There are several types of swarm topologies which can be used depending upon the problem to be solved. Kennedy and Mendes [35] have tested few topologies like pyramid model, star, "small", von Neumann etc. for some function optimization problems.

1.2.2 Ant Colonies Optimization

Ant Colonies Optimization (ACO) algorithms were introduced around 1990 [15], [16], [17]. These algorithms were inspired by the behavior of ant colonies. Ants are social insects, being interested mainly in the colony survival rather than individual survival. Of interests is ants' ability to find the shortest path from their nest to food. This idea was the source of the proposed algorithms inspired from ants' behavior.

When searching for food, ants initially explore the area surrounding their nest in a random manner. While moving, ants leave a chemical pheromone trail on the ground. Ants are guided by pheromone smell. Ants tend to choose the paths marked by the strongest pheromone concentration . When an ant finds a food source, it evaluates the quantity and the quality of the food and carries some of it back to the nest. During the return trip, the quantity of pheromone that an ant leaves on the ground may depend on the quantity and quality of the food. The pheromone trails will guide other ants to the food source. The indirect communication between the ants via pheromone trails enables them to find shortest paths between their nest and food sources. As given by Dorigo et al. [18] the main steps of the ACO algorithm are given below:

1. *pheromone trail initialization*
2. *solution construction using pheromone trail*
   Each ant constructs a complete solution to the problem according to a probabilistic
3. *state transition rule.* The state transition rule depends mainly on the state of the pheromone [64].
4. *pheromone trail update.*

A global pheromone updating rule is applied in two phases. First, an evaporation phase where a fraction of the pheromone evaporates, and then a reinforcement phase where each ant deposits an amount of pheromone which is proportional to the fitness of its solution [64]. This process is iterated until a termination condition is reached.

1.3 Data mining

Historically the notion of finding useful patterns in data has been given a variety of names including data mining, knowledge extraction, information discovery, and data pattern processing. Data mining is the application of specific algorithms for extracting patterns from data [22]. The additional steps in the KDD process, such as data selection, data cleaning, incorporating appropriate prior knowledge, and proper interpretation of the results are essential to ensure that useful knowledge is derived form the data.

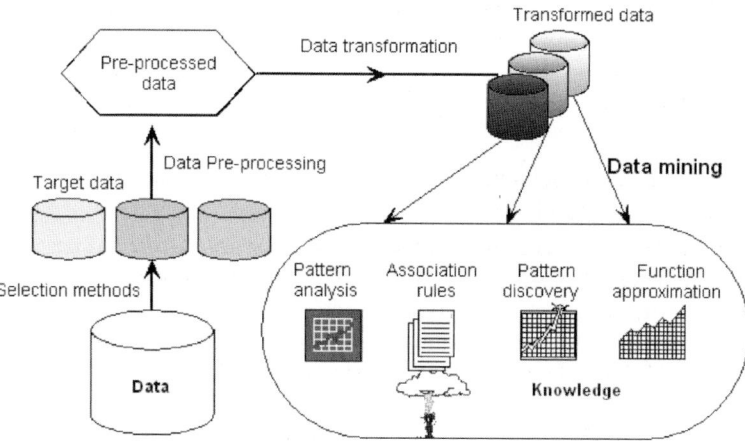

Fig. 1.7. Steps of the knowledge discovery process

1.3.1 Steps of Knowledge Discovery

Here we broadly outline some of its basic steps of the data mining process as illustrated in Figure 1.7 [22], [2].

1. *Developing and understanding the application domain, the relevant prior knowledge, and identifying the goal of the KDD process.*
2. *Creating target data set.*

3. *Data cleaning and preprocessing*: basic operations such as the removal of noise, handling missing data fields.
4. *Data reduction and projection*: finding useful features to represent the data depending on the goal of the task. Using dimensionality reduction or transformation methods to reduce the effective number of variables under consideration or to find invariant representation of data.
5. *Matching the goals of the KDD process to a particular* **data mining method**: Although the boundaries between prediction and description are not sharp, the distinction is useful for understanding the overall discovery goal.

   The goals of knowledge discovery are achieved via the following data mining methods:

   - **Clustering**: identification of a finite set of categories or clusters to describe the data.
   - **Summation**: finding a compact description for subset of data, e.g. the derivation of summary for association of rules and the use of multivariate visualization techniques.
   - **Dependency modeling**: finding a model which describes significant dependencies between variables.
   - **Regression**: learning a function which maps a data item to a real-valued prediction variable and the discovery of functional relationships between variables.
   - **Classification**: learning a function that classifies a data item into one of several predefined classes.
   - **Change and Deviation Detection:** discovering the most significant changes in the data from previously measured or normative values.

1.4 Swarm Intelligence and Knowledge Discovery

Data mining and particle swarm optimization may seem that they do not have many properties in common. However, they can be used together to form a method which often leads to the result, even when other methods would be too expensive or difficult to implement. Omran [47], [48] has used particle swarm optimization methods for pattern recognition and image processing. A new clustering method based on PSO is proposed and is applied to unsupervised classification and image segmentation. The PSO-based approaches are proposed to tackle the color image quantization and spectral unmixing problems.

Visual data mining via the construction of virtual reality spaces for the representation of data and knowledge, involves the solution of optimization problems. Valdes [68] introduced a hybrid technique based on particle swarm optimization (PSO) combined with classical optimization methods. This approach is applied to very high dimensional data from microarray gene expression experiments in order to understand the structure of both raw and processed data. Experiments with data sets corresponding to Alzheimer's disease show that high quality visual representation can be obtained by combining PSO with classical optimization methods. The behavior of some of the parameters controlling the swarm evolution was also studied.

Sousa et al. [61], [62] have proposed the use of PSO as a tool for data mining. In order to evaluate the usefulness of PSO for data mining, an empirical comparison of the performance of three variants of PSO with another evolutionary algorithm (Genetic Algorithm), in rule discovery for classification tasks is used. Such tasks are considered core tools for decision support systems in a widespread area, ranging from the industry, commerce, military and scientific fields. The data sources used here for experimental testing are commonly used and considered as a de facto standard for rule discovery algorithms reliability ranking. The results obtained in these domains seem to indicate that PSO algorithms are competitive with other evolutionary techniques, and can be successfully applied to more demanding problem domains.

Recommender systems are new types of internet-based software tools, designed to help users to find their way through today's complex on-line shops and entertainment websites. Ujjin and Bentley [66], [67] have described a new recommender system, which employs a particle swarm optimization (PSO) algorithm to learn personal preferences of users and provide tailored suggestions. Experiments are carried out to observe the performance of the system and results are compared to those obtained from the genetic algorithm (GA) recommender system and a standard, non-adaptive system based on the Pearson algorithm [7].

Another very important application of PSO is in the domain of cascading classifiers. *Cascading classifiers* have been used to solve pattern recognition problems in the last years. The main motivations behind such a strategy are the improvement of classification accuracy and the reduction of the complexity. The issue of class-related reject thresholds for cascading classifier systems is an important problem. It has been demonstrated in the literature that class-related reject thresholds provide an error-reject trade-off better than a single global threshold. Oliveira, Britto and Sabourin [46] proposed the using of the PSO for finding thresholds in order to improving the error-reject trade-off yielded by class-related reject thresholds. It has been proved to be very effective in solving real valued global optimization problems. In order to show the benefits of such an algorithm, they have applied it to optimize the thresholds of a cascading classifier system devoted to recognize handwritten digits. In a cascading classifier the inputs rejected by the first stage are handled by the next ones using costlier features or classifiers.

Settles and Rylander [56] have proposed a PSO method for neural network training. Chen and Abraham [8] investigated how the seemingly chaotic behavior of stock markets could be well represented using several soft computing techniques. Authors considered the flexible neural tree algorithm, a wavelet neural network, local linear wavelet neural network and finally a feed-forward artificial neural network. The parameters of the different learning techniques are optimized by the PSO approach. Experiment results reveal that PSO could play an important role to fine tune the parameters for optimal performance.

Breast cancer is one of the major tumor related cause of death in women. Various artificial intelligence techniques have been used to improve the diagnoses procedures and to aid the physician's efforts. Chen and Abraham [9] reported a preliminary study to detect breast cancer using a Flexible Neural Tree (FNT), Neural Network (NN), Wavelet Neural Network (WNN) and their ensemble combination. For the FNT model, a tree-structure based evolutionary algorithm and PSO are used to find an optimal FNT. For the NN and WNN, the PSO is employed to optimize the free parameters. The performance of each approach is evaluated using the breast cancer data set. Simulation results show that the obtained FNT model has a fewer number of variables with reduced number of input features and without significant reduction in the detection accuracy. The overall accuracy could be improved by using an ensemble approach by a voting method.

Chen et al. [10] proposed an evolutionary procedure to design hierarchical or multilevel fuzzy system Takagi-Sugeno Fuzzy Systems (TS-FS). The hierarchical structure is evolved using Probabilistic Incremental Program Evolution (PIPE) with specific instructions. The fine tuning of the if-then rules parameters encoded in the structure is accomplished using PSO. The proposed method interleaves both PIPE and PSO optimizations. The new method results in a smaller rule-base and good learning ability. The proposed hierarchical TS-FS is evaluated by using some forecasting problems. When compared to other hierarchical TS-FS, the proposed hybrid approach exhibits competing results with high accuracy and smaller size of the hierarchical architecture.

Skopos et al. [60] have proposed a PSO method for locating periodic orbits in a three-dimensional (3D) model of barred galaxies. Method developed an appropriate scheme that transforms the problem of finding periodic orbits into the problem of detecting global minimizers of a function, which is defined on the Poincaré surface section of the Hamiltonian system. By combining the PSO method with deflection techniques, they succeeded in tracing systematically several periodic orbits of the system.

Cluster analysis has become an important technique in exploratory data analysis, pattern recognition, machine learning, neural computing, and other engineering. The clustering aims at identifying and extracting significant groups in underlying data. The four main classes of clustering algorithms are partitioning methods, hierarchical methods, density based clustering and grid-based clustering. *Document clustering* is a fundamental operation used in unsupervised document organization, automatic topic extraction, and information retrieval. Fast and high-quality document clustering algorithms play an important role in effectively navigating, summarizing, and organizing information. Recent studies have shown that partitional clustering algorithms are more suitable for clustering large datasets due to their relatively low computational requirements [63], [73]. In the field of clustering, K-means algorithm is the most popularly used algorithm to find a partition that minimizes mean square error (MSE) measure. Although K-means is an extensively useful clustering algorithm, it suffers from several drawbacks. The objective function of the K-means is not convex and hence it may contain local minima. Consequently, while minimizing the objective function, there is possibility of getting stuck at local minima (also at local maxima and saddle point) [55]. The performance of the K-means algorithm depends on the initial choice of the cluster centers. Besides, the Euclidean norm is sensitive to noise or outliers. Hence K-means algorithm should be affected by noise and outliers [72], [28]. In addition to the K-means algorithm, several algorithms, such as Genetic Algorithm (GA) [28], [53] and Self-Organizing Maps (SOM) [43], have been used for document clustering. Cui et al. [12] proposed a PSO based hybrid document clustering algorithm. The PSO clustering algorithm performs a globalized search in the entire solution space. In the experiments, they applied the PSO, K-means and a hybrid PSO clustering algorithm on four different text document datasets. The results illustrate that the hybrid PSO algorithm can generate more compact clustering results than the K-means algorithm.

Swarming agents in networks of physically distributed processing nodes may be used for data acquisition, data fusion, and control applications. An architecture for active surveillance systems in which simple mobile agents collectively process real time data from heterogeneous sources at or near the origin of the data is used. The system requirements are motivated with the needs of a surveillance system for the early detection of large-scale bioterrorist attacks on a civilian population, but the same architecture is applicable to a wide range of other domains. The pattern detection and classification processes executed by the proposed system emerge from the coordinated activities of agents of two populations in a shared computational environment. Detector agents draw each other's attention to significant spatiotemporal patterns in the observed data stream. Classifier agents rank the detected patterns according to their respective criterion. The resulting system-level behavior is adaptive and robust.

Ye and Chen [24] introduced an evolutionary PSO learning-based method to optimally cluster N data points into K clusters. The hybrid PSO and K-means, with a novel alternative metric algorithm is called Alternative KPSO-clustering (AKPSO)method. This is developed to automatically detect the cluster centers of geometrical structure data sets. In AKPSO algorithm, the special alternative metric is considered to improve the traditional K-means clustering algorithm to deal with various structure data sets. Simulation results compared with some well-known clustering methods demonstrate the robustness and efficiency of the novel AKPSO method.

In the literature, there are some works related to co-evolutionary Particle Swarm Optimization (Co-PSO) [40], [58], [1]. According to Shi and Krohling [58], [37] each population is run using the standard PSO algorithm, using the other population as its environment [1]. Preliminary results demonstrated that Co-PSO constitutes a promising approach to solve constrained optimization problems. The problem is the difficulty to obtain fine tuning of the solution using a uniform distribution.

1.5 Ant Colony Optimization and Data mining

Ant colony based clustering algorithms have been first introduced by Deneubourg et al. [13] by mimicking different types of naturally-occurring emergent phenomena. Ants gather items to form heaps (clustering of dead corpses or cemeteries) observed in the species of *Pheidole Pallidula* and *Lasius Niger*. The basic mechanism underlying this type of aggregation phenomenon is an attraction between dead items mediated by the ant workers: small clusters of items grow by attracting workers to deposit more items. It is this positive and auto-catalytic feedback that leads to the formation of larger and larger clusters.

The general idea for data clustering is that isolated items should be picked up and dropped at some other location where more items of that type are present. Ramos et al. [54] proposed *ACLUSTER* algorithm to follow real ant-like behaviors as much as possible. In that sense, bio-inspired spatial transition probabilities are incorporated into the system, avoiding randomly moving agents, which encourage the distributed algorithm to explore regions manifestly without interest. The strategy allows guiding ants to find clusters of objects in an adaptive way.

In order to model the behavior of ants associated with different tasks (dropping and picking up objects), the use of combinations of different response thresholds was proposed. There are two major factors that should influence any local action taken by the ant-like agent: the number of objects in its neighborhood, and their similarity. Lumer and Faieta [41] used an average similarity, mixing distances between objects with their number, incorporating it simultaneously into a response threshold function like the algorithm proposed by Deneubourg et al. [13].

Admane et al. [4], presented *AntPart*, which is an exclusive unsupervised classification technique inspired by the behavior of a particular species of ants called *Pachycondyla apicalis*. The performances of this method were compared with those of three other ones, also inspired by the social behavior of ants: *AntClass, AntTree* and *AntClust*.

Kuo et al. [4], [38] proposed ant K-means (AK) clustering method. AK algorithm modifies the K-means as locating the objects in a cluster with the probability, which is updated by the pheromone, while the rule of updating pheromone is according to total within cluster variance (TWCV).

Tsai et al. [65] proposed a novel clustering method called ant colony optimization with different favor algorithm which performed better than the fast self-organizing map (SOM) K-means approach and genetic K-means algorithm.

Weng et al. [69] proposed a time series segmentation algorithm based on the ant colony optimization algorithm to exhibit the changeability of the time series data. Authors used the Bottom-Up method, which has been reported to give good results for time series segmentation. The research result shows that time series segmentation run by the ACO algorithm not only automatically identifies the number of segments, but its segmentation cost was lower than that of the time series segmentation using the Bottom-Up method.

Shelokar et al. [57] developed an ant colony optimization metaheuristic as a rule based machine learning method, called as ant colony classifier system, and applied to three process engineering examples. The learning algorithm addresses the problem of knowledge acquisition in terms of rules from example cases by developing and maintaining the knowledge base through the use of simple mechanism, pheromone trail information matrix and use of available heuristic information. The performance of an ant colony classifier is compared with the well-known decision tree based C4.5 algorithm in terms of the predictive accuracy on test cases and the simplicity of rules discovered.

Handl et al. [10] proposed a novel ant based clustering method by incorporating adaptive, heterogeneous ants, a time-dependent transporting activity, and a method that transforms the spatial embedding produced by the algorithm into an explicit partitioning. Empirical results demonstrate the ability of ant-based clustering and sorting to automatically identify the number of clusters inherent to a data collection, and to produce high quality solutions. However, the performance of the algorithm for topographic mapping was not really very good.

Web usage mining attempts to discover useful knowledge from the secondary data obtained from the interactions of the users with the Web. Web usage mining has become very critical for effective Web site management, creating adaptive Web sites, business and support services, personalization, network traffic flow analysis and so on. Abraham and Ramos [3] proposed an ant clustering algorithm to discover Web usage patterns (data clusters) and a linear genetic programming approach to analyze the visitor trends. Empirical results clearly show that ant colony clustering performs well when compared to a self organizing map (for clustering Web usage patterns).

1.6 Conclusions

In this Chapter, we introduced some of the preliminary concepts of swarm intelligence with an emphasis on particle swarm optimization and ant colony optimization algorithms. We then described the basic data mining terminologies and also illustrated some of the past and ongoing works of swarm intelligence in data mining.

References

1. Abdelbar AM, Ragab S, Mitri S (2003) Applying Co-Evolutionary Particle Swam Optimization to the Egyptian Board Game Seega. In Proceedings of The First Asian-Pacific Workshop on Genetic Programming, (S.B. Cho, N. X. Hoai and Y. Shan editors), 9-15, Canberra, Australia
2. Abonyi J., Feil B. and Abraham A. (2005), Computational Intelligence in Data Mining', Informatica: An International Journal of Computing and Informatics, Vol. 29, No. 1, pp. 3-12

3. Abraham A, Ramos V (2003) Web Usage Mining Using Artificial Ant Colony Clustering and Genetic Programming, 2003 IEEE Congress on Evolutionary Computation (CEC2003), Australia, IEEE Press, ISBN 0780378040, 1384-1391
4. Admane L, Benatchba K, Koudil M, Siad L, Maziz S (2006) AntPart: an algorithm for the unsupervised classification problem using ants, Applied Mathematics and Computation (http://dx.doi.org/10.1016/j.amc.2005.11.130)
5. Barrat A, Weight M (2000) On the properties of small-world network models. The European Physical Journal, 13:547-560
6. Blum C (2005) Ant colony optimization: Introduction and recent trends. Physics of Life Reviews, 2, 353–373
7. Breese, J.S., Heckerman, D., Kadie, C. Empirical analysis of predictive algorithms for collaborative filtering. In Proceedings of the 14th Conference on Uncertainty in Artificial Intelligence, pp. 43-52, 1998
8. Chen Y, Abraham A, (2006) Hybrid Learning Methods for Stock Index Modeling, Artificial Neural Networks in Finance, Health and Manufacturing: Potential and Challenges, J. Kamruzzaman, R. K. Begg and R. A. Sarker (Eds.), Idea Group Inc. Publishers, USA
9. Chen Y, Abraham A (2005) Hybrid Neurocomputing for Detection of Breast Cancer, The Fourth IEEE International Workshop on Soft Computing as Transdisciplinary Science and Technology (WSTST'05), Japan, Springer Verlag, Germany, pp. 884-892
10. Chen Y, Peng L, Abraham A (2006) Programming Hierarchical Takagi Sugeno Fuzzy Systems, The 2nd International Symposium on Evolving Fuzzy Systems (EFS2006), IEEE Press
11. Couzin ID, Krause J, James R, Ruxton GD, Franks NR (2002) Collective Memory and Spatial Sorting in Animal Groups, Journal of Theoretical Biology, 218, 1-11
12. Cui X, Potok TE (2005) Document Clustering Analysis Based on Hybrid PSO+K-means Algorithm, Journal of Computer Sciences (Special Issue), ISSN 1549-3636, pp. 27-33
13. Deneubourg JL, Goss S, Franks N, Franks AS, Detrain C, Chretien L (1991) The dynamics of collective sorting: Robot-like ants and ant-like robots. Proceedings of the First International Conference on Simulation of Adaptive Behaviour: From Animals to Animats, Cambridge, MA: MIT Press, 1, 356-365
14. Dall'Asta L, Baronchelli A, Barrat A, Loreto V (2006) Agreement dynamics on small-world networks. Europhysics Letters
15. Dorigo M, Blum C (2005) Ant colony optimization theory: A survey. Theoretical Computer Science, 344(2-3), 243-278
16. Dorigo M, Di Caro G, Gambardella LM (1999) Ant algorithms for discrete optimization. Artificial Life, 5(2), 137-72
17. Dorigo M, Gambardella LM (1997) Ant colony system: A cooperative learning approach to the traveling salesman problem. IEEE Transaction on Evolutionary Computation, 1(1), 53-66
18. Dorigo M, Bonaneau E, Theraulaz G (2000) Ant algorithms and stigmergy, Future Generation Computer Systems, 16, 851-871
19. Eberhart RC, Kennedy J (1995) A new optimizer using particle swarm theory. In Proceedings of the Sixth International Symposium on Micromachine and Human Science, Nagoya, Japan, 39-43
20. Eberhart RC, Shi Y (2001) Particle swarm optimization: developments, applications and resources. In Proceedings of the IEEE Congress on Evolutionary Computation (CEC), Seoul, Korea
21. Eberhart RC, Simpson PK, Dobbins RW (1996) Computational Intelligence PC Tools, Boston, MA: Academic Press Professional

22. Fayyad U, Piatestku-Shapio G, Smyth P, Uthurusamy R (1996) Advances in Knowledge Discovery and Data Mining, AAAI/MIT Press
23. Flake G (1999) The Computational Beauty of Nature. Cambridge, MA: MIT Press
24. Fun Y, Chen CY (2005) Alternative KPSO-Clustering Algorithm, Tamkang Journal of Science and Engineering, 8(2), 165-174
25. Handl J, Knowles J, Dorigo M (2006) Ant-based clustering and topographic mapping. Artificial Life 12(1) (in press)
26. Hu X, Shi Y, Eberhart RC (2004) Recent Advences in Particle Swarm, In Proceedings of Congress on evolutionary Computation (CEC), Portland, Oregon, 90-97
27. Jasch F, Blumen A (2001) Trapping of random walks on small-world networks. Physical Review E 64, 066104
28. Jones G, Robertson A, Santimetvirul C, Willett P (1995) Non-hierarchic document clustering using a genetic algorithm. Information Research, 1(1)
29. Kennedy J, Eberhart RC (1995) Particle Swarm Optimization. In Proceedings of IEEE International Conference on Neural Networks, Perth, Australia, IEEE Service Center, Piscataway, NJ, Vol.IV, 1942-1948
30. Kennedy J (1997) Minds and cultures: Particle swarm implications. Socially Intelligent Agents. Papers from the 1997 AAAI Fall Symposium. Technical Report FS-97-02, Menlo Park, CA: AAAI Press, 67-72
31. Kennedy J (1998) The Behavior of Particles, In Proceedings of 7th Annual Conference on Evolutionary Programming, San Diego, USA
32. Kennedy J (1997) The Particle Swarm: Social Adaptation of Knowledge. In Proceedings of IEEE International Conference on Evolutionary Computation, Indianapolis, Indiana, IEEE Service Center, Piscataway, NJ, 303-308
33. Kennedy J (1992) Thinking is social: Experiments with the adaptive culture model. Journal of Conflict Resolution, 42, 56-76
34. Kennedy J, Eberhart R (2001) Swarm Intelligence, Morgan Kaufmann Academic Press
35. Kennedy J, Mendes R (2002) Population structure and particle swarm performance. In Proceedings of the IEEE Congress on Evolutionary Computation (CEC), 1671-1676
36. Krause J, Ruxton GD (2002) Living in Groups. Oxford: Oxford University Press
37. Krohling RA, Hoffmann F, Coelho LS (2004) Co-evolutionary Particle Swarm Optimization for Min-Max Problems using Gaussian Distribution. In Proceedings of the Congress on Evolutionary Computation 2004 (CEC'2004), Portland, USA, volume 1, 959-964
38. Kuo RJ, Wang HS, Hu TL, Chou SH (2005) Application of ant K-means on clustering analysis, Computers & Mathematics with Applications, Volume 50, Issues 10-12, 1709-1724
39. Liu Y, Passino KM (2000) Swarm Intelligence: Literature Overview, http://www.ece.osu.edu/ passino/swarms.pdf
40. Lovbjerg M, Rasmussen TK, Krink T (2001) Hybrid Particle Swarm Optimiser with Breeding and Subpopulations. Proc. of the third Genetic and Evolutionary Computation Conference (GECCO-2001), volume 1, 469-476
41. Lumer ED, Faieta B (1994) Diversity and Adaptation in Populations of Clustering Ants. Clio D, Husbands P, Meyer J and Wilson S (Eds.), Proceedings of the Third International Conference on Simulation of Adaptive Behaviour: From Animals to Animats 3, Cambridge, MA: MIT Press, 501-508
42. Major PF, Dill LM (1978) The three-dimensional structure of airborne bird flocks. Behavioral Ecology and Sociobiology, 4, 111-122
43. Merkl D (2002) Text mining with self-organizing maps. Handbook of data mining and knowledge, Oxford University Press, Inc. New York, 903-910

44. Moore C, Newman MEJ (2000) Epidemics and percolation in small-world networks. Physics. Review. E 61, 5678-5682
45. Newman MEJ, Jensen I, Ziff RM (2002) Percolation and epidemics in a two-dimensional small world, Physics Review, E 65, 021904
46. Oliveira LS, Britto AS Jr., Sabourin R (2005) Improving Cascading Classifiers with Particle Swarm Optimization, International Conference on Document Analysis and Recognition (ICDAR 2005), Seoul, South Korea, 570-574
47. Omran, M. Particle Swarm optimization methods for pattern Recognition and Image Processing, Ph.D. Thesis, University of Pretoria, 2005
48. Omran, M., Salman, A. and Engelbrecht, A. P. Image classification using particle swarm optimization. Proceedings of the 4th Asia-Pacific Conference on Simulated Evolution and Learning 2002 (SEAL 2002), Singapore. pp. 370-374, 2002
49. Paredis J (1994) Steps towards coevolutionary classification neural networks, Artificial Life IV, MIT Press, 359-365
50. Partridge BL, Pitcher TJ (1980) The sensory basis of fish schools: relative role of lateral line and vision. Journal of Comparative Physiology, 135, 315-325
51. Partridge BL (1982) The structure and function of fish schools. Science American, 245, 90-99
52. Pomeroy P (2003) An Introduction to Particle Swarm Optimization, http://www.adaptiveview.com/articles/ipsop1.html
53. Raghavan VV, Birchand K (1979) A clustering strategy based on a formalism of the reproductive process in a natural system. Proceedings of the Second International Conference on Information Storage and Retrieval, 10-22
54. Ramos V, Muge, F, Pina, P (2002) Self-organized data and image retrieval as a consequence of inter-dynamic synergistic relationships in artificial ant colonies. Soft Computing Systems - Design, Management and Applications, Proceedings of the 2nd International Conference on Hybrid Intelligent Systems, IOS Press, 500-509
55. Selim SZ, Ismail MA (1984) K-means Type Algorithms: A Generalized Convergence Theorem and Characterization of Local Optimality, IEEE Transaction on Pattern Analysis and Machine Intelligence, 6, 81-87
56. Settles M, Rylander B (2002) Neural network learning using particle swarm optimizers. Advances in Information Science and Soft Computing, 224-226
57. Shelokar PS, Jayaraman VK, Kulkarni BD (2004) An ant colony classifier system: application to some process engineering problems, Computers & Chemical Engineering, 28(9), 1577-1584
58. Shi Y, Krohling RA (2002) Co-evolutionary particle swarm optimization to solving min-max problems. In Proceedings of the IEEE Conference on Evolutionary Computation, Hawai, 1682-1687
59. Shi Y, Eberhart RC (1998) A modified particle swarm optimizer. In Proceedings of the IEEE Congress on Evolutionary Computation (CEC), Piscataway, NJ. 69-73
60. Skopos C, Parsopoulus KE, Patsis PA, Vrahatis MN (2005) Particle swarm optimization: an efficient method for tracing periodic orbits in three-dimensional galactic potential, Mon. Not. R. Astron. Soc. 359, 251-260
61. Sousa T, Neves A, Silva A (2003) Swarm Optimisation as a New Tool for Data Mining, International Parallel and Distributed Processing Symposium (IPDPS'03), 144b
62. Sousa T, Silva A, Neves A (2004) Particle Swarm based Data Mining Algorithms for classification tasks, Parallel Computing, Volume 30, Issues 5-6, 767-783
63. Steinbach M, Karypis G, Kumar V, (2000) A Comparison of Document Clustering Techniques. TextMining Workshop, KDD

64. Toksari MD (2006) Ant colony optimization for finding the global minimum. Applied Mathematics and Computation, (in press)
65. Tsai CF, Tsai CW, Wu HC, Yang T (2004) ACODF: a novel data clustering approach for data mining in large databases, Journal of Systems and Software, Volume 73, Issue 1, 133-145
66. Ujjin S, Bentley PJ (2002) Learning User Preferences Using Evolution. In Proceedings of the 4th Asia-Pacific Conference on Simulated Evolution and Learning, Singapore
67. Ujjin S, Bentley PJ (2003) Particle swarm optimization recommender system. Proceedings of the IEEE Swarm Intelligence Symposium 2003 (SIS 2003), Indianapolis, Indiana, USA, 124-131
68. Valdes J (2004) Building Virtual Reality Spaces for Visual Data Mining with Hybrid Evolutionary-Classical Optimization: Application to Microarray Gene Expression Data. Proceedings of the IASTED International Joint Conference on Artificial Intelligence and Soft Computing (ASC'2004), 713-720
69. Weng SS, Liu YH (2006) Mining time series data for segmentation by using Ant Colony Optimization, European Journal of Operational Research, (http://dx.doi.org/10.1016/j.ejor.2005.09.001)
70. Watts DJ (1999) Small Worlds: The Dynamics of Networkds Between Order and Randomness. Princeton University Press
71. Watts DJ, Strogatz SH (1998) Collective dynamics of small-world networks. Nature, 393, 440-442
72. Wu KL, Yang MS (2002) Alternative C-means Clustering Algorithms. Pattern Recognition, 35, 2267-2278
73. Zhao Y, Karypis G (2004) Empirical and Theoretical Comparisons of Selected Criterion Functions for Document Clustering, Machine Learning, 55(3), 311-331

2

Ants Constructing Rule-Based Classifiers

David Martens[1], Manu De Backer[1], Raf Haesen[1],
Bart Baesens[2,1], Tom Holvoet[3]

[1] Department of Applied Economic Sciences, K.U.Leuven
Naamsestraat 69, B-3000 Leuven, Belgium
{David.Martens;Manu.Debacker;Raf.Haesen;Bart.Baesens}@econ.kuleuven.be

[2] University of Southampton, School of Management
Highfield Southampton, SO17 1BJ, United Kingdom
Bart@soton.ac.uk

[3] Department of Computer Science, K.U.Leuven
Celestijnenlaan 200A, B-3001 Leuven, Belgium
Tom.Holvoet@cs.kuleuven.be

Summary. This chapter introduces a new algorithm for classification, named AntMiner+, based on an artificial ant system with inherent self-organizing capabilities. The usage of ant systems generates scalable data mining solutions that are easily distributed and robust to failure. The introduced approach differs from the previously proposed AntMiner classification technique in three aspects. Firstly, AntMiner+ uses a $\mathcal{MAX}$-$\mathcal{MIN}$ ant system which is an improved version of the originally proposed ant system, yielding better performing classifiers. Secondly, the complexity of the environment in which the ants operate has substantially decreased. This simplification results in more effective decision making by the ants. Finally, by making a distinction between ordinal and nominal variables, AntMiner+ is able to include intervals in the rules which leads to fewer and better performing rules. The conducted experiments benchmark AntMiner+ with several state-of-the-art classification techniques on a variety of datasets. It is concluded that AntMiner+ builds accurate, comprehensible classifiers that outperform C4.5 inferred classifiers and are competitive with the included black-box techniques.

2.1 Introduction

In recent decades, innovative storage technologies and the success of the Internet have caused a true explosion of data. This data is typically distributed, continuously updated and contains valuable, yet hidden knowledge. Data mining is the overall process of extracting knowledge from this raw data. Although many techniques have been proposed and successfully implemented, few take into account the importance of the comprehensibility aspect of the generated models or the ability to deal with distributed data. Artificial ant systems are inspired on real ant colonies and are specifically designed to provide robust, scalable and distributed solutions. By performing local actions and indirect communication only, ants are able to achieve complex overall behavior. The approach described in this chapter, named AntMiner+, takes advantage of the inherent benefits of ant systems and puts them in a data mining context. Comprehensible, accurate classifiers in the form of simple **if-then-else** rules are extracted from data by the ants. The environment of the ants is defined as a directed acyclic graph (DAG) where an ant, walking from start to end, gradually constructs a rule. AntMiner+ uses a $\mathcal{MAX}$-$\mathcal{MIN}$ ant system, which is an improved version of the originally proposed ant system [41] and enhances the performance by a stronger exploitation of the best solutions.

The remainder of this chapter is structured as follows. In Sect. 2.2 we shortly explain the basics of ant systems, data mining and introduce the use of ant systems for data mining. This is further elaborated on in Section 2.3 where we explain the workings of our approach: AntMiner+. The final sections report on the results of our experiments on various datasets.

2.2 Ant Systems and Data Mining

2.2.1 Ant Systems

Artificial ant systems are inspired on the behavior of real ant colonies and are part of a relatively new concept in artificial intelligence, called swarm intelligence [5]. Swarm Intelligence is the property of a system whereby the collective behaviors of (unsophisticated) agents interacting locally with their environment cause coherent functional global patterns to emerge. A biological ant is a simple insect with limited capabilities but an ant colony is able to behave in complex manners and come to intelligent solutions for problems such as the transportation of heavy items and finding the shortest path between the food source and the nest. This complex behavior emerges from self-organization and indirect communication between the ants. The indirect way of communication, through the environment rather than directly between the individuals, is also known as stigmergy [18]. More specifically, ants communicate through a chemical substance called pheromone that each ant drops on its path. When an ant finds a pheromone trail it is likely to follow this path and reinforce the pheromone. The pheromone trail intensity is increased and the path will become more likely to be followed by other ants. In turn, when no ants follow the same path the pheromone trail intensity decreases, this process is called evaporation.

The same ideas are used for artificial ant systems [11]: a number of computational concurrent and asynchronous agents move through their environment and by doing so incrementally construct a solution for the problem at hand. Ants move by applying a stochastic local decision policy based on two parameters, the pheromone and heuristic values. The pheromone amount of a trail is a measure for the number of ants that recently have passed the trail and the heuristic value is a problem dependent value. When an ant comes at a crossroad, it is more likely to choose the trail with the higher pheromone and heuristic values. When an ant arrives at its destination, the ant's solution is evaluated and the trail followed by the ant is updated according to its quality. Updating the trails entails two phenomena: evaporation and reinforcement. Evaporation means that the pheromone level of the trails are diminished gradually. In this way less accurate trails will disappear. Reinforcement means that the pheromone level is increased proportionally to the quality of the corresponding candidate solution for the target problem. As a result, the solution provided by the ants will converge to a (sub)optimal solution of the problem.

In essence, the design of an ant system implies the specification of the following aspects:

- An environment that represents the problem domain in such a way that it lends itself to incrementally building a solution for the problem;
- A problem dependent heuristic evaluation function (η), which represents a quality factor for the different solutions;
- A rule for pheromone updating (τ), which takes into account the evaporation and the reinforcement of the trails;
- A probabilistic transition rule based on the value of the heuristic function (η) and on the strength of the pheromone trail (τ) that is used to iteratively construct a solution;
- A clear specification of when the algorithm converges to a solution.

Ant systems have shown to be a viable method for tackling hard combinatorial optimization problems [10]. A short overview of the literature, though not exhaustive, is provided in Table 2.1.

The performance of traditional ant systems, however, is rather poor on larger problems [37]. Stützle et al. [41] advocate that improved performance can be obtained by a stronger exploitation of the best solutions, combined with an effective mechanism for avoiding early search stagnation[4]. The authors propose a $\mathcal{MAX}$-$\mathcal{MIN}$ ant system ($\mathcal{MM}$AS) that differs from a normal ant system in three aspects:

- After each iteration only the best ant is allowed to add pheromone to its trail. This allows for a better exploitation of the best solution found;
- To avoid stagnation of the search, the range of possible pheromone trails is limited to an interval $[\tau_{min}, \tau_{max}]$;

---

[4]The situation where all ants take the same path and thus describe the same solution.

- Each trail is initialized with a pheromone value of τ_{max}, as such the algorithm achieves a higher exploration at the beginning of the algorithm.

Table 2.1. Literature Overview

\multicolumn{3}{c}{Overview of the applications of Ant Systems}
Data Mining	Clustering	Abraham et al. [1] Handle et al. [20] Schockaert et al. [33]
	Classification	Parpinelli et al. [28, 29] Liu et al. [22] Ramos et al. [31]
Operations Research	Traveling Salesman Problem	Dorigo et al. [12] Gambardella et al. [16, 17] Eyckelhof et al. [13] Stützle et al. [38, 39, 40]
	Vehicle Routing Problem	Wade et al. [45] Bullnheimer [6] Cicirello et al. [8]
	Quadratic Assignment Problem	Maniezzo et al. [24, 25] Gambardella et al. [15] Stützle et al. [36]
	Scheduling Problems	Colorni et al. [9] Socha et al. [35] Forsyth et al. [14]
	Telecommunications	Schoonderwoerd et al. [34] Di Caro et al. [7]

2.2.2 Data Mining

Over the past decades we have witnessed an explosion of data. Although much information is available in this data, it is hidden in the vast collection of raw data. Data mining entails the overall process of extracting knowledge from this data and addresses the rightly expressed concern of Naisbitt [27]:

> *"We are drowning in information but starving for knowledge."*
> *– John Naisbitt*

Different types of data mining are discussed in the literature [2], such as regression, classification and clustering. The task of interest here is classification, which is the task of assigning a datapoint to a predefined class or group according to its predictive characteristics. The classification problem and accompanying data mining techniques are relevant in a wide variety of domains such as financial engineering, medical diagnostic and marketing. The result of a classification technique is a model which makes it possible to classify future data points based on a set of specific characteristics in an automated way. In the literature, there is a myriad of different techniques proposed for this classification task, some of the most commonly used being C4.5, logistic regression, linear and quadratic discriminant analysis, k-nearest neighbor, artificial neural networks and support vector machines [19].

The performance of the classifier is typically determined by its accuracy on an independent test set. Benchmarking studies [3] have shown that the non-linear classifiers generated by neural networks and support vector machines score best on this performance measure. However, comprehensibility can be a key requirement as well, demanding that the user can understand the motivations behind the model's prediction. In some domains, such as credit scoring and medical diagnostic, the lack of comprehensibility is a major issue and causes a reluctance to use the classifier or even complete rejection of the model. In a credit scoring context, when credit has been denied the Equal Credit Opportunity Act of the U.S. requires that the financial institution provides specific reasons why the customer's application was rejected, whereby vague reasons for denial are illegal. In the medical diagnostic domain as well, clarity and explainability are major constraints. The most suited classifiers for this type of problem are of course rules and trees. C4.5 is one of the techniques that constructs such comprehensible classifiers, but other techniques, such as rule extraction from neural network and support vector machine classifiers, have been proposed as well [4].

Our approach focuses on building accurate, though comprehensible classifiers, fit for dynamic, distributed environments.

2.2.3 Data Mining with Ant Systems

The first application of ant systems for data mining was reported in [28], where the authors introduce the AntMiner algorithm for the discovery of classification rules. Extensions and optimizations of the AntMiner are described in AntMiner2 and AntMiner3 [22]. The aim is to extract simple **if-then-else** rules from data, where the condition part of the rule is a conjunction of terms. All attributes are assumed to be categorical since the terms are of the form $< Variable = Value >$, e.g. $< Sex = male >$.

The original AntMiner works as follows. Each ant starts with an empty rule and chooses a term $< V_i = Value_k >$ to add to it's rule. The choice of the term to add is dependent on the pheromone function ($\tau(t)$) and the heuristic value (η) associated with each term. This choice is furthermore constrained since each variable can occur at most once in a rule to avoid inconsistencies such as $< Sex = male >$ **and** $< Sex = female >$. The ant keeps adding terms to its partial rule until either all variables have been used in the rule or if adding any term would make the rule cover less cases than a user-defined minimum. The class predicted by this rule is determined by the majority class of the training cases covered by the rule. Afterwards the rule is pruned in order to remove irrelevant terms and the pheromone levels are adjusted, increasing the pheromone of the trail followed by the ant and evaporating the others. Another ant starts its search with the new pheromone trails to guide its search. This process is repeated until all ants have constructed a rule or when ants have already converged to the same constructed rule. The best rule among these constructed rules is added to the list of discovered rules and the training cases covered by this rule are removed from the training set. This overall process is repeated until the number of uncovered training cases is lower than a specific threshold.

The heuristic value in AntMiner is defined as an information theoretic measure in terms of the entropy, which can be seen as an impurity measure. AntMiner2 on the other hand uses a much simpler, though less accurate, density estimation equation as the heuristic value with the assumption that the small induced errors are compensated by the pheromone level. This makes AntMiner2 computationally less expensive without a degradation of the performance. Two key changes have been proposed in AntMiner3 [22], resulting in an increased accuracy. A different update rule is used and more exploration is incorporated with a different transition rule that increases the probability of choosing terms not yet used in previously constructed rules.

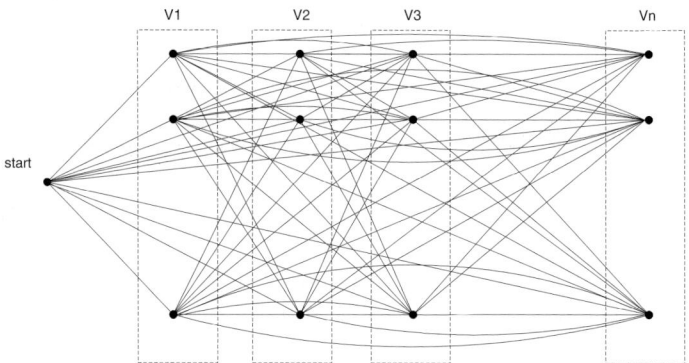

Fig. 2.1. Construction Graph of AntMiner

In these AntMiner versions, an ant can add terms corresponding to any of the variables that are not yet present in the partially constructed rule, with any of its values. This approach is illustrated in Fig. 2.1 which shows a graph representation of the ants' environment. Each 'column' or node group corresponds to a variable and every 'row' corresponds to a value. Each ant going to node $n_{i,k}$ (node in column i and row k) adds the term $< V_i = Value_k >$ to its rule. All ants begin in the start node and then start adding terms by walking through the construction graph representing the problem domain. As shown by Fig. 2.1, the complexity of the construction graph, measured by the number of edges, is $O(\frac{avg^2}{2} \cdot n^2)$ with n the number of variables and avg the average number of values per variable.

$$n \cdot avg + (n-1) \cdot avg^2 + (n-2) \cdot avg^2 + \ldots + avg^2 \approx avg^2 \cdot \frac{n(n+1)}{2} \quad (2.1)$$

2.3 AntMiner+

We build further on the work introduced in the previous AntMiner versions and try to resolve some issues. First of all, we define the environment as a directed acyclic construction graph which allows a clear representation of the problem domain and considerably improves the performance of the ant system. Furthermore, we introduce the better performing $\mathcal{MAX}$-$\mathcal{MIN}$ ant system for mining rules. To the best of our knowledge, there is no use of the $\mathcal{MAX}$-$\mathcal{MIN}$ ant system technique for the discovery of classification rules.

The main working of our algorithm is described in pseudo-code 2.3.1 below. First, a directed acyclic construction graph is created that acts as the environment of the ants. All ants begin in the start node and walk through their environment to the end node, gradually constructing a rule. Only the ant that describes the best rule, i.e. covers the most training points, will have the pheromone of its followed trail increased. Evaporation decreases the pheromone of all edges. Supplementary modifications of the pheromone levels may be needed since the $\mathcal{MAX}$-$\mathcal{MIN}$ approach additionally requires the pheromone levels to lie within a given interval. Since the probabilities are the same for all ants in the same iteration, these values are calculated in advance. When all the edges of one path have a pheromone level τ_{max} and all others edges have pheromone level τ_{min}, the rule corresponding to the path with τ_{max} will be extracted and training data covered by this rule removed from the training set. This iterative process will be repeated until enough training points have been covered or when early stopping occurs (cf. Sect. 2.3.5). Details of AntMiner+ are provided in the next sections.

Pseudo-code 2.3.1 AntMiner+

```
construct graph
while (not min. percentage of training data covered or early stopping)
   initialize heuristics, pheromones and probabilities of edges
   while (not converged)
      create ants
      let ants run from source to sink
      evaporize edges
      update path of best ant
      adjust pheromone levels if outside boundaries
      kill ants
      update probabilities of edges
   end
   extract rule
   flag the data points covered by the extracted rule
end
evaluate performance on test set
```

2.3.1 The Construction Graph

The AntMiner+ construction graph is defined as a simple DAG which provides a comprehensible view of the solution space. Ants in a node of variable V_i are only allowed to go to nodes of variable V_{i+1}. Consequently, each path from start to end node represents a rule. Similarly as before, each ant going from node $n_{i,j}$ to node $n_{i+1,k}$ adds the term $< V_{i+1} = Value_k >$ to its rule. Since binary classification is performed, at the end node the rule consequent $< class = 1 >$ is added to the rule. So during the walk from start to stop, an ant gradually constructs a complete rule. To allow for rules where not all variables are involved, hence shorter rules, an extra *dummy node* is added to each variable whose value is undetermined, meaning it can take any of the values available. This fits well in the construction graph and makes the need for pruning superfluous.

Although only categorical variables can be used in our implementation, we make a distinction between nominal and ordinal variables. Each nominal variable has one node group, but for the ordinal however, we build two node groups to allow for intervals to be chosen by the ants. The first node group corresponds to the lower bound of the interval and should thus be interpreted as $< V_{i+1} \geq Value_k >$, the second node group determines the upper bound, giving $< V_{i+2} \leq Value_l >$. This allows to have less, shorter and actually better rules. Note that in the ordinal case V_{i+1} is equal to V_{i+2}. Figure 2.2 gives a general view of the construction graph with the first variable being nominal and the second one ordinal, hence having two node groups. The complexity of this construction graph is $O(n \cdot avg^2)$, far below the complexity of the construction graph defined by previous

AntMiner versions[5]. The lower complexity of the AntMiner+ construction graph reduces the number of probability computations and makes the best possible term to add more obvious.

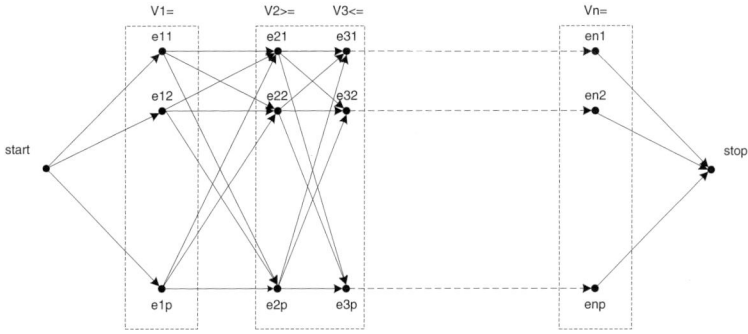

Fig. 2.2. Construction Graph of AntMiner+

2.3.2 Edge Probabilities

The probability $P_{(n_{i,j},n_{i+1,k})}$ is the probability that an ant which is in node $n_{i,j}$ (node for which variable V_i is equal to its jth value) will go to node $n_{i+1,k}$. This probability is commonly defined as follows, with $|condition|$ denoting the number of datapoints fulfilling $condition$:

$$P_{(n_{i,j},n_{i+1,k})} = \frac{[\tau_{(n_{i,j},n_{i+1,k})}]^\alpha \cdot [\eta_{n_{i+1,k}}]^\beta}{\sum_{l=1}^{|V_{i+1}|}[\tau_{(n_{i,j},n_{i+1,l})}]^\alpha \cdot [\eta_{n_{i+1,l}}]^\beta} \quad (2.2)$$

Notice that this probability is dependent on two values: the heuristic value η and the pheromone value τ. The relative weights of these values are determined by the α and β parameters.

2.3.3 Heuristic Value

The heuristic value gives for each node in the construction graph a notion of its quality in the problem domain. For data mining, the quality is usually measured by the number of data points that are covered (described) by a value. Since we extract rules for the $class = 1$ case, we define the heuristic value for the node $n_{i,k}$ as follows:

---

[5] Note that n is now a value slightly higher than the number of variables since ordinal variables have two node groups.

$$\eta_{n_{i,k}} = \frac{|V_i = Value_k \ \& \ class = 1|}{|V_i = Value_k|} \quad (2.3)$$

As mentioned before, more complex and accurate formulas can be provided here, e.g. information entropy measures. This extra information however is computationally demanding and is already included (implicitly) in the pheromone value: small potential errors in the heuristic value will be compensated by the pheromone levels. The same argument goes for the fact that we do not take into account the history of the ant in the heuristic value, that is we do not look at how many data points are already covered by the rule so far. This implies that the sequence of the variables in the construction graph is irrelevant.

2.3.4 Pheromone Updating

Generally, updating the pheromone trail of an ant system is accomplished in two phases, being evaporation and reinforcement. Applying the ideas of $\mathcal{MAX\text{-}MIN}$ ant systems has direct consequences for the pheromone updating rule in the following way:

Evaporation

Evaporation in an ant system is accomplished by diminishing the pheromone level of each trail according to the following rule:

$$\tau_{(n_{i,j}, n_{i+1,k})}(t+1) = \rho \cdot \tau_{(n_{i,j}, n_{i+1,k})}(t), \quad (2.4)$$

where ρ is the evaporation factor. Typical values for ρ lie in the range $[0.8, 0.99]$ [41].

Reinforcement

In a $\mathcal{MAX\text{-}MIN}$ ant system reinforcement of the pheromone trail is only applied to the best ant's path. The best ant can be chosen as either the iteration best ant, or the global best ant. Results described in the literature bias our choice towards the iteration best ant [41]. This means that, taking into account the evaporation factor as well, the update rule for the best ant's path can be described by:

$$\tau_{(n_{i,j}, n_{i+1,k})}(t+1) = \rho \cdot \tau_{(n_{i,j}, n_{i+1,k})}(t) + \Delta^{best} \quad (2.5)$$

Clearly, the reinforcement of the best ant's path, Δ^{best}, should be proportional to the quality of the path. For data mining, we define the quality of a rule by the sum of its *confidence* and its *coverage*. Confidence is an indication of the number of correctly classified data points by a rule compared to the total number of data points covered by that rule. The coverage gives an indication of the overall importance of the specific rule by looking at the number of correctly classified data points over the total number of uncovered data points. More formally, the pheromone amount to add to the path of the iteration best ant is given by the benefit of the path of the iteration best ant, as indicated by 2.6, with $rule_{ib-ant}$ being the rule described by the iteration best ant, and Cov a boolean expressing whether a datapoint is already covered by one of the extracted rules or not.

$$\Delta^{best} = \underbrace{\frac{|rule_{ib-ant} \ \& \ Class = 1|}{|rule_{ib-ant}|}}_{\text{confidence}} + \underbrace{\frac{|rule_{ib-ant} \ \& \ Class = 1|}{|Cov = 0|}}_{\text{coverage}} \qquad (2.6)$$

Let us consider a dataset with 1000 data points and 2 ants extracting rules. Ant 1 describes rule R_1 that covers 100 data points, of which 90 are correctly described. Ant 2 extracts rule R_2, that covers 300 data points, of which 240 are correctly classified. As Table 2.2 shows rule R_2 has a higher quality and therefore ant 2 will be the iteration best ant.

Table 2.2. Example on calculation of rule quality

Rule	Confidence	Coverage	Quality
R_1	$\frac{90}{100} = 0.9$	$\frac{100}{1000} = 0.1$	1.0
R_2	$\frac{240}{300} = 0.8$	$\frac{300}{1000} = 0.3$	1.1

τ boundaries

An additional restriction imposed by the $\mathcal{MAX\text{-}MIN}$ ant systems is that the pheromone level of the edges is restricted by an upper-bound (τ_{max}) and a lower-bound (τ_{min}). Furthermore, these bounds are dynamically altered during the execution of the ant system, so the values converge to the exact maximum value of τ_{max}. Every time an *iteration best* ant improves the results of the current *global best* ant, the boundaries should be updated in the following way [41]:

$$\tau_{max} = \frac{1}{1-\rho} \cdot \Delta^{best} \qquad (2.7)$$

The lower-bound (τ_{min}) for the pheromone level is derived from the probability that the best solution is found upon convergence p_{best}, the maximum pheromone level τ_{max} and the average number of values per variable. The exact derivation is given in [41].

$$\tau_{min} = \frac{\tau_{max} \cdot (1 - \sqrt[n]{p_{best}})}{(avg - 1) \cdot \sqrt[n]{p_{best}}} \qquad (2.8)$$

2.3.5 Early Stopping

AntMiner+ stops constructing rules when either a predefined percentage of training points has been covered by the inferred rules or when early stopping occurs. The last criterion is often used in data mining and is explained in more detail.

The ultimate goal of AntMiner+ is to produce a model which performs well on new, unseen test instances. If this is the case, we say that the classifier generalises well. To do so, we basically have to avoid the classifier from fitting the noise or idiosyncrasies in the training data. This can be realized by monitoring the error on a separate validation set during the training session. When the error measure on the validation set starts to increase, training is stopped, thus effectively preventing the rule base from fitting the noise in the training data. This stopping technique is known as *early stopping*. Note that this causes loss of data that cannot be used for constructing the rules and hence, this method may not be appropriate for small data sets.

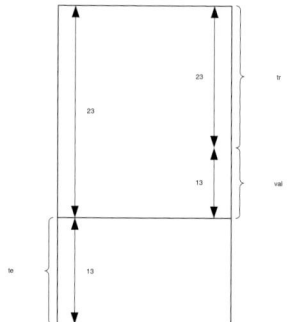

Fig. 2.3. Dataset splitup

The dataset is split up in three sets, as indicated in Fig. 2.3: a training set used by AntMiner+ to infer rules from, a validation set to implement the early stopping rule, and a test set to calculate an unbiased performance estimate. As a rule of thumb, two third of the complete dataset is typically used for training and validation and the remaining third for testing. Of the data points set aside for training and validation, two third is training data and one third validation data. A typical plot of the accuracy for the three types of data is shown in Fig. 2.4. The data used here is the german credit scoring dataset described further on in this chapter. As the figure shows, early stopping makes sure the rules generalise well. Going beyond the induced stop increases the training accuracy, yet lowers the validation (and also test) accuracy because overfitting occurs.

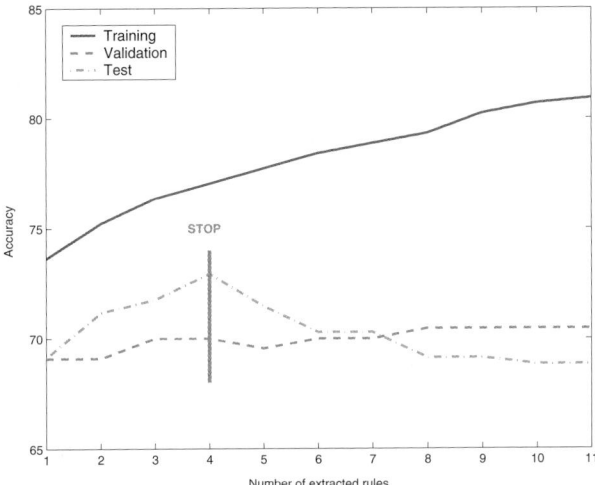

Fig. 2.4. Illustration of the early stopping rule

2.4 Distributed Data Mining With AntMiner+: a Credit Scoring Case

To demonstrate the possible use of AntMiner+ in a dynamic, distributed environment we developed the following credit scoring case, see Fig. 2.5. This case has not yet been put in practice and should only be viewed as an illustration of how to distribute AntMiner+.

One of the key decisions financial institutions have to make is to decide whether or not to grant a loan to an applicant. This basically comes down to a binary classification problem which aims at distinguishing the good payers from the bad payers. Suppose a bank wants to make use of AntMiner+ to build comprehensible classifiers on which to base its credit approval decisions. The bank has some customer data at its disposal such as amount on checking and savings account, possibly distributed over several branch offices. It also has access to data from a credit bureau (such as Experian or Equifax) which consolidates information from a variety of financial institutions. Further available data comes from the Tax and Housing Offices. The externally provided data is continuously updated. A centralized data mining approach would require continuous importing of data across the network, a rather cumbersome task. AntMiner+ on the other hand, sends the ants through the network to the relevant databases. Now the construction graph is dynamic and distributed over different sites, where each database contains a number of node groups. In practice, of course, support for the use of AntMiner+, common customer IDs and routing knowledge needs to be incorporated.

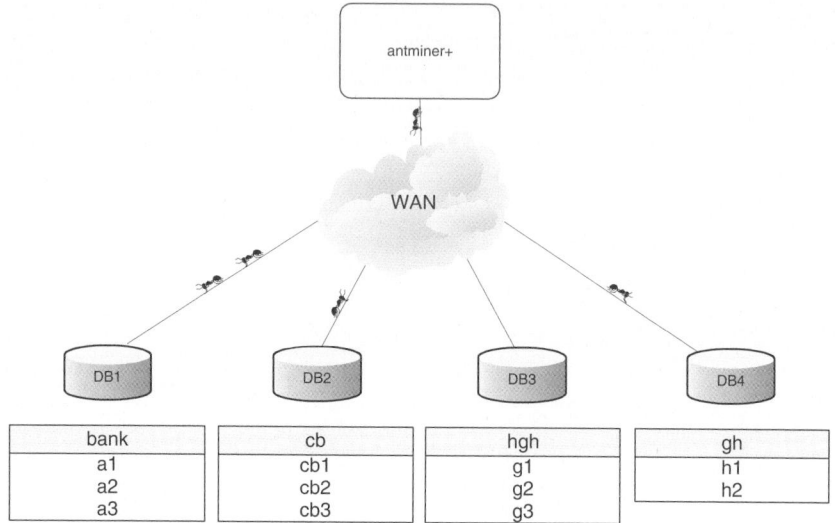

Fig. 2.5. AntMiner+ in distributed environment

The decentralized approach described here which operates in a dynamic, distributed environment allows to detect new customer default profiles quickly, is easily scalable, and robust to failure: ants going lost over the network, or even losing the connection to databases do not severely degrade the performance and AntMiner+ will still be able to work properly.

2.5 Experiments and Results

2.5.1 Experimental Set-Up

We applied AntMiner+ to several publicly available datasets. Training, validation and test set are determined in the manner described before. To eliminate any chance of having unusually good or bad training and test sets, 10 runs are conducted where the data is first randomized before the training, validation and test set are chosen. Experiments are conducted with 1000 ants, the evaporation rate ρ set at 0.85, and the relative weights α and β set at respectively 2 and 1.

A wide range of techniques are chosen to benchmark the AntMiner+ classifier against. C4.5 is the popular decision tree builder [34] where each leaf assigns class labels to observations. Each of these leaves can be represented by a rule; therefore C4.5 is also a truly comprehensible classifier. K-nearest neighbor classifiers (kNN) classify a data instance by considering only the k most similar data points in the training set. We apply kNN to our datasets for k = 1 and k = 10. This technique is called lazy since it does not involve creating a classification model, but rather defers the decisions on how to classify new data points beyond the training session. For both C4.5 and kNN we use the Weka workbench [34]. Also included is the commonly used logistic regression (logit) classifier and the non-linear support vector machine (SVM) classifier [44]. We take the Least Squares SVM variant [43] with RBF kernel using the LS-SVM Matlab toolbox [42].

2.5.2 Datasets

Different datasets are chosen to infer rules from by AntMiner+. Two datasets concern medical diagnosis of breast cancer and one dataset concerns credit scoring; these datasets are chosen for the critical importance of comprehensibility of the generated classifiers. Other datasets used concern the tic-tac-toe toy problem and the two-dimensional ripley dataset which allows for visualization of the extracted rules. Both the tic-tac-toe, the credit scoring and the breast cancer datasets come from the publicly available UCI data repository [21].

Breast Cancer Diagnosis

The task at hand for the breast cancer diagnosis datasets consists of classifying breast masses as being either benign or malignant. For this, attributes of a sample are listed that are deemed relevant. Two different datasets are used, one obtained from the University of Wisconsin Hospitals, Madison from Dr. William H. Wolberg [23] and the other from the University Medical Centre, Institute of Oncology, Ljubljana, Yugoslavia and is provided by M. Zwitter and M. Soklic [26]. This is one of the domains where comprehensibility is a key issue and thus rules are very much preferred. The following rule, with 97.6% training accuracy and 95.7% test accuracy, was extracted by AntMiner+ for the Breast Cancer Wisconsin dataset:

Table 2.3. Example rule on Breast Cancer Wisconsin dataset

if (Clump Thickness $\in [1,8]$ **and** Uniformity of Cell Size $\in [1,9]$ **and** Uniformity of Cell Shape $\in [1,8]$ **and** Marginal Adhesion $\in [1,6]$ **and** Single Epithelial Cell Size $\in [1,9]$ **and** Bare Nuclei $\in [1,5]$ **and** Bland Chromatin $\in [1,8]$ **and** Normal Nucleoli $\in [1,9]$ **and** Mitoses $\in [1,2]$)
then class = benign
else class = malignant

Credit Scoring

As mentioned before, credit scoring involves the discrimination between good payers and bad ones. We used the german credit scoring dataset offered by prof.

Hofmann. An AntMiner+ example rule set is described in Table 2.4 which has a training accuracy of 75.5% and a test accuracy of 72.1%.

Table 2.4. Example rule set on German Credit Scoring Dataset

if (Checking Account < 200 DM **and** Duration > 15 m **and**
     Credit History = no credits taken **and** Savings Account < 1000 DM)
 then class = bad
 else if (Purpose = new car/repairs/education/others **and**
     Credit History = no credits taken/all credits paid back duly at this bank **and**
     Savings Account < 1000 DM)
 then class = bad
 else if (Checking Account < 0 DM **and**
     Purpose = furniture/domestic appliances/business **and**
     Credit History = no credits taken/all credits paid back duly at this bank **and**
     Savings Account < 500 DM)
 then class = bad
 else if (Checking Account < 0 DM **and** Duration > 15 m **and**
     Credit History = delay in paying off in the past **and**
     Savings Account < 500 DM)
 then class = bad
 else class = good

Toy Problems

Also included in our experiments is the *tic-tac-toe* dataset, which encodes the complete set of possible board configurations at the end of tic-tac-toe games where X is assumed to have played first. The target concept is 'win for X' (i.e., true when X has one of 8 possible ways to create a 'three-in-a-row'). The extracted rules can easily be verified in Table 2.5 that shows the board of the game and the 9 variables which can take value X, O or B (blank). An example of a rule base extracted by AntMiner+ is provided in Table 2.6.

Table 2.5. tic-tac-toe game

A1	A2	A3
A4	A5	A6
A7	A8	A9

Table 2.6. Example rule on tic-tac-toe dataset

if (A7 = 1 **and** A8 = 1 **and** A9 = 1) **then** class = X
else if (A3 = 1 **and** A5 = 1 **and** A7 = 1) **then** class = X
else if (A1 = 1 **and** A5 = 1 **and** A9 = 1) **then** class = X
else if (A3 = 1 **and** A6 = 1 **and** A9 = 1) **then** class = X
else if (A4 = 1 **and** A5 = 1 **and** A6 = 1) **then** class = X
else if (A2 = 1 **and** A5 = 1 **and** A8 = 1) **then** class = X
else if (A1 = 1 **and** A4 = 1 **and** A7 = 1) **then** class = X
else if (A1 = 1 **and** A2 = 1 **and** A3 = 1) **then** class = X
else class = O

Ripley's dataset [32] has two variables and two classes, where the classes are drawn from two normal distributions with a high degree of overlap. This two-dimensional dataset allows for visualization of the classifiers. Since AntMiner+ can only deal with categorical variables, the continuous values of the two variables are divided into 50 intervals of equal length. The Ripley dataset is shown in Fig. 2.6, together with the decision boundary defined by the rules extracted by AntMiner+ (accuracy 90.8%), and a support vector machine model (accuracy 91.4%).

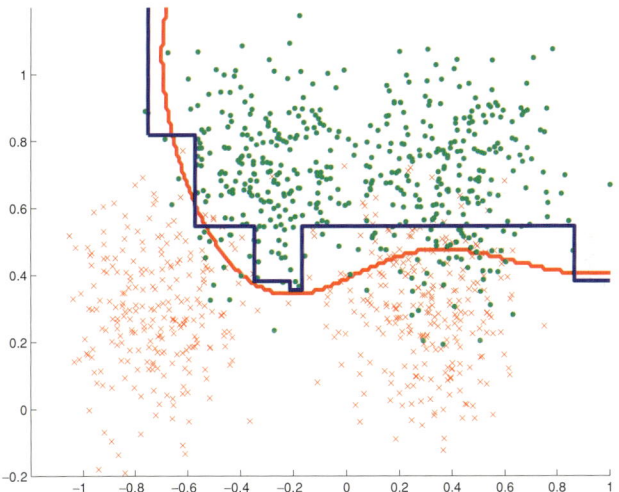

Fig. 2.6. AntMiner rules (**❙**) and SVM decision boundary (**❙**) for Ripley's dataset

2.5.3 Software Implementation

AntMiner+ is implemented in the platform-independent, object-oriented Java programming environment, with usage of the MySQL open source database server. Several screenshots of the Graphical User Interface (GUI) of AntMiner+, applied to the breast cancer wisconsin and tic-tac-toe datasets, are provided in Fig. 2.7. The GUI shows the construction graph with the width of the edges being proportional to their pheromone value. Extracted rules, with their training, validation and test accuracy, are displayed in the bottom box.

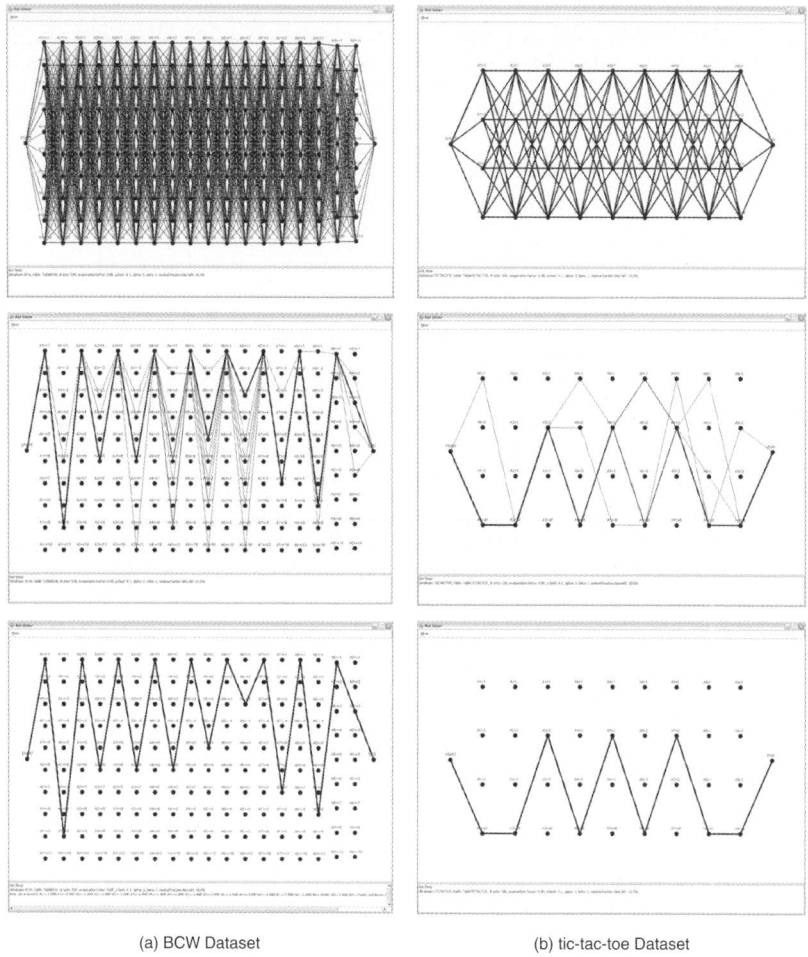

(a) BCW Dataset   (b) tic-tac-toe Dataset

Fig. 2.7. Screenshots of AntMiner+ during different stages of execution: from initialization (top) to convergence (bottom)

2.5.4 Discussion

The results of our experiments are shown in Table 2.7. For each dataset, the number of data instances (inst) and attributes (attr) as well as the accuracy and number of generated rules are displayed. The techniques are ranked according to their accuracy and the average ranking (AR) over the different datasets of each technique is included in the table, hence a low AR indicates good performance. The best average test set performance over 10 runs is underlined and denoted in bold face for each data set. We then use a paired t-test to test the performance differences. Performances that are not significantly different at the 5% level from the top performance with respect to a one-tailed paired t-test are tabulated in bold face. Statistically significant underperformances at the 1% level are emphasized in italics. Performances significantly different at the 5% level but not at the 1% level are reported in normal script. Results published for AntMiner1 and 3, reported in [29, 22], are also listed in the table but not included in the comparison as just described since we did not conduct experiments with these AntMiner versions ourselves.

Table 2.7. Average out-of-sample performance

		bcw		bcl		ger		ttt		rip	
		inst	attr	inst	attr	inst	attr	inst	attr	inst	attr
		683	9	277	9	1000	19	958	9	1250	2
Technique	AR	Acc	#R	Acc	#R	Acc	#R	Acc	#R	Acc	#R
AntMiner		92.63	10.1	75.28	7.1			70.99	16.5		
AntMiner3		94.32	13.2					76.58	18.6		
AntMiner+	3.2	95.79	1	**77.05**	3.9	72.29	3.9	**99.76**	8	89.41	3.9
C4.5	4.6	**94.38**	11	75.68	21	**72.91**	36	*84.17*	95	89.08	6
1NN	4.4	95.84		*74.74*		72.48		97.98		88.66	
10NN	1.8	96.48		<u>**78.42**</u>		74.26		95.18		<u>**90.80**</u>	
SVM	3.8	92.81		76.56		73.68		91.06		89.78	
logit	3.2	**96.54**		76.77		<u>**75.24**</u>		*65.56*		88.92	

The best performance is achieved by 10NN with an average ranking of 1.8. However, the nearest neighbor techniques are lazy in the sense that there is no actual classifier. Comprehensibility of such decisions, based on the similarity with training data, is limited. The SVM models perform consistently well, but the non-linear, complex nature of the generated classifiers makes them rather incomprehensible for humans. Logistic regression achieves good results as well but is troubled with similar opacity issues. Equations 2.9 and 2.10 describe the form of respectively the SVM and logistic regression classifiers and clearly indicate the opacity of these models.

$$y_{SVM}(\mathbf{x}) = \text{sign}[\sum_{i=1}^{N} \alpha_i y_i \exp\{-\|\mathbf{x}-\mathbf{x}_i\|_2^2/\sigma^2\} + b] \qquad (2.9)$$

$$y_{logit}(\mathbf{x}) = 1/(1+\exp\{-(w_0+\mathbf{w}^T\mathbf{x})\}) \qquad (2.10)$$

The only techniques that deal with the comprehensibility aspect are C4.5 and Antminer+. With an overall average ranking of 3.2 AntMiner+ holds a top three place among the included state-of-the-art classification techniques. AntMiner+ outperforms C4.5 on all but one dataset and consistently does so with fewer rules, making AntMiner+ the best performing technique when considering both accuracy and comprehensibility.

The better results can be attributed to our $\mathcal{MAX}$-$\mathcal{MIN}$ approach, our simple construction graph with the inclusion of dummy nodes, as well as our ability to include intervals in our rules. The $\mathcal{MAX}$-$\mathcal{MIN}$ ant system is better able to combine exploration of the search space and exploitation of the best solutions found, and has been shown to perform better than the usual ant system [9, 37, 41]. The construction graph modeled as a DAG reduces the complexity of the problem, yet the presence of dummy nodes enables AntMiner+ to infer short rules. Surely, the intervals play a crucial role as well in attaining fewer and shorter rules. This is best demonstrated with the breast cancer wisconsin dataset. The only possible way to achieve an accurate classifier with only 1 rule is when intervals are allowed. Several weaknesses need to be kept in mind as well. AntMiner+ requires more computational time than C4.5 to achieve its results[6] and is only able to deal with categorical variables. Parallelization of the inherently distributed AntMiner+ system could decrease the computation time needed.

2.6 Conclusion and Future Research

AntMiner+ is a technique that successfully incorporates swarm intelligence in data mining. Using a $\mathcal{MAX}$-$\mathcal{MIN}$ system, AntMiner+ builds comprehensible, accurate rule-based classifiers that perform competitively with state-of-the-art classification techniques. Although ants have a limited memory and perform actions based on local information only, the ants come to complex behavior due to self-organization and indirect communication. The intrinsic properties of ant systems allow us to easily distribute our approach to provide robust and scalable data mining solutions.

Still, several challenges lie ahead. Most real-life datasets contain various continuous variables. One approach to deal with this is by categorizing these variables in a pre-processing step. Incorporating the variables in a dynamically

---

[6]Time order of minutes up to one hour on Xeon 2.4GHz, 1GB RAM for the datasets analyzed.

changing construction graph is another option and focus of current research. An issue faced by ant systems, is the necessity to instantiate various parameters, such as the weight parameters α and β. These parameters are typically determined by trial and error or with more fancy techniques such as genetic algorithms or local search techniques. We are currently investigating the possibility of including these two variables in the construction graph, which would have the supplementary benefit of generating dynamically changing parameters as the environment changes. Once again, the ants will take over the work from the user.

Acknowledgment

We would like to thank the Flemish Research Council (FWO, Grant G.0615.05), and the Microsoft and KBC-Vlekho-K.U.Leuven Research Chairs for financial support to the authors.

References

1. Abraham A, Ramos V (2003) Web usage mining using artificial ant colony clustering. In Proceedings of Congress on Evolutionary Computation (CEC2003), Australia, IEEE Press, ISBN 0780378040, 1384-1391
2. Baesens B (2003) Developing intelligent systems for credit scoring using machine learning techniques. PhD thesis, K.U.Leuven
3. Baesens B, Van Gestel T, Viaene S, Stepanova M, Suykens J, Vanthienen J (2003) Benchmarking state-of-the-art classification algorithms for credit scoring. Journal of the Operational Research Society, 54(6):627–635
4. Baesens B, Setiono R, Mues C, Vanthienen J (2003) Using neural network rule extraction and decision tables for credit-risk evaluation. Management Science, 49(3):312–329
5. Bonabeau E, Dorigo M, Theraulaz G (2001) Swarm intelligence: From natural to artificial systems. Journal of Artificial Societies and Social Simulation, 4(1)
6. Bullnheimer B, Hartl RF, Strauss C (1999) Applying the ant system to the vehicle routing problem. In: Osman IH, Roucairol C, Voss S, Martello S (eds) Meta-Heuristics: Advances and Trends in Local Search Paradigms for Optimization
7. Di Caro G, Dorigo M (1998) Antnet: Distributed stigmergetic control for communications networks. Journal of Artificial Intelligence Research, 9:317–365
8. Cicirello VA, Smith SF (2001) Ant colony control for autonomous decentralized shop floor routing. In: the Fifth International Symposium on Autonomous Decentralized Systems, pages 383–390
9. Colorni A, Dorigo M, Maniezzo V, Trubian M (1994) Ant system for jobshop scheduling. Journal of Operations Research, Statistics and Computer Science, 34(1):39–53
10. Dorigo M Ant colony optimization [http://iridia.ulb.ac.be/ mdorigo/aco/aco.html].

11. Dorigo M, Maniezzo V, Colorni A (1991) Positive feedback as a search strategy. Technical Report 91016, Dipartimento di Elettronica e Informatica, Politecnico di Milano, IT
12. Dorigo M, Maniezzo V, Colorni A (1996) The Ant System: Optimization by a colony of cooperating agents. IEEE Transactions on Systems, Man, and Cybernetics Part B: Cybernetics, 26(1):29–41
13. Eyckelhof CJ, Snoek M (2002) Ant systems for a dynamic tsp. In: ANTS '02: Proceedings of the Third International Workshop on Ant Algorithms, pages 88–99, London, UK. Springer-Verlag
14. Forsyth P, Wren A (1997) An ant system for bus driver scheduling. Research Report 97.25, University of Leeds School of Computer Studies
15. Gambardella LM, Taillard E, Dorigo M (1999) Ant colonies for the quadratic assignment problem. Journal of the Operational Research Society, (50):167–176
16. Gambardella LM, Dorigo M (1995) Ant-q: A reinforcement learning approach to the traveling salesman problem. In: Proceedings of the Eleventh International Conference on Machine Learning, pages 252–260
17. Gambardella LM, Dorigo M (1996) Solving symmetric and asymmetric tsps by ant colonies. In: Proceedings of the IEEE International Conference on Evolutionary Computation (ICEC'96), pages 622–627
18. Grassé PP (1959) La reconstruction du nid et les coordination inter-individuelles chez bellicositermes natalensis et cubitermes sp. la thérie de la stigmergie: Essai d'interprétation du comportement des termites constructeurs. Insect. Soc., 6:41–80
19. Hand D (2002) Pattern detection and discovery. In: Hand D, Adams N, Bolton R (eds) Pattern Detection and Discovery, volume 2447 of Lecture Notes in Computer Science, pages 1–12. Springer
20. Handl J, Knowles J, Dorigo M (2003) Ant-based clustering: a comparative study of its relative performance with respect to k-means, average link and 1d-som. Technical Report TR/IRIDIA/2003-24, Universite Libre de Bruxelles
21. Hettich S, Bay SD (1996) The uci kdd archive [http://kdd.ics.uci.edu]
22. Liu B, Abbass HA, McKay B (2004) Classification rule discovery with ant colony optimization. IEEE Computational Intelligence Bulletin, 3(1):31–35
23. Mangasarian OL, Wolberg WH (1990) Cancer diagnosis via linear programming. SIAM News, 23(5):1–18
24. Maniezzo V (1998) Exact and approximate nondeterministic tree-search procedures for the quadratic assignment problem. Research CSR 98-1, Scienze dell'Informazione, Università di Bologna, Sede di Cesena, Italy
25. Maniezzo V, Colorni A (1999) The ant system applied to the quadratic assignment problem. IEEE Transactions on Knowledge and Data Engineering
26. Michalski RS, Mozetic I, Hong J, Lavrac N (1986) The multi-purpose incremental learning system aq15 and its testing application to three medical domains. In: AAAI, pages 1041–1047
27. Naisbitt J (1988) Megatrends : Ten New Directions Transforming Our Lives. Warner Books
28. Parpinelli RS, Lopes HS, Freitas AA (2001) An ant colony based system for data mining: Applications to medical data. In: Lee Spector, Goodman E, Wu A, Langdon WB, Voigt H, Gen M, Sen S, Dorigo M, Pezeshk S, Garzon M, Burke E (eds) Proceedings of the Genetic and Evolutionary Computation Conference (GECCO-2001), pages 791–797, San Francisco, California, USA, 7-11. Morgan Kaufmann

29. Parpinelli RS, Lopes HS, Freitas AA (2002) Data mining with an ant colony optimization algorithm. IEEE Transactions on Evolutionary Computation, 6(4):321–332
30. Quinlan JR (1993) C4.5: Programs for Machine Learning. Morgan Kaufmann Publishers Inc., San Francisco, CA, USA
31. Ramos V, Abraham A (2003) Swarms on continuous data. In: Proceedings of the Congress on Evolutionary Computation, IEEE Press, pages 1370–1375
32. Ripley BD (1994) Neural networks and related methods for classification. Journal of the Royal Statistical Society B, 56:409–456
33. Schockaert S, De Cock M, Cornelis C, Kerre EE (2004) Efficient clustering with fuzzy ants. Applied Computational Intelligence
34. Schoonderwoerd R, Holland OE, Bruten JL, Rothkrantz LJM (1996) Ant-based load balancing in telecommunications networks. Adaptive Behavior, (2):169–207
35. Socha K, Knowles J, Sampels M (2002) A $\mathcal{MAX}$-$\mathcal{MIN}$ ant system for the university timetabling problem. In: Dorigo M, Di Caro G, Sampels M (eds) Proceedings of ANTS 2002 – Third International Workshop on Ant Algorithms, volume 2463 of Lecture Notes in Computer Science, pages 1–13. Springer-Verlag, Berlin, Germany
36. StützleT, Dorigo M (1999) Aco algorithms for the quadratic assignment problem. In: Dorigo M, Corne D, Glover F (eds) New Ideas in Optimization
37. Stützle T, Hoos HH (1996) Improving the ant-system: A detailed report on the $\mathcal{MAX}$-$\mathcal{MIN}$ ant system. Technical Report AIDA 96-12, FG Intellektik, TU Darmstadt, Germany
38. Stützle T, Hoos HH (1997) The $\mathcal{MAX}$-$\mathcal{MIN}$ ant system and local search for the traveling salesman problem. In: Proceedings of the IEEE International Conference on Evolutionary Computation (ICEC'97), pages 309–314
39. Stützle T, Hoos HH (1998) Improvements on the ant system: Introducing the $\mathcal{MAX}$-$\mathcal{MIN}$ ant system. In: Steele NC, Albrecht RF, Smith GD (eds) Artificial Neural Networks and Genetic Algorithms, pages 245–249
40. Stützle T, Hoos HH (1999) $\mathcal{MAX}$-$\mathcal{MIN}$ ant system and local search for combinatorial optimization problems. In: Osman IH, Voss S, Martello S, Roucairol C (eds) Meta-Heuristics: Advances and Trends in Local Search Paradigms for Optimization, pages 313–329
41. Stützle, Hoos HH (2000) $\mathcal{MAX}$-$\mathcal{MIN}$ ant system. Future Generation Computer Systems, 16(8):889–914
42. Suykens JAK, Van Gestel T, De Brabanter J, De Moor B, Vandewalle J (2002) Least Squares Support Vector Machines. World Scientific, Singapore
43. Suykens JAK, Vandewalle J (1999) Least squares support vector machine classifiers. Neural Process. Lett., 9(3):293–300
44. Vapnik VN (1995) The nature of statistical learning theory. Springer-Verlag, New York, NY, USA
45. Wade A, Salhi S (2004) An ant system algorithm for the mixed vehicle routing problem with backhauls. In: Metaheuristics: computer decision-making, pages 699–719, Norwell, MA, USA, 2004. Kluwer Academic Publishers
46. Witten IH, Frank E (2000) Data mining: practical machine learning tools and techniques with Java implementations. Morgan Kaufmann Publishers Inc., San Francisco, CA, USA

3

Performing Feature Selection with ACO

Richard Jensen

Department of Computer Science, The University of Wales, Aberystwyth, UK
rkj@aber.ac.uk

Summary. The main aim of feature selection is to determine a minimal feature subset from a problem domain while retaining a suitably high accuracy in representing the original features. In real world problems FS is a must due to the abundance of noisy, irrelevant or misleading features. However, current methods are inadequate at finding optimal reductions. This chapter presents a feature selection mechanism based on Ant Colony Optimization in an attempt to combat this. The method is then applied to the problem of finding optimal feature subsets in the fuzzy-rough data reduction process. The present work is applied to two very different challenging tasks, namely web classification and complex systems monitoring.

3.1 Introduction

Many problems in machine learning involve high dimensional descriptions of input features. It might be expected that the inclusion of an increasing number of features would increase the likelihood of including enough information to distinguish between classes. Unfortunately, this is not true if the size of the training dataset does not also increase rapidly with each additional feature included. This is the so-called curse of dimensionality. A high-dimensional dataset increases the chances that a data-mining algorithm will find spurious patterns that are not valid in general. It is therefore not surprising that much research has been carried out on dimensionality reduction [6, 18]. However, existing work tends to destroy the underlying semantics of the features after reduction.

The task of feature selection is to significantly reduce dimensionality by locating minimal subsets of features, at the same time retaining data semantics. The use of rough set theory (RST) [21] to achieve such data reduction has proved very successful. Over the past twenty years, rough set theory has become a topic of great interest to researchers and has been applied to many domains (e.g. classification [8], systems monitoring [29], clustering [12], expert systems [32]). This success is due in part to the following aspects of the theory: only the facts hidden in data are analysed, no additional information about the data is required (such as thresholds or expert knowledge), and it finds a minimal knowledge representation. Given a dataset with discretized attribute values, it is possible to find a subset (termed a *reduct*) of the original attributes using RST that are the most informative; all other attributes can be removed from the dataset with minimal information loss.

However, it is most often the case that the values of attributes may be both crisp and *real-valued*, and this is where traditional rough set theory encounters a problem. It is not possible in the theory to say whether two attribute values are similar and to what extent they are the same; for example, two close values may only differ as a result of noise, but in RST they are considered to be as different as two values of a different order of magnitude.

It is, therefore, desirable to develop these techniques to provide the means of data reduction for crisp and real-value attributed datasets which utilises the extent to which values are similar. This could be achieved through the use of *fuzzy-rough* sets. Fuzzy-rough set theory is an extension of crisp rough set theory, allowing all memberships to take values in the range [0,1]. This permits a higher degree of flexibility compared to the strict requirements of crisp rough sets that only deal with full or zero set membership. They encapsulate the related but distinct concepts of vagueness (for fuzzy sets [37]) and indiscernibility (for rough sets [21]), both of which occur as a result of imprecision, incompleteness and/or uncertainty in knowledge [9].

Ant Colony Optimization (ACO) techniques are based on the behaviour of real ant colonies used to solve discrete optimization problems [2]. These have been successfully applied to a large number of difficult combinatorial problems such as the quadratic assignment and the traveling salesman problems. This method is particularly attractive for feature selection as there seems to be no heuristic that can guide search to the optimal minimal subset (of features) every time. Additionally, it can be the case that ants discover the best feature combinations as they proceed throughout the search space. This chapter investigates how ant colony optimization may be applied to the difficult problem of finding optimal feature subsets, using fuzzy-rough sets, within web classification and systems monitoring programs.

The rest of this chapter is structured as follows. The second section describes the theory of rough sets and particularly focuses on its role as a feature selection tool. The extension to this approach, fuzzy-rough set feature selection, is detailed in the third section. Section 4 introduces the main concepts in ACO and details how this may be applied to the problem of feature selection in general, and fuzzy-rough feature selection in particular. The fifth section describes the experimentation carried out using the crisp ACO-based feature selector. The application of the fuzzy-rough techniques to web content classification and complex system monitoring is detailed in section 6. Section 7 concludes the chapter, and proposes further work in this area.

3.2 Rough Feature Selection

Rough set theory [10, 20, 21] is an extension of conventional set theory that supports approximations in decision making. It possesses many features in common (to a certain extent) with the Dempster-Shafer theory of evidence [30] and fuzzy set theory [35]. The rough set itself is the approximation of a vague concept (set) by a pair of precise concepts, called lower and upper approximations, which are a classification of the domain of interest into disjoint categories. The lower approximation is a description of the domain objects which are known with certainty to belong to the subset of interest, whereas the upper approximation is a description of the objects which possibly belong to the subset.

Rough Set Attribute Reduction (RSAR) [3] provides a filter-based tool by which knowledge may be extracted from a domain in a concise way; retaining the information content whilst reducing the amount of knowledge involved. The main advantage that rough set analysis has is that it requires no additional parameters to operate other than the supplied data [11]. It works by making use of the granularity structure of the data only.

3.2.1 Theoretical Background

Central to RSAR is the concept of indiscernibility. Let $I = (\text{universe}, \mathbb{A})$ be an information system, where universe is a non-empty set of finite objects (the universe) and $\mathbb{A}$ is a non-empty finite set of attributes such that $a : \text{universe} \rightarrow V_a$ for every $a \in \mathbb{A}$. V_a is the set of values that attribute a may take. For a decision table, $\mathbb{A} = \{\mathbb{C} \cup \mathbb{D}\}$ where $\mathbb{C}$ is the set of input features and $\mathbb{D}$ is the set of class indices. Here, a class index $d \in \mathbb{D}$ is itself a variable $d : \text{universe} \rightarrow \{0,1\}$ such that for $a \in \text{universe}, d(a) = 1$ if a has class d and $d(a) = 0$ otherwise.
With any $P \subseteq \mathbb{A}$ there is an associated equivalence relation $IND(P)$:

$$IND(P) = \{(x,y) \in \text{universe}^2 | \forall a \in P, a(x) = a(y)\} \quad (3.1)$$

The partition of universe, generated by $IND(P)$ is denoted universe/$IND(P)$ (or universe/P) and can be calculated as follows:

$$/IND(P) = \otimes \{a \in P : /IND(\{a\})\}, \quad (3.2)$$

where

$$A \otimes B = \{X \cap Y : \forall X \in A, \forall Y \in B, X \cap Y \neq \emptyset\} \quad (3.3)$$

If $(x,y) \in IND(P)$, then x and y are indiscernible by attributes from P. The equivalence classes of the P-indiscernibility relation are denoted $[x]_P$.
Let $X \subseteq$ universe. X can be approximated using only the information contained within P by constructing the P-*lower* and P-*upper* approximations of X:

$$\underline{P}X = \{x \,|\, [x]_P \subseteq X\} \quad (3.4)$$

$$\overline{P}X = \{x \,|\, [x]_P \cap X \neq \emptyset\} \quad (3.5)$$

Let P and Q be equivalence relations over universe, then the positive, negative and boundary regions can be defined as:

$$POS_P(Q) = \bigcup_{X \in \text{universe}/Q} \underline{P}X$$
$$NEG_P(Q) = \text{universe} - \bigcup_{X \in \text{universe}/Q} \overline{P}X$$
$$BND_P(Q) = \bigcup_{X \in \text{universe}/Q} \overline{P}X - \bigcup_{X \in \text{universe}/Q} \underline{P}X$$

The positive region contains all objects of universe that can be classified to classes of universe/Q using the information in attributes P. The boundary region, $BND_P(Q)$, is the set of objects that can possibly, but not certainly, be classified in this way. The negative region, $NEG_P(Q)$, is the set of objects that cannot be classified to classes of universe/Q.

An important issue in data analysis is discovering dependencies between attributes. Intuitively, a set of attributes Q depends totally on a set of attributes P, denoted $P \Rightarrow Q$, if all attribute values from Q are uniquely determined by values of attributes from P. If there exists a functional dependency between values of Q and P, then Q depends totally on P. In rough set theory, dependency is defined in the following way:

For $P, Q \subset \mathbb{A}$, it is said that Q depends on P in a degree k ($0 \le k \le 1$), denoted $P \Rightarrow_k Q$, if

$$k = \gamma_P(Q) = \frac{|POS_P(Q)|}{|\text{universe}|} \quad (3.6)$$

If $k = 1$, Q depends totally on P, if $0 < k < 1$, Q depends partially (in a degree k) on P, and if $k = 0$ then Q does not depend on P.

By calculating the change in dependency when an attribute is removed from the set of considered conditional attributes, a measure of the significance of the attribute can be obtained. The higher the change in dependency, the more significant the attribute is. If the significance is 0, then the attribute is dispensable. More formally, given P, Q and an attribute $a \in P$,

$$\sigma_P(Q, a) = \gamma_P(Q) - \gamma_{P-\{a\}}(Q) \quad (3.7)$$

3.2.2 Reduction Method

The reduction of attributes is achieved by comparing equivalence relations generated by sets of attributes. Attributes are removed so that the reduced set provides the same predictive capability of the decision feature as the original. A *reduct* is defined as a subset of minimal cardinality R_{min} of the conditional attribute set $\mathbb{C}$ such that $\gamma_R(\mathbb{D}) = \gamma_\mathbb{C}(\mathbb{D})$.

$$R = \{X : X \subseteq \mathbb{C}, \gamma_X(\mathbb{D}) = \gamma_\mathbb{C}(\mathbb{D})\} \quad (3.8)$$

$$R_{min} = \{X : X \in R, \forall Y \in R, |X| \le |Y|\} \quad (3.9)$$

The intersection of all the sets in R_{min} is called the *core*, the elements of which are those attributes that cannot be eliminated without introducing more contradictions to the dataset. In RSAR, a subset with minimum cardinality is searched for.

The problem of finding a reduct of an information system has been the subject of much research. The most basic solution to locating such a subset is to simply generate *all* possible subsets and retrieve those with a maximum rough set dependency degree. Obviously, this is an expensive solution to the problem and is only practical for very simple datasets. Most of the time only one reduct is required as, typically, only one subset of features is used to reduce a dataset, so all the calculations involved in discovering the rest are pointless.

To improve the performance of the above method, an element of pruning can be introduced. By noting the cardinality of any pre-discovered reducts, the current possible subset can be ignored if it contains more elements. However, a better approach is needed - one that will avoid wasted computational effort.

QUICKREDUCT($\mathbb{C},\mathbb{D}$).
$\mathbb{C}$, the set of all conditional features;
$\mathbb{D}$, the set of decision features.

(1) $R \leftarrow \{\}$
(2)   **do**
(3) $T \leftarrow R$
(4) $\forall x \in (\mathbb{C} - R)$
(5)   **if** $\gamma_{R \cup \{x\}}(\mathbb{D}) > \gamma_T(\mathbb{D})$
(6) $T \leftarrow R \cup \{x\}$
(7) $R \leftarrow T$
(8)   **until** $\gamma_R(\mathbb{D}) == \gamma_\mathbb{C}(\mathbb{D})$
(9)   **return** R

Fig. 3.1. The QUICKREDUCT Algorithm

The QUICKREDUCT algorithm given in Fig. 3.1 (adapted from [3]), attempts to calculate a reduct without exhaustively generating all possible subsets. It starts off with an empty set and adds in turn, one at a time, those attributes that result in the greatest increase in the rough set dependency metric, until this produces its maximum possible value for the dataset. Other such techniques may be found in [23].

Determining the consistency of the entire dataset is reasonable for most datasets. However, it may be infeasible for very large data, so alternative stopping criteria may have to be used. One such criterion could be to terminate the search when there is no further increase in the dependency measure. This will produce exactly the same path to a reduct due to the monotonicity of the measure [3], without the computational overhead of calculating the dataset consistency.

The QUICKREDUCT algorithm, however, is not guaranteed to find a *minimal* subset as has been shown in [4]. Using the dependency function to discriminate between candidates may lead the search down a non-minimal path. It is impossible to predict which combinations of attributes will lead to an optimal reduct based on changes in dependency with the addition or deletion of single attributes. It does result in a close-to-minimal subset, though, which is still useful in greatly reducing dataset dimensionality. However, when maximal data reductions are required, other search mechanisms must be employed. Although these methods also cannot ensure optimality, they provide a means by which the best feature subsets might be found.

3.3 Fuzzy-Rough Feature Selection

The selection process described previously based on crisp rough sets (RSAR) can only operate effectively with datasets containing discrete values. However, most datasets contain real-valued features and so it is necessary to perform a discretization step beforehand. This is typically implemented by standard fuzzification techniques. As membership degrees of feature values to fuzzy sets are not exploited in the process of dimensionality reduction, important information has been lost. By employing *fuzzy-rough* sets, it is possible to use this information to better guide feature selection.

A fuzzy-rough set is defined by two fuzzy sets, fuzzy lower and upper approximations, obtained by extending the corresponding crisp rough set notions. In the crisp case, elements that belong to the lower approximation (i.e. have a membership of 1) are said to belong to the approximated set with absolute certainty. In the fuzzy-rough case, elements may have a membership in the range [0,1], allowing greater flexibility in handling uncertainty.

3.3.1 Fuzzy Equivalence Classes

Fuzzy equivalence classes [9, 19] are central to the fuzzy-rough set approach in the same way that crisp equivalence classes are central to classical rough sets. For typical applications, this means that the decision values and the conditional values may all be fuzzy. The concept of crisp equivalence classes can be extended by the inclusion of a fuzzy similarity relation S on the universe, which determines the extent to which two elements are similar in S. The usual properties of reflexivity ($\mu_S(x,x) = 1$), symmetry ($\mu_S(x,y) = \mu_S(y,x)$) and transitivity ($\mu_S(x,z) \geq \mu_S(x,y) \wedge \mu_S(y,z)$) hold.

Using the fuzzy similarity relation, the fuzzy equivalence class $[x]_S$ for objects close to x can be defined:

$$\mu_{[x]_S}(y) = \mu_S(x,y) \tag{3.10}$$

The following axioms should hold for a fuzzy equivalence class F:

- $\exists x, \mu_F(x) = 1$
- $\mu_F(x) \wedge \mu_S(x,y) \leq \mu_F(y)$
- $\mu_F(x) \wedge \mu_F(y) \leq \mu_S(x,y)$

The first axiom corresponds to the requirement that an equivalence class is non-empty. The second axiom states that elements in y's neighbourhood are in the equivalence class of y. The final axiom states that any two elements in F are related via the fuzzy similarity relation S. Obviously, this definition degenerates to the normal definition of equivalence classes when S is non-fuzzy. The family of normal fuzzy sets produced by a fuzzy partitioning of the universe of discourse can play the role of fuzzy equivalence classes [9].

3.3.2 Fuzzy Lower and Upper Approximations

The fuzzy lower and upper approximations are fuzzy extensions of their crisp counterparts. Informally, in crisp rough set theory, the lower approximation of a set contains those objects that belong to it with certainty. The upper approximation of a set contains the objects that possibly belong. From the literature, the fuzzy P-lower and P-upper approximations are defined as [9]:

$$\mu_{\underline{P}X}(F_i) = \inf_x \max\{1 - \mu_{F_i}(x), \mu_X(x)\} \quad \forall i \quad (3.11)$$

$$\mu_{\overline{P}X}(F_i) = \sup_x \min\{\mu_{F_i}(x), \mu_X(x)\} \quad \forall i \quad (3.12)$$

where /P stands for the partition of the universe of discourse, universe, with respect to a given subset P of features, and F_i denotes a fuzzy equivalence class belonging to /P. Note that although the universe of discourse in feature reduction is finite, this is not the case in general, hence the use of *sup* and *inf* above. These definitions diverge a little from the crisp upper and lower approximations, as the memberships of individual objects to the approximations are not explicitly available. As a result of this, the fuzzy lower and upper approximations are redefined as [14]:

$$\mu_{\underline{P}X}(x) = \sup_{F \in /P} \min(\mu_F(x), \inf_{y \in \text{universe}} \max\{1 - \mu_F(y), \mu_X(y)\}) \quad (3.13)$$

$$\mu_{\overline{P}X}(x) = \sup_{F \in /P} \min(\mu_F(x), \sup_{y \in \text{universe}} \min\{\mu_F(y), \mu_X(y)\}) \quad (3.14)$$

The tuple $< \underline{P}X, \overline{P}X >$ is called a fuzzy-rough set. For this particular feature selection method, the upper approximation is not used, though this may be useful for other methods.

For an individual feature, a, the partition of the universe by $\{a\}$ (denoted universe/$IND(\{a\})$) is considered to be the set of those fuzzy equivalence classes for that feature. For example, if the two fuzzy sets N_a and Z_a are generated for feature a during fuzzification, the partition universe/$IND(\{a\}) = \{N_a, Z_a\}$. If the fuzzy-rough feature selection process is to be useful, it must be able to deal with multiple features, finding the dependency between various subsets of the original feature set. For instance, it may be necessary to be able to determine the degree of dependency of the decision feature(s) with respect to feature set $P = \{a, b\}$. In the crisp case, /P contains sets of objects grouped together that are indiscernible according to both features a and b. In the fuzzy case, objects may belong to many equivalence classes, so the cartesian product of /$IND(\{a\})$ and /$IND(\{b\})$ must be considered in determining /P. In general,

$$/P = \otimes \{a \in P : /IND(\{a\})\} \quad (3.15)$$

For example, if $P = \{a,b\}$, /$IND(\{a\}) = \{N_a, Z_a\}$ and /$IND(\{b\}) = \{N_b, Z_b\}$, then

$$/P = \{N_a \cap N_b, N_a \cap Z_b, Z_a \cap N_b, Z_a \cap Z_b\}$$

Clearly, each set in $/P$ denotes an equivalence class. The extent to which an object belongs to such an equivalence class is therefore calculated by using the conjunction of constituent fuzzy equivalence classes, say $F_i, i = 1, 2, ..., n$:

$$\mu_{F_1 \cap ... \cap F_n}(x) = min(\mu_{F_1}(x), \mu_{F_2}(x), ..., \mu_{F_n}(x)) \qquad (3.16)$$

3.3.3 Fuzzy-Rough Reduction Method

Fuzzy-Rough Feature Selection (FRFS) [14] builds on the notion of the fuzzy lower approximation to enable reduction of datasets containing real-valued features. The process becomes identical to the crisp approach when dealing with nominal well-defined features.

The crisp positive region in the standard RST is defined as the union of the lower approximations. By the extension principle, the membership of an object $x \in$ universe, belonging to the fuzzy positive region can be defined by

$$\mu_{POS_P(Q)}(x) = \sup_{X \in /Q} \mu_{PX}(x) \qquad (3.17)$$

Object x will not belong to the positive region only if the equivalence class it belongs to is not a constituent of the positive region. This is equivalent to the crisp version where objects belong to the positive region only if their underlying equivalence class does so.

Using the definition of the fuzzy positive region, a new dependency function between a set of features Q and another set P can be defined as follows:

$$\gamma_P(Q) = \frac{|\mu_{POS_P(Q)}(x)|}{||} = \frac{\sum_{x \in \text{universe}} \mu_{POS_P(Q)}(x)}{||} \qquad (3.18)$$

As with crisp rough sets, the dependency of Q on P is the proportion of objects that are discernible out of the entire dataset. In the present approach, this corresponds to determining the fuzzy cardinality of $\mu_{POS_P(Q)}(x)$ divided by the total number of objects in the universe.

A new QUICKREDUCT algorithm, based on the crisp version [3], has been developed as given in Fig. 3.2. It employs the new dependency function γ' to choose which features to add to the current reduct candidate. The algorithm terminates when the addition of any remaining feature does not increase the dependency. As with the original algorithm, for a dimensionality of n, the worst case dataset will result in $(n^2 + n)/2$ evaluations of the dependency function. However, as fuzzy-rough set-based feature selection is used for dimensionality reduction prior to any involvement of the system which will employ those features belonging to the resultant reduct, this operation has no negative impact upon the run-time efficiency of the system.

FRQUICKREDUCT(C,D).
C, the set of all conditional features;
D, the set of decision features.

(1) $R \leftarrow \{\}, \gamma_{best} \leftarrow 0, \gamma_{prev} \leftarrow 0$
(2)   **do**
(3) $T \leftarrow R$
(4) $\gamma_{prev} \leftarrow \gamma_{best}$
(5) $\forall x \in (C - R)$
(6)   **if** $\gamma_{R \cup \{x\}}(D) > \gamma_T(D)$
(7) $T \leftarrow R \cup \{x\}$
(8) $\gamma_{best} \leftarrow \gamma_T(D)$
(9) $R \leftarrow T$
(10) **until** $\gamma_{best} = \gamma_{prev}$
(11) **return** R

Fig. 3.2. The fuzzy-rough QUICKREDUCT algorithm

Object	a	b	c	q
1	−0.4	−0.3	−0.5	no
2	−0.4	0.2	−0.1	yes
3	−0.3	−0.4	−0.3	no
4	0.3	−0.3	0	yes
5	0.2	−0.3	0	yes
6	0.2	0	0	no

Table 3.1. Example dataset: crisp decisions

3.3.4 A Worked Example

Table 3.1 contains three real-valued conditional attributes and a crisp-valued decision attribute. To begin with, the fuzzy-rough QUICKREDUCT algorithm initializes the potential reduct (i.e. the current best set of attributes) to the empty set.

Using the fuzzy sets defined in Fig. 3.3 (for all conditional attributes), and setting $A = \{a\}$, $B = \{b\}$, $C = \{c\}$ and $Q = \{q\}$, the following equivalence classes are obtained:

$$/A = \{N_a, Z_a\}$$
$$/B = \{N_b, Z_b\}$$
$$/C = \{N_c, Z_c\}$$
$$/Q = \{\{1,3,6\},\{2,4,5\}\}$$

The first step is to calculate the lower approximations of the sets A, B and C, using (3.13). To clarify the calculations involved, table 3.2 contains the membership degrees of objects to fuzzy equivalence classes. For simplicity, only A will be

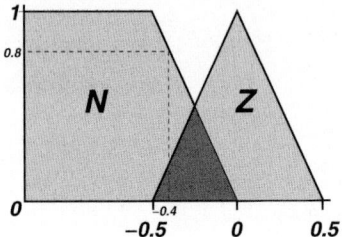

Fig. 3.3. Fuzzifications for conditional features

considered here; that is, using A to approximate Q. For the first decision equivalence class $X = \{1,3,6\}$, $\mu_{\underline{A}\{1,3,6\}}(x)$ needs to be calculated:

$$\mu_{\underline{A}\{1,3,6\}}(x) = \sup_{F \in /A} min(\mu_F(x), \inf_{y \in \text{universe}} max\{1 - \mu_F(y), \mu_{\{1,3,6\}}(y)\})$$

Considering the first fuzzy equivalence class of A, N_a:

$$min(\mu_{N_a}(x), \inf_{y \in \text{universe}} max\{1 - \mu_{N_a}(y), \mu_{\{1,3,6\}}(y)\})$$

Table 3.2. Membership values of objects to corresponding fuzzy sets

Object	a		b		c		q	
	N_a	Z_a	N_b	Z_b	N_c	Z_c	$\{1,3,6\}$	$\{2,4,5\}$
1	0.8	0.2	0.6	0.4	1.0	0.0	1.0	0.0
2	0.8	0.2	0.0	0.6	0.2	0.8	0.0	1.0
3	0.6	0.4	0.8	0.2	0.6	0.4	1.0	0.0
4	0.0	0.4	0.6	0.4	0.0	1.0	0.0	1.0
5	0.0	0.6	0.6	0.4	0.0	1.0	0.0	1.0
6	0.0	0.6	0.0	1.0	0.0	1.0	1.0	0.0

For object 2 this can be calculated as follows. From table 3.2 it can be seen that the membership of object 2 to the fuzzy equivalence class N_a, $\mu_{N_a}(2)$, is 0.8. The remainder of the calculation involves finding the smallest of the following values:

$$max(1-\mu_{N_a}(1), \mu_{\{1,3,6\}}(1)) = max(0.2, 1.0) = 1.0$$
$$max(1-\mu_{N_a}(2), \mu_{\{1,3,6\}}(2)) = max(0.2, 0.0) = 0.2$$
$$max(1-\mu_{N_a}(3), \mu_{\{1,3,6\}}(3)) = max(0.4, 1.0) = 1.0$$
$$max(1-\mu_{N_a}(4), \mu_{\{1,3,6\}}(4)) = max(1.0, 0.0) = 1.0$$
$$max(1-\mu_{N_a}(5), \mu_{\{1,3,6\}}(5)) = max(1.0, 0.0) = 1.0$$
$$max(1-\mu_{N_a}(6), \mu_{\{1,3,6\}}(6)) = max(1.0, 1.0) = 1.0$$

From the calculations above, the smallest value is 0.2, hence:

$$\min(\mu_{N_a}(x), \inf_{y \in \text{universe}} \max\{1-\mu_{N_a}(y), \mu_{\{1,3,6\}}(y)\}) = \min(0.8, \inf\{1, 0.2, 1, 1, 1, 1\})$$
$$= 0.2$$

Similarly for Z_a

$$\min(\mu_{Z_a}(x), \inf_{y \in \text{universe}} \max\{1-\mu_{Z_a}(y), \mu_{\{1,3,6\}}(y)\})$$
$$= \min(0.2, \inf\{1, 0.8, 1, 0.6, 0.4, 1\})$$
$$= 0.2$$

Thus,
$$\mu_{\underline{A}\{1,3,6\}}(2) = 0.2$$

Calculating the A-lower approximation of $X = \{1,3,6\}$ for every object gives

$$\mu_{\underline{A}\{1,3,6\}}(1) = 0.2 \quad \mu_{\underline{A}\{1,3,6\}}(2) = 0.2$$
$$\mu_{\underline{A}\{1,3,6\}}(3) = 0.4 \quad \mu_{\underline{A}\{1,3,6\}}(4) = 0.4$$
$$\mu_{\underline{A}\{1,3,6\}}(5) = 0.4 \quad \mu_{\underline{A}\{1,3,6\}}(6) = 0.4$$

The corresponding values for $X = \{2,4,5\}$ can also be determined:

$$\mu_{\underline{A}\{2,4,5\}}(1) = 0.2 \quad \mu_{\underline{A}\{2,4,5\}}(2) = 0.2$$
$$\mu_{\underline{A}\{2,4,5\}}(3) = 0.4 \quad \mu_{\underline{A}\{2,4,5\}}(4) = 0.4$$
$$\mu_{\underline{A}\{2,4,5\}}(5) = 0.4 \quad \mu_{\underline{A}\{2,4,5\}}(6) = 0.4$$

It is a coincidence here that $\mu_{\underline{A}\{2,4,5\}}(x) = \mu_{\underline{A}\{1,3,6\}}(x)$ for this example. Using these values, the fuzzy positive region for each object can be calculated via using

$$\mu_{POS_A(Q)}(x) = \sup_{X \in \mathbb{Q}} \mu_{\underline{A}X}(x)$$

This results in:

$$\mu_{POS_A(Q)}(1) = 0.2 \quad \mu_{POS_A(Q)}(2) = 0.2$$
$$\mu_{POS_A(Q)}(3) = 0.4 \quad \mu_{POS_A(Q)}(4) = 0.4$$
$$\mu_{POS_A(Q)}(5) = 0.4 \quad \mu_{POS_A(Q)}(6) = 0.4$$

The next step is to determine the degree of dependency of Q on A:

$$\gamma_A(Q) = \frac{\sum_{x \in U} \mu_{POS_A(Q)}(x)}{|U|} = 2/6$$

Calculating for B and C gives:

$$\gamma_B(Q) = \frac{2.4}{6}, \quad \gamma_C(Q) = \frac{1.6}{6}$$

From this it can be seen that attribute b will cause the greatest increase in dependency degree. This attribute is chosen and added to the potential reduct. The process iterates and the two dependency degrees calculated are

$$\gamma_{\{a,b\}}(Q) = \frac{3.4}{6}, \quad \gamma_{\{b,c\}}(Q) = \frac{3.2}{6}$$

Adding attribute a to the reduct candidate causes the larger increase of dependency, so the new candidate becomes $\{a,b\}$. Lastly, attribute c is added to the potential reduct:

$$\gamma'_{\{a,b,c\}}(Q) = \frac{3.4}{6}$$

As this causes no increase in dependency, the algorithm stops and outputs the reduct $\{a,b\}$. The dataset can now be reduced to only those attributes appearing in the reduct. When crisp RSAR is performed on this dataset (after using the same fuzzy sets to discretize the real-valued attributes), the reduct generated is $\{a,b,c\}$, i.e. the full conditional attribute set. Unlike crisp RSAR, the true minimal reduct was found using the information on degrees of membership. It is clear from this example alone that the information lost by using crisp RSAR can be important when trying to discover the smallest reduct from a dataset.

Conventional hill-climbing approaches to feature selection such as the algorithm presented above often fail to find maximal data reductions or minimal reducts. Some guiding heuristics are better than others for this, but as no perfect heuristic exists there can be no guarantee of optimality. When maximal data reductions are required, other search mechanisms must be employed. Although these methods also cannot ensure optimality, they provide a means by which the best feature subsets might be found. This motivates the development of feature selection based on Ant Colony Optimization.

3.4 Ant-based Feature Selection

Swarm Intelligence (SI) is the property of a system whereby the collective behaviours of simple agents interacting locally with their environment cause coherent functional global patterns to emerge [2]. SI provides a basis with which it is possible to explore collective (or distributed) problem solving without centralized control or the provision of a global model. One area of interest in SI is Particle Swarm Optimization [17], a population-based stochastic optimization technique. Here, the system is initialised with a population of random solutions, called particles. Optima are searched for by updating generations, with particles moving through the parameter space towards the current local and global optimum particles. At each time step, the velocities of all particles are changed depending on the current optima.

Ant Colony Optimization (ACO) [2] is another area of interest within SI. In nature, it can be observed that real ants are capable of finding the shortest route between a food source and their nest without the use of visual information and hence possess no global world model, adapting to changes in the environment. The deposition of pheromone is the main factor in enabling real ants to find the shortest routes over a period of time. Each ant probabilistically prefers to follow a direction rich in this chemical. The pheromone decays over time, resulting in much less pheromone on less popular paths. Given that over time the shortest route will have the higher rate of ant traversal, this path will be reinforced and the others diminished until all ants follow the same, shortest path (the "system" has converged to a single solution). It is also possible that there are many equally short paths. In this situation, the rates of ant traversal over the short paths will be roughly the same, resulting in these paths being maintained while others are ignored. Additionally, if a sudden change to the environment occurs (e.g. a large obstacle appears on the shortest path), the ACO system can respond to this and will eventually converge to a new solution. Based on this idea, artificial ants can be deployed to solve complex optimization problems via the use of artificial pheromone deposition.

ACO is particularly attractive for feature selection as there seems to be no heuristic that can guide search to the optimal minimal subset every time. Additionally, it can be the case that ants discover the best feature combinations as they proceed throughout the search space. This section discusses how ACO may be applied to the difficult problem of finding optimal feature subsets and, in particular, fuzzy-rough set-based reducts.

3.4.1 ACO Framework

An ACO algorithm can be applied to any combinatorial problem as far as it is possible to define:

- *Appropriate problem representation.* The problem can be described as a graph with a set of nodes and edges between nodes.
- *Heuristic desirability (η) of edges.* A suitable heuristic measure of the "goodness" of paths from one node to every other connected node in the graph.
- *Construction of feasible solutions.* A mechanism must be in place whereby possible solutions are efficiently created. This requires the definition of a suitable traversal stopping criterion to stop path construction when a solution has been reached.
- *Pheromone updating rule.* A suitable method of updating the pheromone levels on edges is required with a corresponding evaporation rule, typically involving the selection of the n best ants and updating the paths they chose.
- *Probabilistic transition rule.* The rule that determines the probability of an ant traversing from one node in the graph to the next.

Each ant in the artificial colony maintains a memory of its history - remembering the path it has chosen so far in constructing a solution. This history can be used in the evaluation of the resulting created solution and may also contribute to the decision process at each stage of solution construction.

Two types of information are available to ants during their graph traversal, local and global, controlled by the parameters β and α respectively. Local information is obtained through a problem-specific heuristic measure. The extent to which the measure influences an ant's decision to traverse an edge is controlled by the parameter β. This will guide ants towards paths that are likely to result in good solutions. Global knowledge is also available to ants through the deposition of artificial pheromone on the graph edges by their predecessors over time. The impact of this knowledge on an ant's traversal decision is determined by the parameter α. Good paths discovered by past ants will have a higher amount of associated pheromone. How much pheromone is deposited, and when, is dependent on the characteristics of the problem. No other local or global knowledge is available to the ants in the standard ACO model, though the inclusion of such information by extending the ACO framework has been investigated [2].

3.4.2 Feature Selection

The feature selection task may be reformulated into an ACO-suitable problem [13, 16]. ACO requires a problem to be represented as a graph - here nodes represent features, with the edges between them denoting the choice of the next feature. The search for the optimal feature subset is then an ant traversal through the graph where a minimum number of nodes are visited that satisfies the traversal stopping criterion. Figure 3.4 illustrates this setup - the ant is currently at node a and has a choice of which feature to add next to its path (dotted lines). It chooses feature b next based on the transition rule, then c and then d. Upon arrival at d, the current subset $\{a,b,c,d\}$ is determined to satisfy the traversal stopping criteria (e.g. a suitably high classification accuracy has been achieved with this subset, assuming that the selected features are used to classify certain objects). The ant terminates its traversal and outputs this feature subset as a candidate for data reduction.

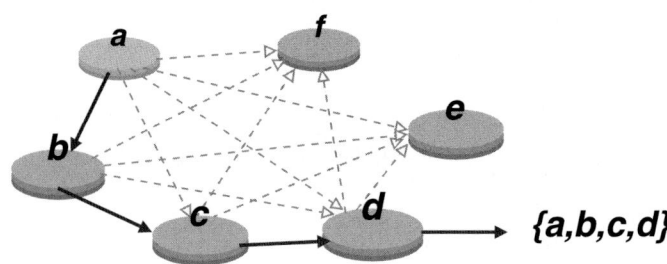

Fig. 3.4. ACO problem representation for feature selection

A suitable heuristic desirability of traversing between features could be any subset evaluation function - for example, an entropy-based measure [24] or the fuzzy-rough set dependency measure. Depending on how optimality is defined for the particular application, the pheromone may be updated accordingly. For instance, subset minimality and "goodness" are two key factors so the pheromone update should be proportional to "goodness" and inversely proportional to size. How "goodness" is determined will also depend on the application. In some cases, this may be a heuristic evaluation of the subset, in others it may be based on the resulting classification accuracy of a classifier produced using the subset.

The heuristic desirability and pheromone factors are combined to form the so-called probabilistic transition rule, denoting the probability of an ant k at feature i choosing to move to feature j at time t:

$$p_{ij}^k(t) = \frac{[\tau_{ij}(t)]^\alpha \cdot [\eta_{ij}]^\beta}{\sum_{l \in J_i^k}[\tau_{il}(t)]^\alpha \cdot [\eta_{il}]^\beta} \qquad (3.19)$$

where J_i^k is the set of ant k's unvisited features, η_{ij} is the heuristic desirability of choosing feature j when at feature i and $\tau_{ij}(t)$ is the amount of virtual pheromone on edge (i, j). The choice of α and β is determined experimentally. Typically, several experiments are performed, varying each parameter and choosing the values that produce the best results.

Selection Process

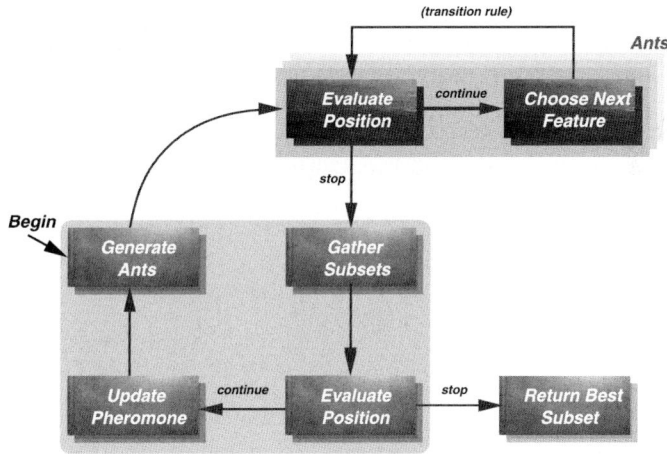

Fig. 3.5. ACO-based feature selection overview

The overall process of ACO feature selection can be seen in Fig. 3.5. It begins by generating a number of ants, k, which are then placed randomly on the graph (i.e. each ant starts with one random feature). Alternatively, the number of ants to place on the graph may be set equal to the number of features within the data; each ant starts path construction at a different feature. From these initial positions, they traverse edges probabilistically until a traversal stopping criterion is satisfied. The resulting subsets are gathered and then evaluated. If an optimal subset has been found or the algorithm has executed a certain number of times, then the process halts and outputs the best feature subset encountered. If neither condition holds, then the pheromone is updated, a new set of ants are created and the process iterates once more.

Complexity Analysis

The time complexity of the ant-based approach to feature selection is $O(IAk)$, where I is the number of iterations, A the number of original features, and k the number of ants. In the worst case, each ant selects all the features. As the heuristic is evaluated after each feature is added to the reduct candidate, this will result in A evaluations per ant. After one iteration in this scenario, Ak evaluations will have been performed. After I iterations, the heuristic will be evaluated IAk times.

Pheromone Update

Depending on how optimality is defined for the particular application, the pheromone may be updated accordingly. To tailor this mechanism to find fuzzy-rough set reducts, it is necessary to use the dependency measure given in (3.18) as the stopping criterion. This means that an ant will stop building its feature subset when the dependency of the subset reaches the maximum for the dataset (the value 1 for consistent datasets). The dependency function may also be chosen as the heuristic desirability measure, but this is not necessary. In fact, it may be of more use to employ a non-rough set related heuristic for this purpose. By using an alternative measure such as an entropy-based heuristic, the method may avoid feature combinations that may mislead the fuzzy-rough set-based heuristic. Again, the time complexity of this fuzzy-rough ant-based method will be the same as that mentioned earlier, $O(IAk)$.

The pheromone on each edge is updated according to the following formula:

$$\tau_{ij}(t+1) = (1-\rho).\tau_{ij}(t) + \Delta\tau_{ij}(t) \qquad (3.20)$$

where

$$\Delta\tau_{ij}(t) = \sum_{k=1}^{n}(\gamma'(S^k)/|S^k|) \qquad (3.21)$$

This is the case if the edge (i,j) has been traversed; $\Delta\tau_{ij}(t)$ is 0 otherwise. The value ρ is a decay constant used to simulate the evaporation of the pheromone, S^k is the feature subset found by ant k. The pheromone is updated according to both the rough (or fuzzy-rough) measure of the "goodness" of the ant's feature subset (γ') and the size of the subset itself. By this definition, all ants update the pheromone. Alternative strategies may be used for this, such as allowing only the ants with the currently best feature subsets to proportionally increase the pheromone.

3.5 Crisp Ant-based Feature Selection Evaluation

In order to compare several mainstream approaches to crisp rough set-based feature selection with ACO-based selection, an investigation into how these methods perform in terms of resulting subset optimality has been carried out here. Several real and artificial datasets are used for this purpose. In particular, it is interesting to compare those methods that employ an incremental-based search strategy with those that adopt a more complex stochastic/probabilistic mechanism. Five techniques for finding crisp rough set reducts are tested here on 13 datasets. These techniques are: RSAR (using QUICKREDUCT), EBR (an entropy-based approach [15]), GenRSAR (genetic algorithm-based), AntRSAR (ant-based) and SimRSAR (simulated annealing-based)[1].

3.5.1 Experimental Setup

Before the experiments are described, a few points must be made about the later three approaches, GenRSAR, AntRSAR and SimRSAR.

GenRSAR employs a genetic search strategy in order to determine rough set reducts. The initial population consists of 100 randomly generated feature subsets, the probabilities of mutation and crossover are set to 0.4 and 0.6 respectively, and the number of generations is set to 100. The fitness function considers both the size of subset and its evaluated suitability, and is defined as follows:

$$fitness(R) = \gamma_R(\mathbb{D}) * \frac{|\mathbb{C}| - |R|}{|\mathbb{C}|} \tag{3.22}$$

AntRSAR follows the mechanism described in section 3.4.2. Here, the precomputed heuristic desirability of edge traversal is the entropy measure, with the subset evaluation performed using the rough set dependency heuristic (to guarantee that true rough set reducts are found). The number of ants used is set to the number of features, with each ant starting on a different feature. For the datasets used here, the performance is not affected significantly using this number of ants. However, for datasets containing thousands of features or more, fewer ants may have to be chosen due to computational limitations. Ants construct possible solutions until they reach a rough set reduct. To avoid fruitless searches, the size of the current best reduct is used to reject those subsets whose cardinality exceed this value. Pheromone levels are set at 0.5 with a small random variation added. Levels are increased by only those ants who have found true reducts. The global search is terminated after 250 iterations, α is set to 1 and β is set to 0.1.

SimRSAR employs a simulated annealing-based feature selection mechanism [15]. The states are feature subsets, with random state mutations set to changing three features (either adding or removing them). The cost function attempts to maximize the rough set dependency (γ) whilst minimizing the subset cardinality. For these experiments, the cost of subset R is defined as:

$$cost(R) = \left[\frac{\gamma_\mathbb{C}(\mathbb{D}) - \gamma_R(\mathbb{D})}{\gamma_\mathbb{C}(\mathbb{D})}\right]^a + \left[\frac{|R|}{|\mathbb{C}|}\right]^b \tag{3.23}$$

where a and b are defined in order to weight the contributions of dependency and subset size to the overall cost measure. In the experiments here, $a = 1$ and $b = 3$. The initial temperature of the system is estimated as $2 * |\mathbb{C}|$ and the cooling schedule is $T(t+1) = 0.93 * T(t)$.

---

[1] These algorithms and datasets (as well as FRFS and antFRFS) can be downloaded from the webpage: http://users.aber.ac.uk/rkj/index.html

The experiments were carried out on 3 datasets from [25], namely *m-of-n*, *exactly* and *exactly2*. The remaining datasets are from the machine learning repository [1]. Those datasets containing real-valued attributes have been discretized to allow all methods to be compared fairly.

3.5.2 Experimental Results

Table 3.3. Subset sizes found for five techniques

Index	Dataset	Features	Optimal	RSAR	EBR	AntRSAR	SimRSAR	GenRSAR
0	M-of-N	13	6	8	6	6	6	6(6) 7(12)
1	Exactly	13	6	9	8	6	6	6(10) 7(10)
2	Exactly2	13	10	13	11	10	10	10(9) 11(11)
3	Heart	13	6	7	7	6(18) 7(2)	6(29) 7(1)	6(18) 7(2)
4	Vote	16	8	9	9	8	8(15) 9(15)	8(2) 9(18)
5	Credit	20	8	9	10	8(12) 9(4) 10(4)	8(18) 9(1) 11(1)	10(6) 11(14)
6	Mushroom	22	4	5	4	4	4	5(1) 6(5) 7(14)
7	LED	24	5	12	5	5(12) 6(4) 7(3)	5	6(1) 7(3) 8(16)
8	Letters	25	8	9	9	8	8	8(8) 9(12)
9	Derm	34	6	7	6	6(17) 7(3)	6(12) 7(8)	10(6) 11(14)
10	Derm2	34	8	10	10	8(3) 9(17)	8(3) 9(7)	10(2) 11(8)
11	WQ	38	12	14	14	12(2) 13(7) 14(11)	13(16) 14(4)	16
12	Lung	56	4	4	4	4	4(7) 5(12) 6(1)	6(8) 7(12)

Table 3.3 presents the results of the five methods on the 13 datasets. It shows the size of reduct found for each method, as well as the size of the optimal (minimal) reduct. RSAR and EBR produced the same subset every time, unlike AntRSAR and SimRSAR that often found different subsets and sometimes different subset cardinalities. On the whole, it appears to be the case that AntRSAR and SimRSAR outperform the other three methods. This is at the expense of the time taken to discover these reducts as can be seen in Fig. 3.6 (results for RSAR and EBR do not appear as they are consistently faster than the other methods). In all experiments the rough ordering of techniques with respect to time is: RSAR < EBR ≤ SimRSAR ≤ AntRSAR ≤ GenRSAR. AntRSAR and SimRSAR perform similarly throughout - for some datasets, AntRSAR is better (e.g. Vote) and for others SimRSAR is best (e.g. LED). The performance of these two methods may well be improved by fine-tuning the parameters to each individual dataset.

From these results it can be seen that even for small and medium-sized datasets, incremental hill-climbing techniques often fail to find minimal subsets. For example, RSAR is misled early in the search for the LED dataset, resulting in it choosing 7 extraneous features. Although this fault is due to the non-optimality of the guiding heuristic, a perfect heuristic does not exist rendering these approaches unsuited to problems where a minimal subset is essential. However, for most real world applications, the extent of reduction achieved via such methods is acceptable. For

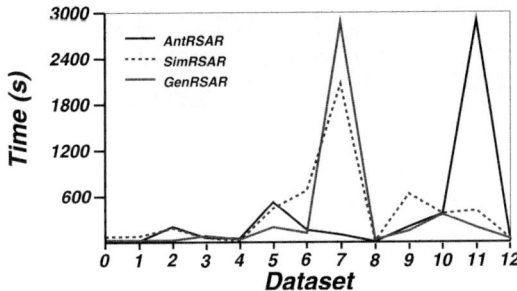

Fig. 3.6. Average runtimes for AntRSAR, SimRSAR and GenRSAR

systems where the minimal subset is required (perhaps due to the cost of feature measurement), stochastic feature selection should be used.

3.6 Fuzzy Ant-based Feature Selection Evaluation

To show the utility of fuzzy-rough feature selection and to compare the hill-climbing and ant-based fuzzy-rough approaches, the two methods are applied as pre-processors to web classification and within a complex systems monitoring application. Both methods preserve the semantics of the surviving features after removing redundant ones. This is essential in satisfying the requirement of user readability of the generated knowledge model, as well as ensuring the understandability of the pattern classification process.

3.6.1 Web Classification

There are an estimated 1 billion web pages available on the world wide web, with around 1.5 million web pages being added every day. The task to find a particular web page, which satisfies a user's requirements by traversing hyper-links, is very difficult. To aid this process, many web directories have been developed - some rely on manual categorization whilst others make decisions automatically. However, as web page content is vast and dynamic, manual categorization is becoming increasingly impractical. Automatic web site categorization is therefore required to deal with these problems.

System Overview

The general overview of the classification system developed here can be seen in Fig. 3.7. A key issue in the design of the system was that of modularity; it should be able to integrate with existing (or new) techniques. The current implementations allow this flexibility by dividing the overall process into several independent sub-modules:

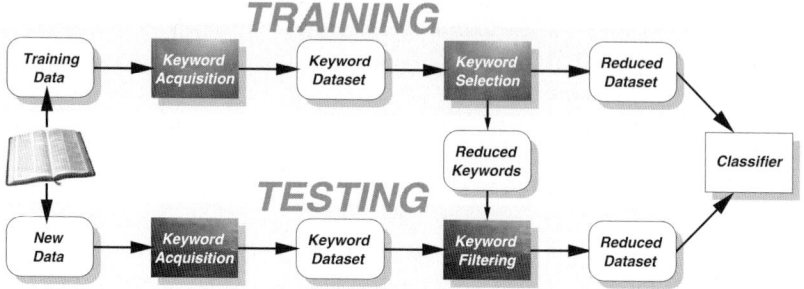

Fig. 3.7. Modular decomposition of the classification system

- *Keyword Acquisition.* From the collection of web documents, only the natural language terms are extracted and considered to be keywords. These are then weighted according to their perceived importance in the document, resulting in a new dataset of weight-term pairs. These weights are almost always real-valued, hence the problem serves well to test the present work. For this, the TF-IDF metric [27] is used which assigns higher weights to those keywords that occur frequently in the current document but not in most others. Note that in this work, no sophisticated keyword acquisition techniques methods are used as the current focus of attention is on the evaluation of attribute reduction. However, the use of more effective keyword acquisition techniques recently built in the area of information retrieval would help improve the system's overall classification performance further.

- *Keyword Selection.* As the newly generated datasets are too large, mainly due to keyword redundancy, a dimensionality reduction step is carried out using the techniques described previously.

- *Keyword Filtering.* Employed only in testing, this simple module filters the keywords obtained during acquisition, using the reduct generated in the keyword selection module.

- *Classification.* This final module uses the reduced dataset to perform the actual categorization of the test data. Four classifiers were used for comparison, namely C4.5 [24], JRip [5], PART [33] and a fuzzy rule inducer, QSBA [26]. Both JRip and PART are available from [34].
  C4.5 creates decision trees by choosing the most informative features and recursively partitioning the data into subtables based on their values. Each node in the tree represents a feature with branches from a node representing the alternative values this feature can take according to the current subtable. Partitioning stops when all data items in the subtable have the same classification. A leaf node is then created, and this classification assigned.

JRip learns propositional rules by repeatedly growing rules and pruning them. During the growth phase, antecedents are added greedily until a termination condition is satisfied. Antecedents are then pruned in the next phase subject to a pruning metric. Once the ruleset is generated, a further optimization is performed where rules are evaluated and deleted based on their performance on randomized data.

PART generates rules by means of repeatedly creating partial decision trees from data. The algorithm adopts a separate-and-conquer strategy in that it removes instances covered by the current ruleset during processing. Essentially, a rule is created by building a pruned tree for the current set of instances; the leaf with the highest coverage is made into a rule.

QSBA induces fuzzy rules by calculating the fuzzy subsethood of linguistic terms and the corresponding decision variables. These values are also weighted by the use of fuzzy quantifiers. This method utilises the same fuzzy sets as those involved in the fuzzy-rough reduction methods.

Experimentation and Results

Initially, datasets were generated from large textual corpora collected from Yahoo [36] and separated randomly into training and testing sets. Each dataset is a collection of web documents. Five classification categories were used, namely Art & Humanity, Entertainment, Computers & Internet, Health, Business & Economy. A total of 280 web sites were collected from Yahoo categories and classified into these categories. From this collection of data, the keywords, weights and corresponding classifications were collated into a single dataset.

Table 3.4 shows the resulting degree of dimensionality reduction, performed via selecting informative keywords, by the standard fuzzy-rough method (FRFS) and the ACO-based approach (AntFRFS). AntFRFS is run several times, and the results averaged both for classification accuracy and number of features selected. It can be seen that both methods drastically reduce the number of original features. AntFRFS performs the highest degree of reduction, with an average of 14.1 features occurring in the reducts it locates.

Table 3.4. Extent of feature reduction

Original	FRFS	AntFRFS
2557	17	14.10

To see the effect of dimensionality reduction on classification accuracy, the system was tested on the original training data and a test dataset. The results are summarised in table 3.5. Clearly, the fuzzy-rough methods exhibit better resultant accuracies for the test data than the unreduced method for all classifiers. This demonstrates that feature selection using either FRFS or AntFRFS can greatly aid classification tasks. It is of additional benefit to rule inducers as the induction time is decreased and the generated rules involve significantly fewer features. AntFRFS improves on FRFS in terms of the size of subsets found and resulting testing accuracy for QSBA and PART, but not for C4.5 and JRip.

Table 3.5. Classification performance

Classifier	Original (%) Train Test	FRFS (%) Train Test	AntFRFS (%) Train Test
C4.5	95.89 44.74	86.30 57.89	81.27 48.39
QSBA	100.0 39.47	82.19 46.05	69.86 50.44
JRip	72.60 56.58	78.08 60.53	64.84 51.75
PART	95.89 42.11	86.30 48.68	82.65 48.83

The challenging nature of this particular task can be seen in the overall low accuracies produced by the classifiers, though improved somewhat after feature selection. Both fuzzy-rough approaches require a reasonable fuzzification of the input data, whilst the fuzzy sets are herein generated by simple statistical analysis of the dataset with no attempt made at optimizing these sets. A fine-tuned fuzzification will certainly improve the performance of FRFS-based systems. Finally, it is worth noting that the classifications were checked automatically. Many websites can be classified to more than one category, however only the designated category is considered to be correct here.

3.6.2 Systems Monitoring

In order to further evaluate the fuzzy-rough approaches and to illustrate its domain-independence, another challenging test dataset was chosen, namely the Water Treatment Plant Database [1]. The dataset itself is a set of historical data charted over 521 days, with 38 different input features measured daily. Each day is classified into one of thirteen categories depending on the operational status of the plant. However, these can be collapsed into just two or three categories (i.e. *Normal* and *Faulty*, or *OK*, *Good* and *Faulty*) for plant monitoring purposes as many classifications reflect similar performance. Because of the efficiency of the actual plant the measurements were taken from, all faults appear for short periods (usually single days) and are dealt with immediately. This does not allow for a lot of training examples of faults, which is a clear drawback if a monitoring system is to be produced. Collapsing 13 categories into 2 or 3 classes helps reduce this difficulty for the present application. Note that this dataset has been utilised in many previous studies, including that reported in [29] (to illustrate the effectiveness of applying crisp RSAR as a pre-processing step to rule induction).

It is likely that not all of the 38 input features are required to determine the status of the plant, hence the dimensionality reduction step. However, choosing the most informative features is a difficult task as there will be many dependencies between subsets of features. There is also a monetary cost involved in monitoring these inputs, so it is desirable to reduce this number.

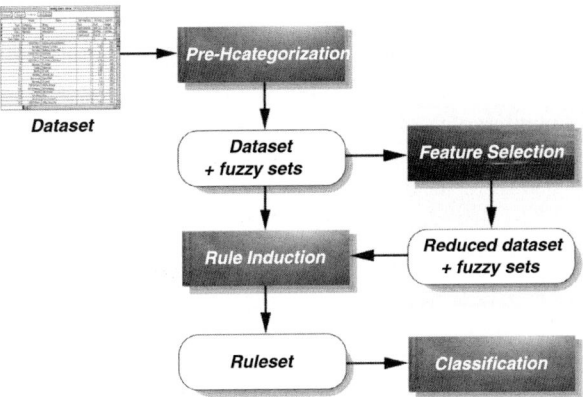

Fig. 3.8. Modular decomposition of the implemented system

Note that the original monitoring system (Fig. 3.8) developed in [29] consisted of several modules; it is this modular structure that allows the FRFS techniques to replace the existing crisp method. Originally, a precategorization step preceded feature selection where feature values were quantized. To reduce potential loss of information, the original use of just the dominant symbolic labels of the discretized fuzzy terms is now replaced by a fuzzification procedure. This leaves the underlying feature values unchanged but generates a series of fuzzy sets for each feature. These sets are generated entirely from the data while exploiting the statistical properties attached to the dataset (in keeping with the rough set ideology in that the dependence of learning upon information provided outside of the training dataset is minimized). This module may be replaced by alternative fuzzifiers, or expert-defined fuzzification if available. Based on these fuzzy sets and the original real-valued dataset, the feature selection module calculates a reduct and reduces the dataset accordingly. Finally, rule induction is performed on the reduced dataset. For this set of experiments, the decision tree method C4.5 [24] is used for induction and the learned rules for classification.

The first set of experiments compares the hill-climbing and ACO-based fuzzy-rough methods. An investigation into another feature selector based on the entropy measure is then presented. This is followed by comparisons with a transformation-based dimensionality reduction approach, PCA [7] and a support vector classifier [22].

Comparison of Fuzzy-Rough Methods

Three sets of experiments were carried out on both the (collapsed) 2-class and 3-class datasets. The first bypasses the feature selection part of the system, using the original water treatment dataset as input to C4.5, with all 38 conditional attributes. The second method employs FRFS to perform the feature selection before induction is carried out. The third uses the ACO-based method, AntFRFS, to perform feature selection over a number of runs, and the results averaged.

Table 3.6. Results for the 2-class dataset

Method	Attributes	γ' value	Training accuracy (%)	Testing accuracy (%)
Unreduced	38	-	98.5	80.9
FRFS	10	0.58783	89.2	74.8
AntFRFS	9.55	0.58899	93.5	77.9

The results for the 2-class dataset can be seen in table 3.6. Both FRFS and AntFRFS significantly reduce the number of original attributes with AntFRFS producing the greatest data reduction on average. As well as generating smaller reducts, AntFRFS finds reducts of a higher quality according to the fuzzy-rough dependency measure. This higher quality is reflected in the resulting classification accuracies for both the training and testing datasets, with AntFRFS outperforming FRFS.

Table 3.7. Results for the 3-class dataset

Method	Attributes	γ' value	Training accuracy (%)	Testing accuracy (%)
Unreduced	38	-	97.9	83.2
FRFS	11	0.59479	97.2	80.9
AntFRFS	9.09	0.58931	94.8	80.2

Table 3.7 shows the results for the 3-class dataset experimentation. The hill-climbing fuzzy-rough method chooses 11 out of the original 38 features. The ACO-based method chooses fewer attributes on average, however this is at the cost of a lower dependency measure for the generated reducts. Again the effect of this can be seen in the classification accuracies, with FRFS performing slightly better than AntFRFS. For both fuzzy methods, the small drop in accuracy as a result of feature selection is acceptable.

By selecting a good feature subset from data it is usually expected that the applied learning method should benefit, producing an improvement in results. For some applications, less features may result in a better classification performance due to the removal of heavy noise attached to those features removed. The ant-based approach should improve upon C4.5 in these situations. However, when the original training (and test) data is very noisy, selected features may not necessarily be able to reflect all the information contained within the original entire feature set. As a result of removing less informative features, partial useful information may be lost. The goal of selection methods in this situation is to minimise this loss, while reducing the number of features to the greatest extent. Therefore, it is not surprising that the classification performance for this challenging dataset can decrease upon data reduction, as shown in table 3.7. However, the impact of feature selection can have different effects on different classifiers. With the use of an alternative classifier in section 3.6.2, performance can be seen to improve for the test data.

The results here also show a marked drop in classification accuracy for the test data. This could be due to the problems encountered when dealing with datasets of small sample size. Overfitting can occur, where a learning algorithm adapts so well to a training set, that the random disturbances present are included in the model as being meaningful. Consequently, as these disturbances do not reflect the underlying distribution, the performance on the test data will suffer. Although such techniques as cross-validation and bootstrapping have been proposed as a way of countering this, these still often exhibit high variance in error estimation.

Comparison with Entropy-based Feature Selection

To support the study of the performance of the fuzzy-rough methods for use as pre-processors to rule induction, a conventional entropy-based technique is used for comparison. This approach utilizes the entropy heuristic typically employed by machine learning techniques such as C4.5 [24]. Those features that provide the most gain in information are selected. A summary of the results of this comparison can be seen in table 3.8.

Table 3.8. Results for the three selection methods

Approach	No. of Classes	No. of Features	Training Accuracy (%)	Testing Accuracy (%)
FRFS	2	10	89.2	74.8
AntFRFS	2	9.55	93.5	77.9
Entropy	2	13	97.7	80.2
FRFS	3	11	97.2	80.9
AntFRFS	3	9.09	94.8	80.2
Entropy	3	14	98.2	80.9

For both the 2-class and 3-class datasets, FRFS and AntFRFS select at least three fewer features than the entropy-based method. However, the entropy-based method outperforms the other two feature selectors with the resulting C4.5 classification accuracies. This is probably due to the fact that C4.5 uses exactly the same entropy measure in generating decision trees. In this case, the entropy-based measure will favour those attributes that will be the most influential in the decision tree generation process. The use of more features here may also contribute to the slightly better classification performance.

Comparison with the use of PCA

The effect of using a different dimensionality reduction technique, namely Principal Components Analysis (PCA) [7], is also investigated. PCA transforms the original features of a dataset with a (typically) reduced number of uncorrelated ones, termed principal components. It works on the assumption that a large feature variance corresponds to useful information, with small variance equating to information that is less useful. The first principle component indicates the direction of maximum data variance. Data is transformed in such a way as to allow the removal of those transformed features with small variance. This is achieved by finding the eigenvectors of the covariance matrix of data points (objects), constructing a transformation matrix from the ordered eigenvectors, and transforming the original data by matrix multiplication.

Here, PCA is applied to the dataset and the first n principal components are used. A range of values is chosen for n to investigate how the performance varies with dimensionality. As PCA irreversibly destroys the underlying dataset semantics, the resulting decision trees are not human-comprehensible nor directly measurable but may still provide useful automatic classifications of new data. Table 3.9 shows the results from applying PCA to the datasets.

Table 3.9. Results for the 2-class and 3-class datasets using PCA

Accuracy	Class	No. of Features								
		5	6	7	8	9	**10**	11	12	13
Training (%)	2	80.0	80.0	80.0	80.0	80.3	**80.3**	80.3	80.8	82.1
Testing (%)	2	72.5	72.5	72.5	72.5	73.3	**73.3**	73.3	35.1	34.4
Training (%)	3	73.6	73.6	73.6	73.6	73.6	**75.9**	75.9	75.9	76.4
Testing (%)	3	80.9	80.9	80.9	80.9	80.9	**80.9**	80.9	80.9	80.2

Both AntFRFS and FRFS significantly outperform PCA on the 2-class dataset. Of particular interest is when 10 principal components are used as this is roughly the same number chosen by AntFRFS and FRFS. The resulting accuracy for PCA is 80.3% for the training data and 73.3% for the test data. For AntFRFS the accuracies were 93.5% (training) and 77.9% (testing), and for FRFS 89.2% (training) and 74.8% (testing). In the 3-class dataset experimentation, both fuzzy-rough methods produce much higher classification accuracies than PCA for the training data. For the test data, the performance is about the same, with PCA producing a slightly higher accuracy than AntFRFS on the whole. It is worth reiterating, however, that PCA does not carry out feature selection but transformation. Hence, the classifier built with such transformed features is hard for human users to understand.

Comparison with the use of a Support Vector Classifier

A possible limitation of employing C4.5 in this context is that it performs a degree of feature selection itself during the induction process. The resulting decision trees do not necessarily contain all the features present in the original training data. As a result of this, it is beneficial to evaluate the use of an alternative classifier that uses all the given features. For this purpose, a support vector classifier [28] is employed, trained by the sequential minimal optimization (SMO) algorithm [22]. The results of the application of this classifier can be found in table 3.10.

Table 3.10. Results for the 2-class and 3-class datasets using SMO

Approach	No. of Classes	No. of Features	Training Accuracy (%)	Testing Accuracy (%)
Unreduced	2	38	80.0	71.8
FRFS	2	10	80.0	72.5
AntFRFS	2	9.55	80.0	72.5
Unreduced	3	38	74.6	80.9
FRFS	3	11	73.6	80.2
AntFRFS	3	9.09	73.6	80.9

For the 2-class dataset, the training accuracy for both FRFS and AntFRFS is the same as that of the unreduced approach. However, this is with significantly fewer attributes. Additionally, the resulting testing accuracy *is* increased with these feature selection methods. With the more challenging 3-class problem, the training accuracies are slightly worse (as seen with the C4.5 analysis). The AntFRFS method performs better than FRFS for the test data and is equal to the unreduced method, again using fewer features.

3.7 Conclusion

This chapter has presented an ACO-based method for feature selection, with particular emphasis on fuzzy-rough feature selection. This novel approach has been applied to aid classification of web content and to complex systems monitoring, with very promising results. In all experimental studies there has been no attempt to optimize the fuzzifications or the classifiers employed. It can be expected that the results obtained with such optimization would be even better than those already observed.

The techniques presented here focus mainly on the use of ACO for rough and fuzzy-rough feature selection. However, many alternative selection measures exist that are used within incremental hill-climbing search strategies to locate minimal subsets. Such measures could be easily incorporated into the existing ACO-framework. For AntFRFS, it can be expected that it is best suited for the optimization of fuzzy classifiers, as the feature significance measure utilizes the fuzzy sets required by these techniques.

There are many issues to be explored in the area of ACO-based feature selection. The impact of parameter settings should be investigated - how the values of α, β and others influence the search process. Other important factors to be considered include how the pheromone is updated and how it decays. There is also the possibility of using different static heuristic measures to determine the desirability of edges. A further extension would be the use of dynamic heuristic measures which would change over the course of feature selection to provide more search information.

Acknowledgement. The author would like to thank Qiang Shen for his support during the development of the ideas presented in this chapter.

References

1. Blake CL, Merz CJ (1998) UCI Repository of machine learning databases. Irvine, University of California http://www.ics.uci.edu/~mlearn/
2. Bonabeau E, Dorigo M, Theraulez G (1999) Swarm Intelligence: From Natural to Artificial Systems. Oxford University Press Inc., New York, NY, USA
3. Chouchoulas A, Shen Q (2001) Rough set-aided keyword reduction for text categorisation. Applied Artificial Intelligence, Vol. 15, No. 9, pp. 843-873
4. Chouchoulas A, Halliwell J, Shen Q (2002) On the Implementation of Rough Set Attribute Reduction. Proceedings of the 2002 UK Workshop on Computational Intelligence, pp. 18-23
5. Cohen WW (1995) Fast effective rule induction. In Machine Learning: Proceedings of the 12th International Conference, pp. 115-123
6. Dash M, Liu H (1997) Feature Selection for Classification. Intelligent Data Analysis, Vol. 1, No. 3, pp. 131-156
7. Devijver P, Kittler J (1982) Pattern Recognition: A Statistical Approach. Prentice Hall
8. Drwal G (2000) Rough and fuzzy-rough classification methods implemented in RClass system. In Proceedings of the 2nd International Conference on Rough Sets and Current Trends in Computing (RSCTC 2000), pp 152-159
9. Dubois D, Prade H (1992) Putting rough sets and fuzzy sets together. In [31], pp. 203-232
10. Düntsch I, Gediga G (2000) Rough Set Data Analysis. In: A. Kent & J. G. Williams (Eds.) Encyclopedia of Computer Science and Technology, Vol. 43, No. 28, pp. 281–301
11. Düntsch I, Gediga G (2000) Rough Set Data Analysis: A road to non-invasive knowledge discovery. Bangor: Methodos
12. Ho TB, Kawasaki S, Nguyen NB (2003) Documents clustering using tolerance rough set model and its application to information retrieval. Studies In Fuzziness And Soft Computing, Intelligent Exploration of the Web, pp. 181-196
13. Jensen R, Shen Q (2003) Finding Rough Set Reducts with Ant Colony Optimization. In Proceedings of the 2003 UK Workshop on Computational Intelligence, pp 15-22
14. Jensen R, Shen Q (2004) Fuzzy-rough attribute eduction with application to web categorization. Fuzzy Sets and Systems, Vol. 141, No. 3, pp. 469-485
15. Jensen R, Shen Q (2004) Semantics-Preserving Dimensionality Reduction: Rough and Fuzzy-Rough Based Approaches. IEEE Transactions on Knowledge and Data Engineering, Vol. 16, No. 12, pp. 1457-1471
16. Jensen R, Shen Q (2005) Fuzzy-Rough Data Reduction with Ant Colony Optimization. Fuzzy Sets and Systems, Vol. 149, No. 1, pp. 5-20
17. Kennedy J, Eberhart RC (1995) Particle swarm optimization. Proceedings of IEEE International Conference on Neura l Networks, pp. 1942-1948

18. Langley P (1994) Selection of relevant features in machine learning. In Proceedings of the AAAI Fall Symposium on Relevance, pp. 1-5
19. Pal SK, Skowron A (eds.) (1999) Rough-Fuzzy Hybridization: A New Trend in Decision Making. Springer Verlag, Singapore
20. Pawlak Z (1982) Rough Sets. International Journal of Computer and Information Sciences, Vol. 11, No. 5, pp. 341-356
21. Pawlak Z (1991) Rough Sets: Theoretical Aspects of Reasoning About Data. Kluwer Academic Publishing, Dordrecht.
22. Platt J (1998) Fast Training of Support Vector Machines using Sequential Minimal Optimization. Advances in Kernel Methods - Support Vector Learning, B. Schölkopf, C. Burges, and A. Smola, eds., MIT Press
23. Polkowski L, Lin TY, Tsumoto S (eds.) (2000) Rough Set Methods and Applications: New Developments in Knowledge Discovery in Information Systems, Vol. 56 Studies in Fuzziness and Soft Computing, Physica-Verlag, Heidelberg, Germany
24. Quinlan JR (1993) C4.5: Programs for Machine Learning. The Morgan Kaufmann Series in Machine Learning. Morgan Kaufmann Publishers, San Mateo, CA
25. Raman B, Ioerger TR (2002) Instance-based filter for feature selection. Journal of Machine Learning Research, Vol. 1, pp. 1-23
26. Rasmani K, Shen Q (2004) Modifying weighted fuzzy subsethood-based rule models with fuzzy quantifiers. In Proceedings of the 13th International Conference on Fuzzy Systems, pp. 1687-1694
27. Salton G, Buckley C (1988) Term Weighting Approaches in Automatic Text Retrieval. Information Processing and Management, Vol. 24, No. 5, pp. 513-523
28. Schölkopf B (1997) Support Vector Learning. R. Oldenbourg Verlag, Munich
29. Shen Q, Chouchoulas A (2000) A modular approach to generating fuzzy rules with reduced attributes for the monitoring of complex systems. Engineering Applications of Artificial Intelligence, Vol. 13, No. 3, pp. 263-278
30. Skowron A, Grzymala-Busse JW (1994) From rough set theory to evidence theory. In Advances in the Dempster-Shafer Theory of Evidence, (R. Yager, M. Fedrizzi, and J. Kasprzyk eds.), John Wiley & Sons, Inc.
31. Slowinski R (ed.) (1992) Intelligent Decision Support. Kluwer Academic Publishers, Dordrecht
32. Swiniarski RW (1996) Rough set expert system for online prediction of volleyball game progress for US olympic team. In Proceedings of the 3rd Biennial European Joint Conference on Engineering Systems Design Analysis, pp. 15-20
33. Witten IH, Frank E (1998) Generating Accurate Rule Sets Without Global Optimization. In Machine Learning: Proceedings of the 15th International Conference, Morgan Kaufmann Publishers, San Francisco
34. Witten IH, Frank E (2000) Data Mining: Practical machine learning tools with Java implementations. Morgan Kaufmann, San Francisco
35. Wygralak M (1989) Rough sets and fuzzy sets - some remarks on interrelations. Fuzzy Sets and Systems, Vol. 29, No. 2, pp. 241-243
36. Yahoo. www.yahoo.com
37. Zadeh LA (1965) Fuzzy sets. Information and Control, 8, pp. 338-353

4

Simultaneous Ant Colony Optimization Algorithms for Learning Linguistic Fuzzy Rules

Michelle Galea[1] and Qiang Shen[2]

[1] Centre for Intelligent Systems and their Application
School of Informatics
University of Edinburgh
Edinburgh EH8 9LE, UK
m.galea@sms.ed.ac.uk

[2] Department of Computer Science
University of Wales
Aberystwyth SY23 3DB, UK
qqs@aber.ac.uk

Summary. An approach based on Ant Colony Optimization for the induction of fuzzy rules is presented. Several Ant Colony Optimization algorithms are run simultaneously, with each focusing on finding descriptive rules for a specific class. The final outcome is a fuzzy rulebase that has been evolved so that individual rules complement each other during the classification process. This novel approach to fuzzy rule induction is compared against several other fuzzy rule induction algorithms, including a fuzzy genetic algorithm and a fuzzy decision tree. The initial findings indicate comparable or better classification accuracy, and superior comprehensibility. This is attributed to both the strategy of evolving fuzzy rules simultaneously, and to the individual rule discovery mechanism, the Ant Colony Optimization heuristic. The strengths and potential of the approach, and its current limitations, are discussed in detail.

4.1 Introduction

Many fuzzy rule induction algorithms are adaptations of crisp rule induction algorithms that fail to take into account a fundamental difference between crisp and fuzzy rules, which is how they interact during the inference or classification process. This chapter presents a strategy based on the simultaneous running of several Ant Colony Optimization (ACO) algorithms, designed specifically with the induction of a complete fuzzy rulebase in mind.

Due to their very nature, fuzzy rules will match or cover all cases within a training set, but to varying degrees. Having a final rulebase of complementary fuzzy rules is therefore essential to the inference process – i.e. it is necessary to avoid

M. Galea and Q. Shen: *Simultaneous Ant Colony Optimization Algorithms for Learning Linguistic Fuzzy Rules*, Studies in Computational Intelligence (SCI) **34**, 75–99 (2006)
www.springerlink.com © Springer-Verlag Berlin Heidelberg 2006

a situation where a case requiring classification is closely matched by two or more rules that have different conclusions. To encourage complementary rules an approach is adopted that allows fuzzy rules describing different classes to be evolved and evaluated simultaneously.

The mechanism adopted for discovering the individual fuzzy rules is based on the ACO heuristic, so that several ACOs are run in parallel, with each constructing rules that describe a specific class. The constructionist nature of the adapted ACO algorithm itself affords several additional advantages, such as providing inbuilt mechanisms for preventing over-fitting to the training data, and dealing with imbalanced datasets, a common occurrence in real-world datasets.

The next section introduces fuzzy rules and rule-based systems, in so far as it is necessary to understand the work presented here. For a more comprehensive exposition the reader is directed to [1] for fuzzy set theory and logic in general, and to [2] for classification and modeling with linguistic fuzzy rules in particular. This section also describes ACO in the context of rule induction, and reviews the limited existing literature on the topic. Section 4.3 highlights the potential advantage provided by simultaneous rule learning, and describes the implemented system. Section 4.4 then presents experiments and an analysis of results. The final section highlights the advantages and limitations of the current research, which in turn suggest several avenues for future work.

4.2 Background

4.2.1 Fuzzy Rules and Rule-Based Systems

There are several different approaches for reasoning with imperfect or imprecise knowledge [3], including fuzzy rule-based systems (FRBSs) that are based on fuzzy set theory and fuzzy logic [4]. FRBSs capture and reason with imprecise or inexact knowledge (in fuzzy logic everything is a measure of degree [5]), and since many real-world problems contain a measure of imprecision and noise, the application of such approximate reasoning systems in these situations is often not only a viable but a necessary approach. This is supported by many successful applications in industry and commerce that deal with automated classification, diagnosis, monitoring and control (e.g. [6, 7]).

A simplified view of an FRBS is depicted in Fig. 4.1 on the facing page. At the core of such a system are:

1. A knowledge base that consists of fuzzy production IF-THEN rules (the rulebase – RB) that conceptualise domain knowledge, and the membership functions (the database – DB) defining the fuzzy sets associated with conditions and conclusions in the rules.
2. An inference procedure that uses this stored knowledge to formulate a mapping from a given input (e.g. in classification, conditions denoted by attribute values) to an output (e.g. in classification, a conclusion denoted by a class label).

The knowledge base has traditionally been determined via discussions with domain experts but this approach has many problems and shortcomings [8] – the interviews are generally long, inefficient and frustrating for both the domain experts and knowledge engineers, especially so in domains where experts make decisions based on incomplete or imprecise information. Data mining for both the fuzzy rules and associated membership functions has therefore been an active research area in the last decade. In this work the membership functions are already determined, and the data mining is applied to the induction of linguistic fuzzy rules.

The following subsections present basic concepts such as fuzzy sets, membership functions and linguistic variables, and describe fuzzy rules and how they are used in the inference process.

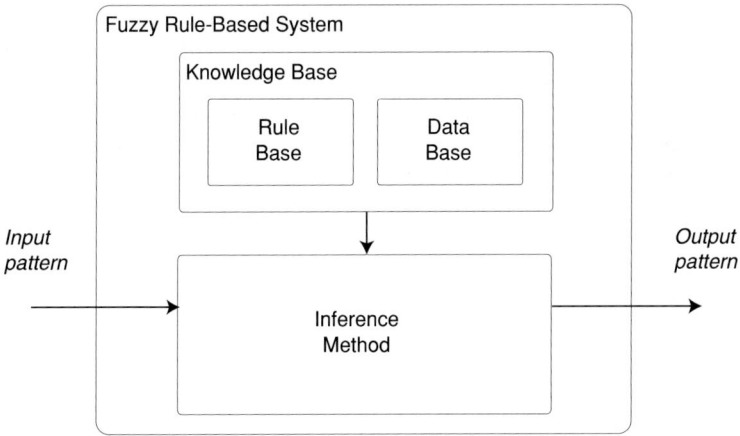

Fig. 4.1. Fuzzy rule-based system

Fuzzy Sets and Operators

A fuzzy set is a generalisation of a classical crisp set. A crisp set has a clearly defined boundary that either fully includes or fully excludes elements. A fuzzy set has a fuzzy boundary and each element u in the universe of discourse U belongs to the fuzzy set, but with a degree of membership in the real interval [0,1]. The closer this value is to 0, the less u may be considered as belonging to the fuzzy set in question, whilst the closer the membership value is to 1, the more u may be considered as belonging.

The degree of membership of the element u for the fuzzy set A is denoted by $\mu_A(u)$, where μ_A is called the membership function of A. This function maps each input $u \in U$ to its appropriate membership value. The fuzzy set A may therefore be denoted by the set of pairs:

$$A = \{(u, \mu_A(u)) \mid u \in U, \mu_A(u) \in [0,1]\} \qquad (4.1)$$

The graph of a membership function may take different shapes, and whether a particular shape is appropriate is generally determined by the application context. Common functions include the triangular, trapezoidal and the Gaussian [7].

Fuzzy sets are associated with each condition in a fuzzy rule and so it is necessary to be able to perform specific operations on single or multiple fuzzy sets. Fuzzy generalisations of the standard set union, intersection and complement are, respectively, *min*, *max* and the additive complement:

$$\mu_{A \cap B}(u) = min(\mu_A(u), \mu_B(u)) \qquad (4.2)$$

$$\mu_{A \cup B}(u) = max(\mu_A(u), \mu_B(u)) \qquad (4.3)$$

$$\mu_{\neg A}(u) = 1 - \mu_A(u) \qquad (4.4)$$

The above three operators are the ones most commonly used for interpreting and combining fuzzy values over the corresponding logical connectives in a fuzzy IF-THEN rule (conjunction, disjunction and negation), but there are several other definitions that may be used instead. In general, an intersection of two fuzzy sets A and B is defined by a binary operator called a *triangular norm* (or *t-norm*), that can aggregate two membership values. Similarly, a union of two fuzzy sets A and B is defined by a binary operator called a *triangular co-norm* (or *s-norm*). Other *t-norms*, *s-norms* and alternative fuzzy complement operators are discussed in [9] in more detail.

Linguistic Variables and Fuzzy Rules

A linguistic variable is a variable that has words or sentences in a natural or synthetic language as its domain values [10]. The domain values are called linguistic terms or labels, and each has associated with it a defining membership function.

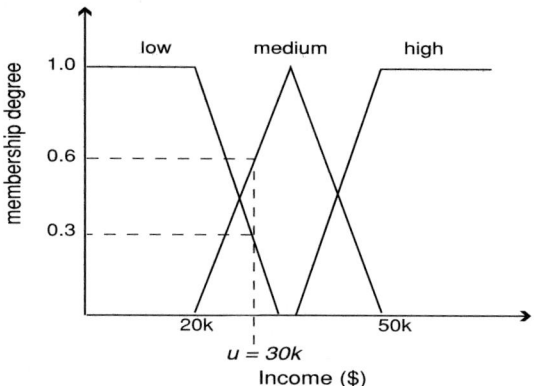

Fig. 4.2. A linguistic fuzzy variable

Figure 4.2 illustrates an example of a linguistic variable called *Income*, that has three linguistic terms in its domain {*low_income, medium_income, high_income*}. It is the overlap between the membership functions defining the linguistic terms that allows fuzzy rules to represent and reason with imprecise or vague information.

When an observation of a linguistic variable is made, or a measurement is taken, the value needs to be 'fuzzified' before it can be used by the FRBS, i.e. its degrees of membership for the different linguistic terms of the variable need to be determined. For instance, consider Fig. 4.2 again. If the income for a person is given as \$30k, this translates to $\mu_{low\_income}(\$30k) = 0.3$, $\mu_{medium\_income}(\$30k) = 0.6$, and $\mu_{high\_income}(\$30k) = 0.0$.

There are different types of fuzzy IF-THEN rules, but the rules induced here are linguistic Mamdani-type rules [11], e.g.:

$$R_1 : \textit{IF } \text{TEMPERATURE } \textit{is Mild } \text{OR } \textit{Cool } \text{AND } \text{WIND } \textit{is Windy}$$
$$\textit{THEN Weightlifting}$$

The underlying dataset which this rule partly describes, with its attributes and domain values, is described in more detail in Sect. 4.4.1.

Linguistic fuzzy rules are a particularly explicit and human-comprehensible form for representing domain knowledge. Comprehensible knowledge, in turn, may help to validate a domain expert's knowledge, refine an incomplete or inaccurate domain theory, provide confidence in the automated system now affecting decisions, highlight previously undiscovered knowledge, and, optimize system performance by highlighting both significant and insignificant features (attributes) of the domain.

Classification using Fuzzy Rules

In an FRBS used for classification purposes, all rules are applied to the input vector, and each will match the input pattern but to varying degrees. How a decision is reached as to what output (classification) should be assigned is handled by the inference method.

There are several different inference methods, with variations on each depending on which *t-norm*, *s-norm* and other aggregation operators are used. References [12, 13] provide several different examples of inference methods. The one used here is a popular one, mainly due to the high transparency of the classification process. It was chosen because it is the one utilised by other works against which the system implemented here is compared. This makes the comparison between the different algorithms more equitable, by ensuring that any difference in performance is due to the rule induction algorithm, and not due to the inference method. To this end, the same fuzzy sets and membership functions that were used by the other algorithms, are also used by the implemented system.

The inference process used is a single winner based-method [13] and the rule that achieves the highest degree of match with the input pattern or vector gets to classify that vector. This is depicted in Fig. 4.3 where $mCond(R_i, u)$ denotes the degree of match between the antecedent part of rule R_i and the input pattern u, and c^{R_i} is the class of R_i.

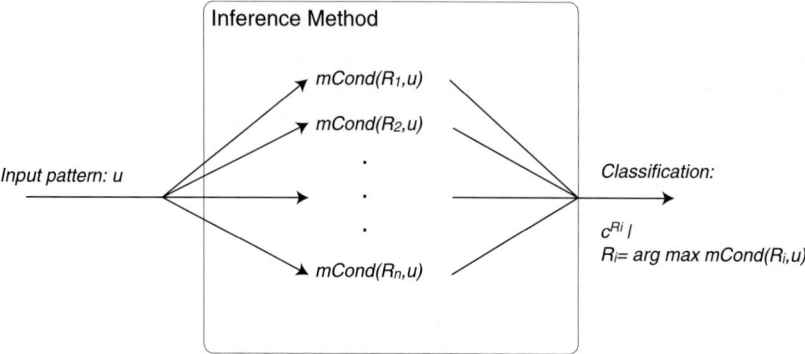

Fig. 4.3. Classification by an FRBS – single winner method

A Rule-Matching Example

Since the process of finding a degree of match between a fuzzy rule antecedent and an input pattern is used not only in classifying the test set for estimating the accuracy of the induced rulebase, but also in constructing the rules and in evaluating them, an example follows.

For illustration purposes a more convenient representation of the rule presented earlier is used: R_1=(0,0,0; 0,1,1; 0,0; 1,0; 0,0,1). This means that there are five attributes, the first four being condition attributes with two or three values (terms) in the domains, and the last representing the class attribute with three possible values (*Volleyball, Swimming* and *Weightlifting* respectively). Terms that are present in the rule are denoted by 1, others by 0. These rules may only classify instances into one class. However, there may be more than one specific attribute value present in a rule (i.e. propositional rules with internal disjunction).

Consider now a fuzzy instance u=(0.9,0.1,0.0; 0.0,0.3,0.7; 0.0,1.0; 0.9,1.0; 0.0,0.3,0.7), i.e. each observation or measurement of each variable has already been fuzzified. The representation is similar as for rule R_1, though the value for each term represents the degree of membership and lies in the range [0,1]. Note that the conclusion attribute values may be greater than 0 for more than one class, but that an instance is considered to belong to the class with the highest degree of membership, and in this case, the class is *Weightlifting*.

The degree of condition match between rule R_1 and instance u is given by

$$mCond(R_1, u) = min_k(mAtt(R_1^k, u^k)) \qquad (4.5)$$

In (4.5) above $mAtt(R_1^k, u^k)$ measures the degree of match between an attribute k in R_1 and the corresponding attribute in u:

$$mAtt(R_1^k, u^k) = \begin{cases} 1 & : R_1^k \ empty \\ max_j(min(\mu_j(R_1^k), \mu_j(u^k))) & : otherwise \end{cases} \qquad (4.6)$$

where R_1^k *empty* indicates that no term from the domain of attribute k is present in rule R_1, and j is a specific term within the domain of attribute k. If the attribute is not represented at all in the rule, the interpretation is that it is irrelevant in making a particular classification. From the rule and instance examples above the attribute matches are: $mAtt(R_1^1, u^1) = 1.0$, $mAtt(R_1^2, u^2) = 0.7$, $mAtt(R_1^3, u^3) = 1.0$ and $mAtt(R_1^4, u^4) = 0.9$, with the degree of match between the rule antecedent of R_1 and the input pattern u therefore being $mCond(R_1, u) = min(1.0, 0.7, 1.0, 0.9) = 0.7$.

If the purpose were classification of the input pattern u, then the degree of condition match between u and all other rule antecedents in the rulebase is determined. For instance, if two other rules were present, say R_2 describing the conditions leading to a decision to go *Swimming*, and R_3 leading to a decision to play *Volleyball*, and their degree of condition matches were $mCond(R_2, u) = 0.2$ and $mCond(R_3, u) = 0.4$, then u would be assigned the same class as that of R_1 – *Weightlifting*. Since the actual class of u is *Weightlifting*, then during training or testing this would be counted as a correct classification. If more than one rule describing different classes obtained the highest degree of condition match with u, this would be considered a misclassification when determining the accuracy of the induced rulebase.

4.2.2 Ant Colony Optimization and Rule Induction

Ant algorithms are heuristics inspired by various behaviours of ants that rely on stigmergy, a form of indirect communication between individual ants that is enabled by effecting changes to a common environment [14]. Examples of such behaviours include cemetery organisation, brood sorting and foraging by real ants. Ant algorithms form a major branch of research in Swarm Intelligence [15], which may be broadly considered as the application of social insect-inspired algorithms to hard problems. They are increasingly being applied to core data mining tasks such as clustering (e.g. [16, 17]), feature selection (e.g. [18, 19]) and rule induction (e.g. [20, 21]).

Ant Colony Optimization (ACO) [22] is a particular instantiation of ant algorithms. It is a population-based algorithm motivated by the foraging strategies of real ants, which have been observed capable of finding the shortest path between their nest and a food source [23]. This is attributed to the fact that ants lay a chemical substance, called a pheromone, along the paths they take, and when presented with a choice between alternative paths, they tend to choose the one with the greatest amount of pheromone. Pheromone, however, evaporates so that over time the shortest path accrues more pheromone as it is traversed more quickly.

In ACO each artificial ant is considered a simple agent, communicating with other ants only indirectly. A high-level description of an ACO-based algorithm is given in Fig. 4.4 on the following page. Following is a brief introduction of the main elements necessary for an implementation of an ACO algorithm [15], set in the context of rule induction. More detail is provided in Sect. 4.3, which describes the implemented system. The first four elements relate to line (2) of Fig. 4.4, the fifth relates to line (3), and the sixth to line (4):

1. An appropriate *problem representation* is required that allows an artificial ant to incrementally build a solution using a *probabilistic transition rule*. The

problem is modelled as a search for a best path through a graph. In the context of rule induction a solution is a rule antecedent and each node of the graph represents a condition that may form part of it, such as OUTLOOK=Sunny, or OUTLOOK=Cloudy.
2. The *probabilistic transition rule* determines which node an ant should visit next. The transition rule is dependent on the *heuristic* value and the *pheromone* level associated with a node. It is biased towards nodes that have higher probabilities, but there is no guarantee that the node with highest probability will get selected. This allows for greater exploration of the solution space.
3. A local *heuristic* provides guidance to an ant in choosing the next node for the path (solution) it is building. This may be similar to criteria used in greedy algorithms, such as information gain for the induction of crisp rules, or fuzzy subsethood values and measurements of vagueness in a fuzzy set, for the induction of fuzzy rules.
4. A *constraint satisfaction method* forces the construction of feasible rules. For instance, if simple propositional IF-THEN rule antecedents are being constructed, then at most only one fuzzy linguistic term from each fuzzy variable may be selected.
5. A *fitness function* determines the quality of the solution built by an ant. This could be a measure based on how well the rule classifies the instances in the training set.
6. The *pheromone update rule* specifies how to modify the pheromone levels of each node in the graph between iterations of an ACO algorithm. For instance, the nodes (conditions) contained in the best rule antecedent created get their pheromone levels increased.

```
(1)    while termination condition false
(2)        each ant constructs a new solution
(3)        evaluate new solutions
(4)        update pheromone levels
(5)    output best solution
```

Fig. 4.4. Basic ACO Algorithm

The application of ant algorithms to classification rule induction is a relatively new research area, but one that is gaining increasing interest. A first attempt is found in [24], where an ACO algorithm is used, however, not for constructing fuzzy rule antecedents, but for assigning rule conclusions. In a graphical representation of the problem, the fixed number of graph nodes are fuzzy rule antecedents found previously by a deterministic method from the training set. An ant traverses the graph, visiting each and every node and probabilistically assigns a rule conclusion to each.

In [20] Parpinelli et al. introduce *Ant-Miner*, a system using ACO algorithms for generating crisp IF-THEN rule antecedents. In the problem graph each node

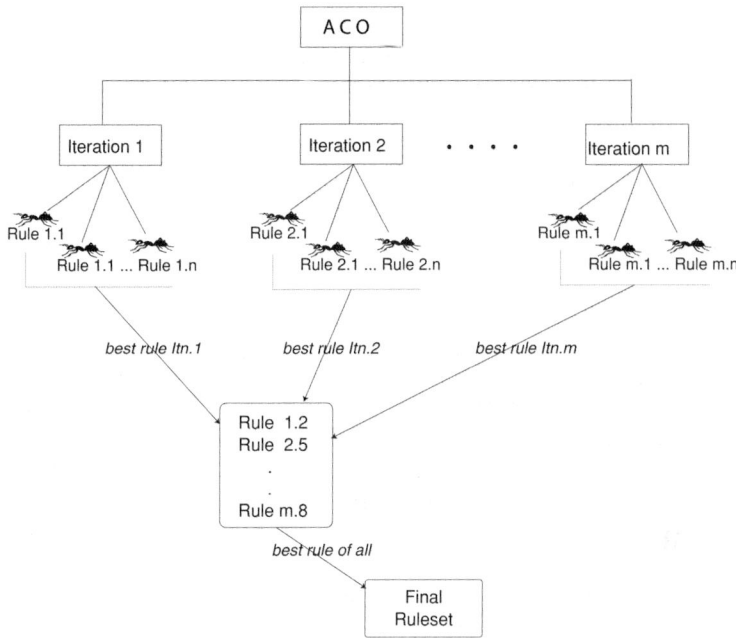

Fig. 4.5. Overview of a basic rule-inducing ACO algorithm

represents a condition that may be selected as part of the crisp rule antecedent being built by an ant. An ant goes round the graph selecting nodes and building its simple propositional rule antecedent. The rule conclusion is assigned afterwards by a deterministic method. Recent interest in *Ant-Miner* has resulted in various modifications to it, and applications to different problem domains. These include changes to the heuristic, transition and pheromone update rules (e.g. [25, 26]), and application to web page classification [27] and handwritten number recognition [28].

The overall strategy *Ant-Miner* uses is one of iterative rule learning – starting with a full training set an ACO algorithm is run and the best rule created by an ant is added to a final ruleset. Instances in the training set that are covered by this best rule are removed before a new ACO algorithm is run, to find another best rule that is added to the final ruleset. This process is re-iterated until only a few instances (as pre-determined by the user) remain in the training set, when a default rule is created to cover them. The final result is an ordered rule list with the rules being applied in the order in which they were created, when classifying a new instance. [29] uses the same iterative strategy as *Ant-Miner* and also produces a decision list. However, Particle Swarm Optimization [30], another major branch of Swarm Intelligence [15], is used instead of ACO as the rule discovery mechanism.

In [21] a different iterative strategy, outlined in Fig. 4.6 on the next page, is used for the induction of fuzzy rules. This system is called *FRANTIC-IRL* and several ACO algorithms are run for each class, with each producing a fuzzy rule that covers a subset of the instances in the training set belonging to that class. Fuzzy rules

```
(1)    for each class
(2)        reinstate full training set
(3)        while classInstRemaining>maxClassInstUncovered
(4)            for numIterations
(5)                each ant constructs rule
(6)                evaluate all rules
(7)                update pheromone levels
(8)            add best rule to final rulebase
(9)            remove covered class instances
(10)   output best rulebase
```

Fig. 4.6. Iterative rule learning for fuzzy rules – *FRANTIC-IRL*

describing one class are produced until only a few class instances remain in the training set, line (3). The system then reinstates the full training set and proceeds to run ACO algorithms to find fuzzy rules describing another class. The process continues until all classes have been adequately described.

This iterative approach to fuzzy rule learning using ACO algorithms compared very favourably against other fuzzy rule induction algorithms, in terms of both accuracy and comprehensibility. The results and analyses also suggested, however, that a simultaneous rule learning strategy would be more appropriate for the induction of fuzzy rules. This chapter presents such a simultaneous strategy for the induction of a cooperative fuzzy rulebase.

4.3 Simultaneous Fuzzy Rule Learning

This section first highlights the potential advantage of simultaneous rule learning over iterative rule learning, and then describes the implemented system – *FRANTIC-SRL*.

4.3.1 Why Simultaneous Rule Learning

As described in the previous section, *FRANTIC-IRL* follows an iterative rule learning approach where the fuzzy rules making up a rulebase are created independently of each other, i.e. without taking into account how they will interact in the final rulebase.

[21] highlights a disadvantage with this approach. *FRANTIC-IRL* was run on the Saturday Morning Problem dataset (also described in detail in Sect. 4.4.1 of this chapter). Rulebase *A* which consists of rules R1–R3 in Table 4.1 on the facing page is one of the rulebases commonly produced, achieving an accuracy of 93.75% on the dataset, while Rulebase *B* which consists of rules R1–R4 is another rulebase and achieves an accuracy of 87.50%. Table 4.2 on the next page helps to illustrate a potential problem with fuzzy rule interaction during classification. The first column is an instance identifier of some of the instances in the dataset, the second provides the actual class of an instance, while columns 3–6 give the degree of match between

a fuzzy rule from Table 4.1 and an instance. The abbreviation in brackets following a rule identifier denotes the class the rule describes: VB–*Volleyball*, SW–*Swimming*, and WL–*Weightlifting*. Column 7 of Table 4.2 gives the classification made by Rulebase *A* (Rb *A*), while the last column gives the classification made by Rulebase *B* (Rb *B*). It should be remembered that an instance is classified by the fuzzy rule with the highest degree of match.

Table 4.1. *FRANTIC-IRL* rulebases for Saturday Morning Problem – Rulebase *A*: R1-R3, 93.75% accuracy; Rulebase *B*: R1-R4, 87.50% accuracy

R1	IF OUTLOOK is NOT_Rain AND TEMPERATURE is NOT_Cool AND HUMIDITY is Normal AND WIND is Not-windy THEN *Volleyball*
R2	IF OUTLOOK is NOT_Rain AND TEMPERATURE is Hot THEN *Swimming*
R3	IF TEMPERATURE is NOT_Hot AND WIND is Windy THEN *Weightlifting*
R4	IF OUTLOOK is NOT_Sunny AND TEMPERATURE is NOT_Mild THEN *Weightlifting*

Table 4.2. Fuzzy rule interaction and classification

| Inst. ID | Actual Class | Degree of match | | | | Classification | |
		R1(VB)	R2(SW)	R3(WL)	R4(WL)	Rb *A*	Rb *B*
5	WL	0.1	0.1	0.3	0.7	WL	WL
6	WL	0.3	0.0	0.4	0.7	WL	WL
7	WL	0.0	0.0	0.1	1.0	WL	WL
10	WL	0.1	0.0	0.9	0.9	WL	WL
13	WL	0.0	0.2	0.8	0.8	WL	WL
14	WL	0.3	0.0	0.7	0.7	WL	WL
15	WL	0.0	0.0	0.8	1.0	WL	WL
8	VB	0.2	0.0	0.0	0.8	VB	WL

Consider now only the instances that actually describe *Weightlifting* (instances in Table 4.2 with 'WL' in column 2) – Rulebase *B* is a closer match to the data than Rulebase *A*, since the additional rule R4 describing *Weightlifting* achieves a very high degree of match with all WL instances. However, R4 also achieves the highest degree of match of all rules with instance 8, and therefore *mis*classifies this instance. Note that Rulebase *A* (rules R1-R3), though occasionally achieving a lower degree of match with WL instances than Rulebase *B*, still manages to correctly classify all WL instances, *and* avoids misclassifying instance 8. This issue arises as a direct consequence of the strategy used to induce the complete fuzzy rulebase – in iterative rule learning the fuzzy rules are added to the final rulebase sequentially, and without taking into account how they may interact with rules describing different classes already in the rulebase, or with other rules that may be added later on.

A strategy is therefore required that encourages optimal fuzzy rule interaction during classification. *FRANTIC-SRL* is a system designed with this requirement in mind – during creation of the rulebase individual rules are not evaluated separately on the training set as in *FRANTIC-IRL*, but are grouped together and evaluated as a potential complete rulebase. *FRANTIC-SRL* is described in detail in the following section.

4.3.2 FRANTIC-SRL

FRANTIC-SRL (Fuzzy Rules from ANT-Inspired Computation – Simultaneous Rule Learning) runs several ACO algorithms in parallel, with each maintaining its own problem graph, pheromone levels and heuristic values. The ACO algorithms are run simultaneously in principle, i.e. this is not as yet a parallel implementation running on multiple processors. An overview of the system is provided in Figure 4.7.

After each class has had its rules created for a particular iteration (Fig.4.7, lines (2)–(3)), all possible combinations of rules (one from each class) are formed into a rulebase and this is tested on the training set (lines (4)–(5)). The rules in the best performing rulebase are used to update the pheromone levels (line (6)), with the rule describing a specific class being used to update the pheromone levels of the associated ACO. The following subsections detail the rule construction and rule evaluation processes.

```
(1)    for numIterations
(2)        for each class
(3)            each ant constructs rule
(4)        for each combined rulebase
(5)            evaluate rulebase
(6)        update pheromone levels
(7)    output best rulebase
```

Fig. 4.7. Simultaneous rule learning for fuzzy rules – *FRANTIC-SRL*

Rule Construction

FRANTIC-SRL has the flexibility to create simple propositional rules (e.g. *IF* TEMPERATURE *is Cool AND* WIND *is Windy THEN Weightlifting*), propositional rules with internal disjunction (e.g. *IF* TEMPERATURE *is Cool OR Mild AND* WIND *is Windy THEN Weightlifting*), and propositional rules that include negated terms (e.g. *IF* TEMPERATURE *is* NOT_*Rain AND* WIND *is Windy THEN Weightlifting*).

When creating a rule antecedent an ant traverses a problem graph where each node represents a term that may be added e.g OUTLOOK=Sunny. In the case of constructing rules with negated terms, the graph has double the number of nodes – one extra for each original linguistic term, e.g. OUTLOOK=NOT_Sunny. The choice of the next node to visit depends on both a heuristic value and the pheromone level associated with the node. It is made probabilistically but is biased towards terms that have relatively higher heuristic and pheromone values.

After selection but before a term is actually added to a rule antecedent, a check is made – this ensures that the resultant rule antecedent covers a minimum number of the appropriate class instances from the training set (set by a parameter called `minInstPerRule`), and is a way of avoiding over-fitting to the training data. As previously mentioned, all fuzzy rules cover all training instances, but to varying degrees, and so what constitutes coverage of an instance by a fuzzy rule needs defining. During rule construction, a fuzzy rule describing a specific class is said to cover or match an instance if:

1. the rule and instance belong to the same class; and,
2. the degree of match between the condition parts of rule and instance is equal to or greater than a pre-defined value, here set by a parameter called `constructionThreshold`.

Consider as an example the fuzzy rule and instance given in Sect. 4.2.1, paragraph *Classification using Fuzzy Rules*. Rule R_1 and instance u belonged to the same class so condition (1.) above is satisfied. The degree of condition match between R and u was found to be $mCond(R,u) = 0.7$. If `constructionThreshold` is set to 0.7 or lower, then R_1 is considered to adequately cover u, while if it is set to higher than 0.7, then R_1 is considered not to cover u.

For simple propositional rules, or rules with negated terms, if an ant does add a term to its rule antecedent then it will not consider other linguistic terms belonging to the same linguistic variable. For example, if the linguistic variable OUTLOOK has terms Sunny, Cloudy, Rain, and the term OUTLOOK=Sunny has just been added to the rule antecedent, then the remaining terms are not considered further. If this restriction is removed, then it is possible for ants to add more than one linguistic term from each variable, with the interpretation being of a disjunctive operator between the terms added, e.g. OUTLOOK=(Sunny OR Mild).

Heuristic

The heuristic used to guide ants when selecting terms is based on fuzzy subsethood values [31], giving a degree to which one fuzzy set A is a subset of another fuzzy set B:

$$S(A,B) = \frac{M(A \cap B)}{M(A)} = \frac{\sum_{u \in U} min(\mu_A(u), \mu_B(u))}{\sum_{u \in U} \mu_A(u)} \quad (4.7)$$

where in this case u is an instance from the training set U, A represents a class label and B a term that may be added to a rule antecedent.

The heuristic value of a term j – η_j – gives a measurement of how important that term is in describing a *specific* class. If there are n class labels in a dataset, j will therefore have n heuristic values associated with it in total. An ACO finding rules to describe a particular class will use the appropriate term heuristic values, i.e. those associated with the class.

The heuristic value for a negated term is the complement of the heuristic value for the non-negated term, i.e. $\eta_{NOT\_j} = 1 - \eta_j$.

Pheromone Updating

Unlike most other ACO implementations, the pheromone here is deposited on the nodes, and not the edges of the graph. This is because it is the actual nodes (terms) themselves that are important in constructing the rule antecedent, and not the order in which they are selected (as opposed to, say, the travelling salesman problem where the order in which the cities are visited is relevant). For instance, the rule:

IF TEMPERATURE *is Mild AND* WIND *is Windy THEN Weightlifting*

is equivalent to the rule

IF WIND *is Windy AND* TEMPERATURE *is Mild THEN Weightlifting*

At the start of an ACO run, all nodes in the graph have an equal amount of pheromone which is set to the inverse of the number of nodes. The pheromone level of individual nodes, however, changes between iterations. At the end of each iteration rules created by all ants are evaluated. The terms in the best rule of an iteration of a particular ACO, say R, get their pheromone levels increased:

$$\tau_j(t+1) = \tau_j(t) + (\tau_j(t) \times Q), \forall j \in R \tag{4.8}$$

i.e. at time $t+1$ each term j that is present in rule R gets its pheromone level increased in proportion to the quality Q of the rule (defined in Sect. 4.3.2).

The pheromone levels of all terms are then normalised (each pheromone level is divided by the sum of all pheromone levels), which results in a decrease of the pheromone levels of terms *not* in R. The pheromone updating process is therefore a reinforcement mechanism – both positive and negative – for ants constructing new rules in successive iterations: terms that have had their pheromone levels increased have a higher chance of being selected, while those that have had their levels decreased have a lower chance.

Transition Rule

Ants select terms while constructing a rule antecedent according to a transition rule that is probabilistic but biased towards terms that have higher heuristic and pheromone levels. The probability that ant m selects term j when building its rule during iteration t is given by:

$$P_j^m(t) = \frac{[\eta_j] \times [\tau_j(t)]}{\sum_{i \in I_m}[\eta_i] \times [\tau_i(t)]} \tag{4.9}$$

where I_m is the set of terms that may still be considered for inclusion in the rule antecedent being built by ant m.

If propositional rules with internal disjunction are being created, then I_m will exclude terms that are already present in the current partial rule antecedent, and terms that have already been considered but found to decrease coverage of the training set below the required number of instances (as set by `minInstPerRule`). If simple propositional rules, or rules that include negated terms are being created, then I_m will further exclude other values within the domain of linguistic variables that already have a term present in the rule antecedent.

The probabilistic nature of the transition rule is a way of introducing exploration into the search for a solution, in the expectation that a more optimal solution may well be found rather than by adhering strictly to terms with the highest values.

Rule Evaluation

Each constructed rule needs to be evaluated and this is done by assessing how accurate it is in classifying the training instances. However, instead of evaluating each rule separately, at the end of each iteration when each class has produced its set of rules, a rule describing one class is combined with one rule describing each of the other classes and together they classify the training set.

The method of classification used during evaluation is the single winner-based method described briefly in Sect. 4.2.1. More specifically, for each instance u:

1. for each rule, calculate the condition match for instance u;
2. assign to instance u the class of the rule with the highest condition match.

The accuracy obtained by a rulebase on the training set is used as a measure of the quality, Q, of each rule within the rulebase. The rules in the rulebase obtaining the highest accuracy are the ones used for updating the pheromone levels in the various ACO algorithms before the next iteration is run.

Currently, all possible rulebases are created and evaluated after an iteration, by combining a rule from one class (ACO), with one rule from each of the other classes. This brings the total number of rulebase evaluations to:

$$numIterations * numAnts^{numClasses} \qquad (4.10)$$

where *numIterations* and *numAnts* are as defined in Table 4.5 on page 92, and *numClasses* is the number of class labels in the training set. It is quite possible, however, that this number may be drastically reduced without its impacting on the quality of the final rulebase. In work using co-operative co-evolution [32] to induce a complete knowledge base (e.g. [33], where one genetic algorithm evolves rulebases, and another evolves membership functions), not all possible combinations are formed. Generally, only a few representatives from each population are used to form different combinations of knowledge bases, and these representatives may be chosen according to fitness, randomly, or a combination of both. This suggests a useful direction for an investigation into the reduction of computational expense relating to rulebase evaluations for *FRANTIC-SRL*.

4.4 Experiments and Analyses

4.4.1 Experiment Setup

This subsection details the datasets, the fuzzy rule induction algorithms against which *FRANTIC-SRL* is compared, and the parameter settings that are used in the empirical study reported below.

The Datasets

Saturday Morning Problem. The first dataset on which *FRANTIC-SRL* is tested is a fuzzified version of the small, artificial Saturday Morning Problem dataset originally used in the induction of decision trees [34]. This dataset was chosen as it has been used by several fuzzy rule inductions algorithm, and so permits a direct comparison between these algorithms and *FRANTIC-SRL*.

The dataset consists of sixteen instances, and four linguistic condition attributes with each having two or three linguistic terms. The class attribute PLAN classifies each instance into *Volleyball*, *Swimming* or *Weightlifting*. The condition attributes are described in Table 4.3.

Table 4.3. Saturday Morning Problem dataset features

Attribute	Linguistic Terms
OUTLOOK	Sunny, Cloudy, Rain
TEMPERATURE	Hot, Cool, Mild
HUMIDITY	Humid, Normal
WIND	Wind, Not-windy

Water Treatment Plant Database. This real-world database [35] is more challenging and contains the daily observations of 38 sensors monitoring the operation of an urban waste water treatment plant at various stages throughout the process, with the objective being to predict faults in the process. Observations were taken over 527 days and are real-valued. The database has 13 possible classifications for each daily set of observations, however, most classifications are assigned to only a few records in the database. Furthermore, when faults are reported these are generally fixed very quickly so that the database is heavily biased by containing a disproportionate number of records indicating normal operation of the plant, versus faulty operation.

The 13 classifications have therefore been collapsed to two: *OK* and *Faulty*, as in [36]. Records that have no assigned classification, and others with missing values have been removed, leaving 377 records for training and testing the rule induction algorithms (289 *OK*, 88 *Faulty*).

Other pre-processing steps include fuzzification of the features using trapezoidal functions, and a feature subset selection process [37] to reduce the number of features

([36] indicated better accuracy results when a reduced water treatment dataset was used). A description of the retained features is shown in Table 4.4. Each feature is described by three linguistic terms (*low, high, normal*), except for the last which uses only two (*low, high*).

Table 4.4. Water Treatment Plant database features

Name	Sensor Description
Q-E	Input to plant – flow
PH-E	Input to plant – pH
DBO-E	Input to plant – biological demand of oxygen
DBO-P	Input to primary settler – biological demand of oxygen
SSV-P	Input to primary settler – volatile suspended solids
PH-D	Input to secondary settler – pH
DQO-D	Input to secondary settler – chemical demand of oxygen
SSV-D	Input to secondary settler – volatile suspended solids
PH-S	Output – pH
SSV-S	Output – volatile suspended solids
RD-SED-G	Global performance, input – sediments

Other Induction Algorithms

The fuzzy rulebases produced by *FRANTIC-SRL* are compared with those produced by a fuzzy decision tree algorithm (*FDT*) [38], a fuzzy genetic algorithm (*FGA*) [39], and two methods based on subsethood values (*FSBA* [40], *WSBA* [41]). Apart from *WSBA*, the algorithm acronyms are not the names given to the algorithms by the original authors, but are introduced here for ease of reference.

FDT is a deterministic algorithm. *FGA* uses randomness to generate rules and so may generate different rulebases achieving different predictive accuracy. The number of fuzzy rules in the rulebases produced by both algorithms is not pre-determined by the user.

FSBA uses subsethood values to select a small number of conditions to formulate one rule for each class in the training set. It is a deterministic algorithm but requires the setting of two parameters α and β. α is a threshold used to determine which linguistic terms should be present in a rule antecedent describing a specific class – terms with a subsethood value equal to or greater than α are selected. If the subsethood values for the linguistic terms associated with a particular class are all lower than α, then an explicit rule can not be created for the class. Instead, an indirect rule is formed and will fire if the membership of the instance to be classified is greater than β (e.g. *IF Membership(OK)* $< \beta$ *THEN OUTCOME is FAULTY*).

WSBA, the second subsethood-based algorithm, uses subsethood values not to determine which linguistic terms should be present in a rule, but to determine a fuzzy quantifier in the range [0,1] for each term, all of which are present. Like *FSBA*, and

FRANTIC-SRL as it is currently implemented, it also generates only one rule per class. Examples of rules generated by all the algorithms are provided in the following subsections.

FRANTIC-SRL Parameters.

FRANTIC-SRL parameters that require setting are listed in Table 4.5, together with a brief description and the values given in order to obtain the results reported here for the two datasets – the Saturday Morning Problem dataset (SM) and the Water Treatment Plant database (WT). Very little parameter tuning has been done and these values are based on a few exploratory runs of the system that indicated reasonable results would be obtained. It is therefore quite possible that different settings of these parameters may lead to even better results.

Table 4.5. *FRANTIC-SRL* Parameters

Parameter Name	Description	SM	WT
numAnts	Number of ants constructing a solution within an iteration	4	10
numIterations	Number of iterations per ACO run	25	30
minInstPerRule	Required during rule construction – minimum number of instances in training set that a rule must cover	4	70%
constructionThreshold	Required during rule construction – sets the value for the threshold below which a rule is considered not to cover an instance in the training set	0.65	0.65

The minInstPerRule parameter is flexible enough so that different values may be given to different classes. This is particularly useful in imbalanced datasets such as the Water Treatment one, where stipulating the same number of instances that a rule must cover for a small class as for a large class is impractical. The value '4' therefore means that for each class a rule must cover at least 4 class instances from the training set, whilst the value '70%' means that a rule should cover at least 70% of the relevant class instances.

Both minInstPerRule and constructionThreshold have been implemented so that their values may change automatically, if necessary, during the running of an experiment. For instance, it is generally the case that the actual number of instances belonging to a particular class in the training set is equal to or greater than the value set by minInstPerRule. On the other hand, constructionThreshold may be set so high the no ant is able to construct a rule that covers the required number of class instances to the specified degree of match. In this case, the values of minInstPerRule and/or constructionThreshold may be automatically and gradually reduced until rules describing the class may be generated.

The adaptive nature of these parameters provides the system with a useful degree of autonomy, and reduces the need for unnecessary user intervention.

4.4.2 Saturday Morning Problem Results

A summary of the results produced on this dataset by various algorithms is provided in Table 4.6 – it gives the percentage classification accuracy on the training set, the number of rules generated, and the average number of conditions in a rule antecedent.

Table 4.6. Saturday Morning Problem – comparison of algorithms

	%Accuracy	#Rules	#Terms
FDT	81.25	6	1.7
FGA	87.50	5	3.2
FSBA	93.75	3	2.3
FRANTIC	93.33	3	2.7

The accuracy of only one rulebase is reported for *FGA* in [39], and reproduced in Table 4.6, and so the assumption here is that it is the best rulebase obtained. The results for *FSBA* reported in [40] are for $\alpha = 0.9$ and $\beta = 0.6$, and the assumption again is that this is the best obtainable. Since *FRANTIC-SRL* is a stochastic-based algorithm the result reported in Table 4.6 is the average of the accuracies obtained from 30 runs of the system with the parameters set as in Table 4.5. The standard deviation of the 30 accuracies is 1.6, i.e. only 2 out of the 30 runs produced accuracies below 93.75%, making the overall accuracy comparable with that obtained by *FSBA*.

Tables 4.7 to 4.10 on the next page provide examples of the rules generated by these algorithms. *FDT* generates simple propositional rules, *FGA* generates propositional rules with internal disjunction, and *FSBA* generates simple propositional rules that however may include negated terms. *FRANTIC-SRL* has the ability to generate all these variations on propositional rules, but a few early runs of the system determined that for this dataset rules that included negated terms were more accurately descriptive. *FRANTIC-SRL* and *FSBA* are also comparable with respect to the number of rules in a rulebase, and the average number of conditions per rule. Note, however, that the final rule produced by *FSBA* has no explanatory power of its own, as it is written in terms of the other rules.

4.4.3 Water Treatment Plant Results

The middle column of Table 4.11 on page 95 indicates the average accuracy obtained by several algorithms after performing stratified ten-fold cross-validation on the Water Treatment dataset. The same folds of the dataset were used for each algorithm, and the stratification ensures that each fold contains approximately the same proportions of instances of the different classes as the original complete dataset

Table 4.7. *FDT* rulebase for Saturday Morning Problem (81.25% accuracy)

R1	IF TEMPERATURE is Hot AND OUTLOOK is Sunny THEN *Swimming*
R2	IF TEMPERATURE is Hot AND OUTLOOK is Cloudy THEN *Swimming*
R3	IF OUTLOOK is Rain THEN *Weightlifting*
R4	IF TEMPERATURE is Mild AND WIND is Windy THEN *Weightlifting*
R5	IF TEMPERATURE is Cool THEN *Weightlifting*
R6	IF TEMPERATURE is Mild AND WIND is Not-windy THEN *Volleyball*

Table 4.8. *FGA* rulebase for the Saturday Morning Problem (87.25% accuracy)

R1	IF OUTLOOK is Sunny OR Cloudy AND TEMPERATURE is Hot THEN *Swimming*
R2	IF OUTLOOK is Rain THEN *Weightlifting*
R3	IF TEMPERATURE is Mild OR Cool AND WIND is Windy THEN *Weightlifting*
R4	IF OUTLOOK is Cloudy OR Rain AND HUMIDITY is Humid THEN *Weightlifting*
R5	IF OUTLOOK is Sunny OR Cloudy AND TEMPERATURE is Mild OR Cool AND HUMIDITY is Normal AND WIND is Not-windy THEN *Volleyball*

Table 4.9. *FSBA* rulebase for Saturday Morning Problem (93.75% accuracy)

R1	IF OUTLOOK is NOT_Rain AND HUMIDITY is Normal AND WIND is Not-windy THEN *Volleyball*
R2	IF OUTLOOK is NOT_Rain AND TEMPERATURE is Hot THEN *Swimming*
R3	IF $MF(R1) < \beta$ AND $MF(R2) < \beta$ THEN *Weightlifting*

Table 4.10. *FRANTIC-SRL* rulebase for Saturday Morning Problem (93.75% accuracy)

R1	IF OUTLOOK is NOT_Rain AND TEMPERATURE is NOT_Cool AND HUMIDITY is Normal AND WIND is Not-windy THEN *Volleyball*
R2	IF OUTLOOK is NOT_Rain AND TEMPERATURE is Hot THEN *Swimming*
R3	IF TEMPERATURE is NOT_Hot AND WIND is Windy THEN *Weightlifting*

did. The figure in brackets is the standard deviation of the accuracies of the ten rulebases produced. The right column gives the average number of terms per rule, with standard deviation in brackets. All these algorithms generate just one rule to describe each class. Note that since *FRANTIC-SRL* is a stochastic algorithm, the *FRANTIC* results presented in Table 4.11 are averages of *ten* ten-fold cross-validations.

As for the Saturday Morning Problem dataset, a few initial runs indicated that *FRANTIC-SRL* rules with negated terms were better descriptors of the Water Treatment Plant database. *FSBA* was run using all combinations of the following values: 0.5,0.55,...,0.85 for α, and 0.5,0.6,...,1.0 for β. The accuracy results reported here are the best obtained with $\alpha = 0.8$ and $\beta = 0.5$. *WSBA* obtained the best results

Table 4.11. Water Treatment Plant – comparison of algorithms

	%Accuracy	#Terms
WSBA	81.74 (7.6)	32.00 (0.0)
FSBA	69.51 (7.0)	4.45 (0.4)
FRANTIC	76.08 (6.6)	2.00 (0.0)

in terms of predictive accuracy, for this particular partitioning of the data. However, it does come at a cost to rule comprehensibility.

There is a considerable difference in the length of the rules produced by each algorithm. *FRANTIC-SRL* produces the most comprehensible rulebases and an example is provided in Table 4.12. An *FSBA* rulebase is given in Table 4.13 – the rules are fairly comprehensible, though not as short as *FRANTIC* rules. It should also be remembered that this algorithm may produce rules that are described in terms of other rules, detracting from the intrinsic comprehensibility of individual rules. Table 4.14 on the next page presents the best rulebase in terms of accuracy produced by *WSBA* – the fuzzy quantifiers attached to each condition allow the rules to be highly accurate, but also results in very long rules.

Table 4.12. *FRANTIC-SRL* rulebase for Water Treatment Plant (84.21% accuracy)

R1	IF SSV-D is NOT Low THEN OUTCOME is *OK*
R2	IF PH-E is NOT High AND SSV-P is Low AND SSV-D is NOT High THEN OUTCOME is *Faulty*

Table 4.13. *FSBA* rulebase for Water Treatment Plant (81.08% accuracy)

R1	IF Q-E is NOT Low AND RD-SED-G is Low THEN OUTCOME is *OK*
R2	IF Q-E is NOT High AND PH-E is NOT Low AND SSV-P is High AND DQO-D is NOT Low AND SSV-D is NOT Low AND SSV-S isNOT Low AND RD-SED-G is Low THEN OUTCOME is *Faulty*

4.5 Conclusions and Future Work

This initial work has demonstrated that *FRANTIC-SRL* is a viable approach to the induction of linguistic fuzzy rules – it appears to achieve a balance between rulebase accuracy and comprehensibility and compares favourably with several other fuzzy rule induction algorithms.

Table 4.14. *WSBA* rulebase for Water Treatment Plant (89.47% accuracy)

R1	IF Q-E is (0.31*Low OR 1.0*Normal OR 0.44*High) AND PH-E is (0.80*Low OR 1.0*Normal OR 0.54*High) AND DBO-E is (0.62*Low OR 0.47*Normal OR 1.0*High) AND DBO-P is (1.0*Low OR 0.84*Normal OR 0.96*High) AND SSV-P is (0.64*Low OR 1.0*Normal OR 0.73*High) AND PH-D is (1.0*Low OR 0.44*Normal OR 0.40*High) AND DBO-D is (1.0*Low OR 0.56*Normal OR 0.68*High) AND SSV-D is (1.0*Low OR 0.68*Normal OR 0.45*High) AND PH-S is (0.63*Low OR 0.91*Normal OR 1.0*High) AND SSV-S is (0.67*Low OR 1.0*Normal OR 0.87*High) AND RD-SED-G is (1.0*Low OR 0.44*High) THEN OUTCOME is *OK*
R2	IF Q-E is (0.51*Low OR 1.0*Normal OR 0.38*High) AND PH-E is (0.31*Low OR 1.0*Normal OR 0.60*High) AND DBO-E is (0.72*Low OR 0.57*Normal OR 1.0*High) AND DBO-P is (1.0*Low OR 0.59*Normal OR 0.71*High) AND SSV-P is (0.00*Low OR 0.08*Normal OR 1.0*High) AND PH-D is (1.0*Low OR 0.60*Normal OR 0.50*High) AND DBO-D is (0.25*Low OR 0.51*Normal OR 1.0*High) AND SSV-D is (0.24*Low OR 0.45*Normal OR 1.0*High) AND PH-S is (0.82*Low OR 1.0*Normal OR 0.87*High) AND SSV-S is (0.16*Low OR 0.36*Normal OR 1.0*High) AND RD-SED-G is (1.0*Low OR 0.35*High) THEN OUTCOME is *Faulty*

A hypothesis driving this work is that fuzzy rules that are evolved and evaluated simultaneously will interact better during the inference process, than fuzzy rules that have been evolved mainly independently of each other. Preliminary findings in [42], comparing *FRANTIC-SRL* with *FRANTIC-IRL* provides some evidence to support this. The results indicate that rule comprehensibility is maintained, that accuracy is maintained or improved, and that faster convergence to solutions, and robustness to value changes in some of the *FRANTIC* parameters may be achieved using the simultaneous approach.

However, *FRANTIC-SRL* as it is currently implemented is limited by the underlying assumption that one rule is sufficient to adequately describe a class, so that n ACO algorithms are run in parallel where n is the number of classes. Though a useful starting point to investigate this simultaneous strategy, this may be a naive assumption when applying *FRANTIC-SRL* to larger and more complex real-world problems. Work will therefore be carried out to extend the system to run as many ACO algorithms as are necessary to adequately describe a class.

One approach to achieving this is to determine beforehand how many rules may be required to describe a class, and then to initiate the appropriate number of ACO algorithms. This may be accomplished by analysing the training data to see whether any subclusters of instances may be found within individual classes – the number of subclusters within a class would then indicate the number of ACO algorithms to be initiated for that class. A more thorough investigation of the potential advantages of the simultaneous approach over the iterative one for the induction of fuzzy rules, may then be accomplished.

Using ACO for individual rule discovery has provided several advantages. The constructionist nature of the algorithm has allowed mechanisms to be built into the rule construction process that allow the user flexibility in determining how general or specific rules should be, as determined by the parameters `minInstPerRule` and `constructionThreshold`. The interdependency and impact of these two parameters require further investigation, but it is clear that together they can work toward prevention of over-fitting to the training data, and hence towards rules with greater generalisation power.

The constraint satisfaction mechanism that ensures that valid fuzzy linguistic rules are built, also permits a useful flexibility – this is seen in *FRANTIC*'s ability to construct simple propositional rules that may also include internal disjunction between attribute values, or negated attribute values. This may be extended to further improve the expressiveness of the knowledge representation used by adding linguistic hedges such as 'very' and 'more or less' [10]. Linguistic hedges are functions that are applied to the linguistic terms of a fuzzy rule in order to increase or decrease the precision of such terms. An example rule might therefore be:

IF TEMPERATURE *is Mild AND* WIND *is Very Windy THEN Weightlifting*

The use of such modifiers enriches the knowledge representation, giving it the flexibility to more accurately describe the underlying dataset, yet maintain the comprehensibility of the induced knowledge.

Another advantage offered by *FRANTIC-SRL* is the obvious and numerous opportunities for a multi-processor implementation. This will provide the necessary speed up in computation when inducing knowledge from very large databases. At a coarse level of granularity, several ACO algorithms may be run truly in parallel. At a finer level of granularity, the numerous ants of each ACO may create their rules simultaneously, for example, and the determination of the rulebase quality at the end of each iteration may also be conducted in parallel.

The potential for the application of ACO to fuzzy rule induction is high, and as yet relatively unexplored.

References

1. Pedrycz W, Gomide F (1998) An introduction to fuzzy sets: analysis and design. A Bradford Book, The MIT Press, Cambridge MA, London
2. Ishibuchi H, Nakashima T, Nii M (2005)Classification and modeling with linguistic information granules: advanced approaches to linguistic data mining. Springer-Verlag, Berlin Heidelberg
3. Parsons S (2001) Qualitative methods for reasoning under uncertainty. The MIT Press, Cambridge MA, London
4. Zadeh L (1965) Fuzzy sets. Information and Control 8:338–353
5. Zadeh L (1988) Fuzzy logic. IEEE Computer 21:83–92
6. Hirota K, Sugeno M (eds) (1995) Industrial applications of fuzzy technology. Advances in fuzzy systems – applications and theory 2. World Scientific

7. Pedrycz W (ed) (1996) Fuzzy modelling: paradigms and practice. Kluwer Academic Publishers, Norwell, MA
8. Buchanan BG, Wilkins DC (eds) (1993) Readings in knowledge acquisition and learning: automating the construction and improvement of expert systems. Morgan Kaufmann Publishers, San Francisco, CA
9. Klir GJ, Yuan B (1998) Operation of fuzzy sets. In: Ruspini EH, Bonisonne PP, Pedrycz W (eds) Handbook of Fuzzy Computation. Institute of Physics Publishing
10. Zadeh L (1975) The concept of a linguistic variable and its application to approximate reasoning – Parts I, II, III. Information Sciences 8:199–249, 8:301–357, 9:43–80
11. Mamdani EH (1976) Advances in the linguistic synthesis of fuzzy controllers. Journal of Man-Machine Studies 8:669–678
12. Cordón O, del Jesus MJ, Herrera F (1999) A proposal on reasoning methods in fuzzy rule-based classification systems. International Journal of Approximate Reasoning 20:21–45
13. Ishibuchi H, Nakashima T, Morisawa T (1999) Voting in fuzzy rule-based systems for pattern classification problems. Fuzzy Sets and Systems 103:223–238
14. Dorigo M, Bonabeau E, Theraulaz G (2000) Ant algorithms and stigmergy. Future Generation Computer Systems 16:851–871
15. Bonabeau E, Dorigo M, Theraulaz G (1999) Swarm intelligence: from natural to artificial systems. Oxford University Press, New York Oxford
16. Abraham A, Ramos V (2003) Web usage mining using artificial ant colony clustering and genetic programming. In: Proceedings of the IEEE Congress on Evolutionary Computation 2:1384–1391
17. Hall L, Kanade P (2005) Swarm based fuzzy clustering with partition validity. In: Proceedings of the IEEE International Conference on Fuzzy Systems 991–995
18. Jensen R, Shen Q (2005) Fuzzy-rough data reduction with ant colony optimization. Fuzzy Sets and Systems 149:5–20
19. Al-Ani A (2005) Feature subset selection using ant colony optimization. International Journal of Computational Intelligence 2:53–58
20. Parpinelli R, Lopes H, Freitas A (2002) Data mining with an ant colony optimization algorithm. IEEE Transactions in Evolutionary Computation 6:321–332
21. Galea M, Shen Q (2004) Fuzzy rules from ant-inspired computation. In: Proceedings of the IEEE International Conference on Fuzzy Systems 3:1691–1696
22. Dorigo M, Stützle T (2004) Ant colony optimization. A Bradford Book, The MIT Press, Cambridge MA, London
23. Goss S, Aron S, Deneubourg J-L, Pasteels JM (1989) Self-organised shortcuts in the Argentine ant. Naturwissenschaften 76:579–581
24. Casillas J, Cordón O, Herrera F (2000) Learning fuzzy rules using ant colony optimization algorithms. In: Proceedings of the 2nd International Workshop on Ant Algorithms 13–21
25. Liu B, Abbas HA, McKay B (2003) Classification rule discovery with ant colony optimization. In: Proceedings of the IEEE/WIC International Conference on Intelligent Agent Technology 83–88
26. Wang Z, Feng B (2004) Classification rule mining with an improved ant colony algorithm. In: Lecture Notes in Artificial Intelligence 3339, Springer-Verlag, 357–367
27. Holden N, Freitas A (2005) Web page classification with an ant colony algorithm. In: Lecture Notes in Computer Science 3242, Springer Verlag, 1092–1102
28. Phokharatkul P, Phaiboon S (2005) Handwritten numerals recognition using an ant-miner algorithm. In: Proceedings of the International Conference on Control, Automation and Systems, Korea

29. Sousa T, Silva A, Neves A (2004) Particle swarm based data mining aglorithms for classification rules. Parallel Computing 30:767–783
30. Kennedy J, Eberhart R (1995) Particle swarm optimization. In: Proceedings of the IEEE International Conference on Neural Networks 4:1942–1948
31. Kosko B (1986) Fuzzy entropy and conditioning. Information Sciences 40:165–174
32. Potter MA, Jong KAD (2000) Cooperative coevolution: an architecture for evolving coadapted subcomponents. Evolutionary Computation 8:1–29
33. Pena-Reyes CA, Sipper M (2001) FuzzyCoCo: a cooperative coevolutionary approach to fuzzy modeling. IEEE Transactions on Fuzzy Systems 9:727–737
34. Quinlan JR (1986) Induction of decision trees. Machine Learning 1:81–106
35. Blake CL, Merz CJ (1998) UCI Repository of Machine Learning Data, Deparatment of Computer Science, University of California, Irvine CA. http://www.ics.uci.edu/~mlearn/MLRepositary.html
36. Shen Q, Chouchoulas A (2002) A rough-fuzzy approach for generating classification rules. Pattern Recognition 35:2425–2438
37. Jensen R, Shen Q (2004) Fuzzy-rough attribute reduction with application to web categorization. Fuzzy Sets and Systems 141:469–485
38. Yuan Y, Shaw MJ (1995) Induction of fuzzy decision trees. Fuzzy Sets and Systems 69:125–139
39. Yuan Y, Zhuang H (1996) A genetic algorithm for generating fuzzy classification rules. Fuzzy Sets and Systems 84:1–19
40. Chen S-M, Lee S-H, Lee C-H (2001) A new method for generating fuzzy rules from numerical data for handling classification problems. Applied Artificial Intelligence 15:645–664
41. Rasmani K, Shen Q (2004) Modifying weighted fuzzy subsethood-based rule models with fuzzy quantifiers. In: Proceedings of the IEEE International Conference on Fuzzy Systems 3:1679–1684
42. Galea M, Shen Q (2005) Iteritive vs simultaneous fuzzy rule induction. In: Proceedings of the IEEE International Conference on Fuzzy Systems 767–772

5

Ant Colony Clustering and Feature Extraction for Anomaly Intrusion Detection

Chi-Ho Tsang[1] and Sam Kwong[2]

[1] Department of Computer Science, City University of Hong Kong, Hong Kong
wilson@cs.cityu.edu.hk
[2] Department of Computer Science, City University of Hong Kong, Hong Kong
cssamk@cityu.edu.hk

Summary. This chapter presents a bio-inspired and efficient clustering model for intrusion detection. The proposed model improves existing ant-based clustering algorithm in searching for near-optimal clustering heuristically, in which the meta-heuristic engages the global optimization principles in the context of swarm intelligence. To further improve the clustering solution and alleviate the curse of dimensionality in network connection data, four unsupervised feature extraction algorithms are studied and evaluated in this work. The experimental results on the real-world benchmark datasets and KDD-Cup99 intrusion detection data demonstrate that our proposed model can provide accurate and robust clustering solution, and its application with the extended infomax independent component analysis algorithm is very effective to detect known and unseen intrusion attacks with high detection rate and recognize normal network traffic with low false positive rate.

Keywords: Ant colony optimization, Clustering; Feature extraction; Intrusion detection; Swarm intelligence

5.1 Introduction

Intrusion detection is one of the important and challenging tasks in network security and data mining. A significant amount of intrusion detection approaches has been proposed over the past two decades. In general, they can be categorized into misuse and anomaly detection approaches in the literature. Misuse detection approach is reliable to identify intrusion attacks in relation to the known signatures of discovered vulnerabilities. However, emergent intervention of security experts is required to define accurate rules or signatures, which limits the application of misuse detection approach to build intelligent Intrusion Detection Systems (IDS). On the other hand, the anomaly detection approach usually deals with statistical analysis and data mining problems. It is able to detect novel attacks without apriori knowledge about them if the classification model has the generalization ability to extract intrusion pattern and knowledge during training.

As it is difficult and costly to obtain bulk of class-labeled network connection records for supervised training, clustering analysis has emerged as an anomaly intrusion detection approach over the past 5 years. Clustering is an unsupervised data exploratory technique that partitions a set of unlabeled data patterns into groups or clusters such that patterns within a cluster are similar to each other but dissimilar from patterns in other clusters. Hence, it is desirable to minimize the intra-cluster distance (better compactness) and maximize the inter-cluster distance (better separateness) in order to abstract the inherent structures and knowledge from data patterns. Clustering algorithms can be broadly categorized into hierarchical and partitional approaches. Hierarchical clustering algorithms construct a dendrogram representing a hierarchy of clusters in which each cluster can be nested within another cluster at the next level in the hierarchy. Partitional clustering algorithms merely construct a single partition for data patterns instead of the nested structure in dendrogram. They usually produce clusters by optimizing a single criterion function, for example, minimizing square error function in K-Means [1] algorithm. Alternatively, clustering algorithms can be classified by (a) agglomerative vs. divisive, (b) hard vs. fuzzy, and (c) deterministic vs. stochastic, etc. The clustering approaches proposed in the IDS literature are often based on hard partitional clustering algorithms using local search strategies such as iterative and hill-climbing methods. These approaches usually suffer from a relative high false positive rate of classifying normal network traffic as they converge to non-optimal clustering, and are sensitive to both the initialization and outliers. In addition, most of the partitional clustering algorithms cannot guarantee the near-optimal solution without specifying the number of clusters k as the input parameter. Since both the size and number of clusters in daily network traffic can be dynamic and not a-prior known in practice, it is desirable for IDS to apply a clustering algorithm that is able to determine the number of clusters automatically, and generate compact as well as separated clusters from the network data containing imbalanced classes.

After the discussion of the clustering problems in IDS, we confine our attention to a swarm intelligent clustering approach with stochastic optimization. Inspired by natural heuristics, ant-based clustering algorithm [2,3] is an unsupervised learning technique that is able to find near-optimal clustering solution without any knowledge of k. As discussed in [4], the algorithm can be regarded as the mixture of hierarchical and partitional clustering approaches. It is similar to a hierarchical algorithm in that all objects are clusters and pairs of similar clusters can be merged together to form larger cluster, while similar to a partitional algorithm in that the initial partition of data can be iteratively merged and split, and the features of object are compared with the average features of other objects in a cluster. The ant-based clustering algorithm has been shown in [5] to be efficient to cluster small and low-dimensional data. However, the performance is highly degraded on clustering large and complex data such as the KDD-Cup99 IDS benchmark data as found in our experiments. We thus propose Ant Colony Clustering Model (ACCM) that improves the algorithm by considering both the global optimization and self-organization issues. In addition, the ACCM is also compared with the well-known K-Means and Expectation-Maximization (E-M) [6] clustering algorithms for performance analysis.

There is a three-fold contribution in this work. First, ACCM empirically improves ant-based clustering algorithms for global optimization. Second, it is efficient to apply ACCM for accurate intrusion detection. Third, four unsupervised feature extraction techniques are evaluated in terms of their effectiveness to improve data clustering on network traffic. The rest of this book chapter is organized as follows. Section 2 discusses the related works of our approach. The basics and problems of ant-based clustering algorithms, as well as our proposed ACCM are described in Section 3 in details. Section 4 presents the experimental results. Finally, we draw the conclusion in Section 5, and highlight our future work in Section 6.

5.2 Related Works

The optimization concept of Ant-based clustering algorithm is similar to that of other evolutionary clustering approaches such as Genetic Algorithm (GA) and Particle Swarm Optimization (PSO) in that they stochastically evolve clustering using meta-heuristics instead of the formal and mathematical optimization theory. The earliest ant-based clustering and sorting model based upon corpse clustering and brood sorting behaviors was proposed by Deneubourg et al. [2] for robotics to cluster and sort physical objects. The basic idea is to pick up the isolated objects in the clustering environment and drop it at some other locations where more objects of that type are present. It encourages small clusters to be destroyed and large clusters to be enlarged. Lumer and Faieta [3] generalize the model for numerical data analysis through projecting the multi-dimensional feature space onto low dimensional (2-D) space for data clustering. They introduce three distinct features that attempt to solve non-optimal clustering problems. These features include the improvements on the population diversity and local adaptation with the use of short-term memory and behavioral switches. Since the work of Lumer and Faieta, many variants of the algorithm have been proposed. Handl et al. [5,7] introduce several clustering strategies for the existing model, such as ℓadaptation method, modified short-term memory and neighborhood function. These modifications attempt to improve both the cluster quality and robustness of the model.

Yang and Kamel [8] propose a clustering ensemble approach using three ant colonies with different moving speed. It aims to enhance the clustering solution by improving the population diversity. On the other hand, Hoe et al. [9] apply a homogeneous ant colony approach for web document clustering, and briefly demonstrate its capability of handling complex data clustering. Ramos and Abraham [10] propose to apply swarm cognitive maps [11] to guide the ants to search clustering from continuous data using pheromone trail. The approach has been applied in some real-world applications such as web usage mining [12]. For intrusion detection domain, the applications of ant-based clustering in IDS are recently proposed by Tsang and Kwong [13], and Ramos and Abraham [14]. In addition, an emotional ant system [15] is also proposed for IDS in sensor network.

There are also many different hybrid approaches proposed in the literature to integrate the ant-based clustering model with other machine learning and soft computing algorithms. They include the cellular automata [16], K-means algorithm [17], self-organizing map [18], fuzzy C-mean algorithm [19] and fuzzy IF-THEN rule system [20]. The issues of hybridization and ensemble approach are not addressed in this work. We mainly focus on the problems and improvements in the basic ant-based clustering model [2,3] that will be discussed in the following sections.

5.3 Ant Colony Clustering Model

5.3.1 Basics and Problems of Ant-based Clustering Approach

The general outline of the ant-based clustering algorithm [3] adopted in this work is given in Fig. 5.1. The algorithm consists of a population of ants that act as autonomous agents and iteratively organize data patterns for optimized clustering. Each data object that represents a multi-dimensional pattern is randomly distributed over the 2-D space initially. At each time step, each ant searches the space through random walking, or jumping with the use of short-term memory. It picks up or drops data objects probabilistically according to the following local density of similarity measure:

$$f(o_i) = \max\left\{0, \frac{1}{s^2}\sum_{o_j \in Neigh_{s \times s}(r)}\left[1 - \frac{d(o_i, o_j)}{\alpha(1 + \frac{v-1}{v_{max}})}\right]\right\} \quad (5.1)$$

where $Neigh_{s \times s}(r)$ denotes the local area of perception (s^2 sites) surrounding the site r where the ant occupies in the 2-D space, and the function $d(o_i, o_j)$ measures the Euclidean distance between two objects in the feature space. The threshold α scales the dissimilarity within each pair of objects, and the moving speed v controls the step-size of the ant searching in the space within one time unit. The probability of unladen ants to pick up data object is measured as:

$$P_{pick}(o_i) = \left(\frac{k_1}{k_1 + f(o_i)}\right)^2 \quad (5.2)$$

and the probability of object-carrying ants to drop an object is measured as:

$$P_{drop}(o_i) = \begin{cases} 2f(o_i) & if\ f(o_i) < k_2 \\ 1 & if\ f(o_i) \geq k_2 \end{cases} \quad (5.3)$$

where the threshold constants k_1 and k_2 adjust the probabilities of picking up and depositing objects. The properties of the local density of similarity measurement and short-term memory are discerned as follows.

```
begin
// --- initialization phase ---
load_initial_parameters()  // e.g. speed v, α-value, radius of perception s, population size
for every data object $o_i$ do
    random_place($o_i$)       // each object is placed randomly on the space (one per cell)
end for
for every ant $a_j$ do
    random_place($a_j$)       // each agent is placed randomly on the space (one per cell)
end for
// --- main loop ---
for iteration $t = 1$ to $t_{max}$ do
    for every ant $a_j$ do
        // --- rule-1 ---
        if ((is_carryingObject($a_j$)) and (is_cellEmpty($r$))) then
            compute $f(o_i)$ and $P_{drop}$              // using (1) and (3), respectively
            draw random number $g$                        // in the interval [0,1]
            if ($P_{drop} > g$) then
                drop($a_j$, $o_i$)
                set_unloading($a_j$)
                memorize_item($o_i$, $r$, $m$)           // store details of $o_i$ and location $r$ into memory $m$
            end if
        end if
        // --- rule-2 ---
        else if ((is_unloading($a_j$)) and (has_object($r$))) then
            compute $f(o_i)$ and $P_{pick}$              // using (1) and (2), respectively
            draw random number $g$                        // in the interval [0,1]
            if ($P_{pick} > g$) then
                pick_up($a_j$, $o_i$)
                set_carryingObject($a_j$)
                search_memory($o_i$, $m$)                // compare properties of memorized items in $m$ with $o_i$
                                                          // and jump to location of the most similar item
            end if
        end if
        // --- rule-3 ---
        wander($a_j$, $v$, $N_{dir}$)                    // move to a new site $r$ where is not occupied by other ants yet
    end for
end for
end
```

Fig. 5.1. High-level description of ant-based clustering algorithm

(a) Perception Area: The number of data objects perceived by each ant in the area s^2 is one of the factors in determining the accuracy of similarity measurement and overall computational time. If s is large, it contributes a faster formation of clusters so that small number of less accurate clusters can be formed initially. If s is small, it contributes a slower formation of clusters so that large number of more accurate clusters can be formed initially.

(b) Similarity Scaling Factor: The scaling value α ranges from the interval (0, 1]. If α is large, then the similarity between objects increases such that it is easier for

ants to drop objects but difficult to pick up objects. Hence, a smaller number of clusters can be formed easily and it contributes the formation of coarse clusters. If α is small, then the similarity between objects decreases such that it is relative easy for ants to pick up objects but difficult for ants to drop objects. Hence, a larger number of clusters can be formed easily, and it contributes the formation of fine-grained clusters. Therefore, the appropriate setting of α value is important and dependent on the statistical pattern of data.

(c) Moving Speed: The moving speed v of an ant can be chosen uniformly in the interval $[1, v_{max}]$. It affects the likelihood of picking up or dropping object. If v is large, then a relative small number of coarse clusters can be formed roughly on large scales initially. If v is small, then a relative large number of compact clusters can be formed accurately on small scales initially. The moving speed is a critical parameter to control the convergence speed. A suitably large v contributes faster convergence rate.

(d) Short-term Memory: Each ant remembers the last m objects it has dropped along with their locations, so it has memory of fixed size m that stores the properties and coordinates of the objects. Whenever the ant picks up a new object, it checks its memory if any most similar object it has dropped. If there is one, it jumps to that location (intends to drop it near that location) and prevents dropping the object in an unvisited place. This feature is shown in [5] that it can reduce the number of statistically equivalent clusters formed in different locations.

The basic ant-based clustering algorithm and a recently improved model [5] are evaluated using KDD-Cup99 data in our experiments. It is found that they suffer from two major problems on clustering large and high-dimensional network data. First, many homogeneous clusters are created and difficult to be merged when they are large in size and spatially separated in a large search space. It is very time-consuming and inefficient to merge these clusters. Second, the density of similarity measure only flavors cluster formation in locally dense regions of similar data objects but cannot discriminate dissimilar objects sensitively. Hence, it is ineffective to split cluster of objects in which their variance is not significantly large. As the probabilities of object picking and dropping are coupled to this measure, impure clusters are formed and large clusters of different classes can be merged if the data objects in their boundaries are similar. These findings show the difficulty in applying ant-based clustering directly for intrusion detection. In the following sections, we adopt the basic algorithm with some modifications in [5], and make further improvements to solve the problems in ACCM. These improvements are discussed as follows.

5.3.2 Measure of Local Regional Entropy

A combination of information entropy and average similarity is proposed as an additional metric to the existing models to identify the spatial regions of coarse clusters, compact clusters and disorder boundary of incorrectly merged clusters. Shannon's information entropy [21] has been widely applied in many fields in the literature to measure the uncertainty concerning an event, or to characterize the impurity of an arbitrary collection of examples. If a discrete-valued random variable X has N outcomes $\{x_1, x_2, \ldots, x_N\}$ which occur with probabilities $\{p(x_1), p(x_2), \ldots, p(x_N)\}$, the entropy of the probability distribution of X is given by:

$$H(X) = -\sum_{i=1}^{N} p(x_i) \log p(x_i) \quad (5.4)$$

The degree of similarity between every pair of data objects can reveal their probability of grouping into the same cluster. Following the principles of self-organization and Shannon's information entropy, each ant can measure the impurity of objects perceived within a local region L (of s^2 sites) and identify if the object o_i at the central site of L is equally likely to group with the other objects o_j using local regional entropy $H(L)$:

$$H(L) = -[g(o_i) \log_2 g(o_i) + (1 - g(o_i)) \log_2 (1 - g(o_i))] \quad (5.5)$$

and average similarity of objects $g(o_i)$ within region L:

$$g(o_i) = \frac{1}{n} \sum_{o_j \in Neigh_{s \times s}(r)} \left[0.5 + \frac{C(o_i, o_j)}{2} \right] \quad (5.6)$$

where n is the number of pair of objects to be measured. The cosine similarity $C(o_i, o_j)$ between every pair of objects is measured as:

$$C(o_i, o_j) = \frac{\sum_{k=1}^{m} o_{ik} \cdot o_{jk}}{\sqrt{\sum_{k=1}^{m} o_{ik}^2} \sqrt{\sum_{k=1}^{m} o_{jk}^2}} \quad (5.7)$$

where o_{ik} represents the kth feature of object o_i. Three examples of the local configurations of data objects over a 9-cell neighborhood in the grid are interpreted in Fig. 5.2a-c in which different classes of objects are presented with different colors. When the data objects in local region are very closely belong to the same cluster in Fig. 5.2a, or very distinct and belong to different clusters in Fig. 5.2b, the uncertainty is low and $H(L)$ is close to 0. We are interested in the disorder configuration of objects in Fig. 5.2c. Either $f(o_i)$ or $g(o_i)$ may give arbitrary or near 0.5 mean value that cannot accurately stimulate the ants to pickup or drop data object at the central site, however high $H(L)$ can identify this complex structure with high uncertainty of grouping the objects into same cluster. Hence, it can be found that compact cluster has properties of high $g(o_i)$ and low $H(L)$, coarse cluster has low $g(o_i)$ and low $H(L)$, and the disorder boundary of incorrectly merged clusters, as depicted in Fig. 5.2c, has high $H(L)$. These properties will be contributed to the following improvements.

5.3.3 Pheromone Infrastructure

The formation of larger clusters and destruction of smaller clusters in ant-based clustering algorithm are based on positive and negative feedbacks in self-organization. The impacts of feedbacks related to the algorithm are discussed in the

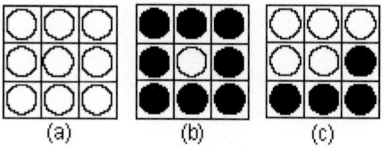

Fig. 5.2. (a-c) Local configuration of objects

work of Theraulaz et al. [22]. In order to build a robust clustering model using these stigmergic effects, it is required to balance their corresponding forces of exploration and exploitation for global optimization. Many species of social insects and animals use some chemical substances called pheromone to communicate indirectly for achieving some collective behaviors such as nest building and food foraging. The global optimality of these swarm intelligent behaviors is the essential criterion for their survival needs. In ACCM, cluster formation and object searching are regarded as nest building and food foraging, respectively. Two types of pheromones are introduced for such searching strategies: cluster-pheromone guides ants to search for compact clusters, while object-pheromone guides ants to search for the objects to be pickup. Their stigmergic interactions are formulated as follows.

(a) Pheromone Deposit Rule: An ant deposits cluster-pheromone on the grid after it drops its carrying object o_i successfully. The intensity of cluster-pheromone at site j, deposited by m ants in colony at time t, is computed as

$$r_{ij}(t) = \sum_{i=1}^{m} [Cj \cdot (1 - H(L)) \cdot g(o_i)] \tag{5.8}$$

where Cj is the quantity of cluster-pheromone deposited by the ant at site j. On the other hand, an ant deposits object-pheromone on the grid after it picks up an object o_i successfully. The intensity of object-pheromone at site j deposited by m ants in colony at time t, is computed as

$$r_{ij}(t) = \sum_{i=1}^{m} [O_j \cdot H(L)] \tag{5.9}$$

where O_j is the quantity of object-pheromone deposited by the ant at site j. A strong cluster-pheromone trail attracts more object-carrying ants to search for compact cluster of similar or identical objects, while a strong object-pheromone trail attracts more unladen ants to search for the cluster of objects with high $H(L)$ and also stimulate them to decrease their similarity scaling α-value in (1) such that they can pick up data objects easily.

(b) Transition Probability: The transition probability of an ant moving from current site i to its neighborhood site j is measured as:

$$P_{ij}^k(t) = \begin{cases} 1/n & if \ \sum_{j=1}^{n}[\tau_{ij}(t)]^\gamma = 0 \ \forall j \in N_{dir} \\ \dfrac{[\tau_{ij}(t)]^\gamma + \beta \sum_{j=1}^{n}[\tau_{ij}(t)]^\gamma}{(\beta+1) \sum_{j=1}^{n}[\tau_{ij}(t)]^\gamma} & otherwise \end{cases} \quad (5.10)$$

where τ is pheromone matrix, $\tau_{ij}(t)$ is the intensity of pheromone on site j at time t, N_{dir} is set of n possible steps, γ and β both control relative dependence on pheromone trails. If an ant senses no intensity of pheromone on its neighborhood sites, it wanders according to its random direction. Otherwise, it probabilistically moves to the site with higher intensity of pheromone. The static scaling parameter γ ranges from $(0, 1]$ and determines the sensitivity to the pheromone intensity. If it is large, the site of high pheromone intensity is more likely to be selected. If it is close to zero, the probability of choosing any one of the neighborhood sites becomes uniformly distributed. Its value is set to 0.7 in our experiments. The scaling parameter β that ranges from $[0, 1]$ is the dynamic rate of successful drop actions of ant in every 2500-iteration that influences its searching behaviors. If the rate is high, which reveals that the ant can search well according to its speed and short-term memory, the influence of pheromone intensity will be relative small. If the rate is low and close to zero, which reveals that the ant needs searching guidance, most likely it will move to a neighborhood site with high pheromone intensity.

(c) Global Pheromone Update Rule: When the model is initialized, all intensity of pheromone in matrix τ is set to 0. At the completion of an iteration t, the pheromone will decay by evaporation and diffusion that balance the deterministic hill climbing and stochastic exploration. The intensity of pheromone at site j observed by an ant at site i at time $(t+1)$ is

$$\tau_{ij}(t+1) = \rho \cdot \tau_{ij}(t) + r_{ij}(t) + q_{ij}(t) \quad (5.11)$$

where $\rho \in (0, 1)$ is evaporation coefficient. The diffusion function proposed by Parunak et al. [23] is adopted as

$$q_{ij}(t+1) = \sum_{j \in N(j')} \frac{F}{|N(j')|} [r_{ij}(t) + q_{ij}(t)] \quad (5.12)$$

where $F \in [0, 1]$ is propagation parameter and $N(j')$ is set of neighbors of site j. The infrastructure guides ants to the best routes with probability in proportional to both intensity and type of pheromone trials for near-optimal cluster merging and splitting. Positive and negative feedbacks are adaptively controlled by the properties of data clusters throughout the clustering process.

5.3.4 Modified Short-term Memory and α-adaptation

In this work, the FIFO based short-term memory in [3] is further modified. Additional properties of object are memorized including local regional entropy and average similarity of successfully dropped object. After an ant successfully picks up new object, it exploits its memory and probabilistically jumps to compact cluster by getting the location of remembered object where has the highest $g(o_i)$ but lowest $H(L)$ of the carrying object therein. It reduces statistically equivalent clusters in a large search space. In addition, since α-value in (5.1) highly scales density of similarity measure, an adaptation scheme proposed in [5] is adopted and adjusted as follows. The parameter α of each ant can be updated using the rule:

$$\alpha \leftarrow \begin{cases} \alpha + 0.01 & if\ r_{fail} > 0.9 \\ \alpha - 0.01 & if\ r_{fail} \leq 0.9 \end{cases} \quad (5.13)$$

where r_{fail} is the rate of failed drop action in every 200-iteration. If r_{fail} is too high, both the alpha value and the similarity between objects will be increased so that the ant can drop the object easily.

5.3.5 Selection Scheme, Parameter Settings and Cluster Retrieval

Balancing the selective pressure and population diversity in sampling mechanism is an important factor in designing evolutionary algorithms. A tournament selection scheme is proposed to counterbalance the population diversity and find the optimal values of control parameters such as α-value, speed and radius of perception over the time. The behavior performance of ant can identify if it is elitist in the colony. At every iteration t, the performance p of each ant can be measured as:

$$p_t = \begin{cases} [g(o_i) + 1 - H(o_i)]/2 & if\ drop\ action\ is\ activated\ by\ P_{drop}\ and\ f(o_i) \\ [1 - g(o_i) + H(o_i)]/2 & if\ pickup\ action\ is\ activated\ by\ P_{pick}\ and\ f(o_i) \\ 0 & if\ no\ drop\ or\ pickup\ action\ is\ activated \end{cases} \quad (5.14)$$

Thus, the average performance P of the ant in every 5000-iteration is measured as:

$$P = \sum_{t=1}^{5000} p_t/n \quad (5.15)$$

which is the score of constructing compact clusters (merging) and destroying incompact clusters (splitting) on different stages of dynamic clustering process, and n is the total number of successful drop and pickup actions. Using 5000-iteration as a generation, all the ants are compared in terms of their average performance. Parameter values of the elitist ants with higher performance remain unchanged in next generation, while the parameter values of the low-fitness ants inherit from (i.e. duplicate) that of the elitist ants during each generation. To avoid a strong selective pressure that produces premature convergence of search and lost of diversity, a small tournament size of 2 is set. This strategy simplifies parameter settings and avoids the robustness of model being droved by their constant values. At the beginning of run, each ant is given different initial parameter values, which are uniformly chosen from bounded intervals. For the sampled KDD-Cup99 IDS data, the initial parameter settings are: grid size: 460*460, population size: 288, speed: $v \in [1, 150]$, radius of perception: $s \in \{3, 5, 7, 9\}$, similarity scaling factor: $\alpha \in (0, 1)$, short-term memory size: $m=20$, maximum number of iterations at each run: $t_{max}=10^6$, and threshold constants for P_{pick} and P_{drop}: $k_1=0.1$, $k_2=0.3$. A weighted single-link agglomerative hierarchical algorithm in [5] is adopted to retrieve concrete clustering from the spatially separated clusters in ACCM.

5.4 Experiments and Results

Our experiments study the characteristics of ACCM by evaluating its cluster validity and classification performance. The ACCM is compared with existing ant-based clustering, K-Means and E-M clustering algorithms for performance analysis. In addition, we examine the effectiveness of applying different feature extraction methods as the preprocessing steps to ACCM for intrusion detection.

5.4.1 Dataset Description and Preprocessing

Six real-world benchmark datasets available from UCI database repository [24] are used to evaluate the performance of our proposed ACCM. Their characteristics are briefly summarized in Table 5.1. The largest class ratio is defined as the ratio of the number of training patterns of the smallest class to that of the largest class. Note that the evaluations on the first five datasets are performed by 10-fold Cross-Validation (10-CV) in our experiments. A brief synopsis of the clustering issues in these datasets is given as follows.

Table 5.1. Characteristics of benchmark datasets. *Note that 16 incomplete patterns with missing values are not included in Wisconsin Breast Cancer dataset

Dataset	# of train patterns	# of test patterns	# of classes	# of features	the largest class ratio
Iris	150	10-CV	3	4	1.0000
Wine	178	10-CV	3	13	0.6761
Wisconsin Breast Cancer*	683	10-CV	2	9	0.5262
Pima Indians Diabetes	768	10-CV	2	8	0.5360
Yeast	1484	10-CV	10	8	0.0108
KDD-Cup99 IDS	494021	311029	23+14	41	7.12e-6

(a) Iris data: The classes "versicolor" and "virginica" are not linearly separable from each other (known to be overlapped) whereas the class "setosa" is linearly separable from them.

(b) Wine data: All three classes are linearly separable. It is commonly used to evaluate a new clustering algorithm.

(c) Wisconsin breast cancer data: The classes "malignant" and "benign" have a small amount of overlapping.

(d) Pima Indians diabetes data: Good sensitivity to the positive diabetes samples is desired.

(e) Yeast data: Dataset has a high degree of class imbalance.

(f) KDD-Cup99 IDS data: The KDD-Cup99 intrusion detection dataset is widely used as the benchmark data to evaluate intrusion detection systems. Comparing with the above datasets, KDD-Cup99 data is relative complex, large-scale, high dimensional and noisy. In addition, it has a very high degree of class imbalance and the classes are highly overlapped. In our experiments, we apply its specified 10% training data that contains 494021 connection records for training. Each connection record represents a sequence of packet transmission starting and ending at a time period, and can be classified as normal, or one of 22 different classes of attacks. All attacks fall into 4 main categories:

(i) Denial-of-Service (DOS) - Denial of the service that is accessed by legitimate users, e.g. SYN flooding (neptune) and land attacks.

(ii) Remote-to-Local (R2L) - Unauthorized access from a remote machine, e.g. password guessing and ftp-write attacks.

(iii) User-to-Root (U2R) - Unauthorized access to gain local super-user (root) privileges, e.g. buffer overflow and perl attacks.

(iv) Probing (Probe) - Surveillance and probing for information gathering, e.g. port scanning and nmap attacks.

To prevent performance deterioration due to class imbalance problem in training, a random sub-sampling method is applied to the three largest classes: 'Normal', 'Neptune' and 'Smurf', which have already included 98% records of the whole training dataset. The new training data contains 10^4 records of Normal class and 10^3 records for each of the Neptune and Smurf classes, while the number of records of other classes remains intact. As a result, total 20752 records are applied for ACCM training. To make the detection task more realistic, the trained model is evaluated using KDD-Cup99 independent test data, which has 311029 records with different class probability distribution and additional 14 unseen attack types. As the network connection records contain both continuous and nominal features, the nominal features such as protocol (TCP/UDP/ICMP), service type (http/ftp/telnet/...) and TCP status flag (SF/REJ/...) are first converted into binary numeric features, hence total 123 numeric features are constructed for numerical computation such as feature extraction.

5.4.2 Metrics of Cluster Validity and Classification Performance

To evaluate the cluster quality delivered by ACCM, three different kinds of cluster validity criteria with different rationales and properties are employed.

(a) *Internal criteria* measure the cluster quality based on the intrinsic characteristics of data. The well-known intra-cluster distance is an internal criterion that measures the compactness of clusters. It is defined as:

$$\text{Intra-cluster distance} = \sum_{C_k \in C} \sum_{o_i \in C_k} d(o_i, c_k)^2 \tag{5.16}$$

Minimization of the intra-cluster distance is desired. The internal criterion is measured based on a specific distance function d, hence it is insufficient to use this criterion alone to compare different kinds of clustering algorithms.

(b) *External criteria* measure the cluster quality by using a priori known cluster information such as the actual class labels. Two external criteria – cluster entropy [25] and mutual information [26] are used. They are briefly described as follows. Cluster entropy measures the homogeneity of data in cluster using the distribution of all classes within each cluster. It is defined as:

$$\text{Cluster entropy} = -\sum_i (\frac{c(i,j)}{\sum_i c(i,j)}) \cdot \log(\frac{c(i,j)}{\sum_i c(i,j)}) \tag{5.17}$$

The overall cluster entropy is computed by averaging over the set of all clusters. Minimization of the overall cluster entropy is desired. As it does not measure the compactness of clusters in terms of the number of clusters, it is biased to favor large number of small clusters. Mutual information takes into account higher-order information dependencies across all clusters and classes. It can be measured as:

$$\text{Mutual information} = \frac{2}{n}\sum_{l=1}^{k}\sum_{h=1}^{g} n_l^{(h)} \frac{\log(n_l^{(h)} n / \sum_{i=1}^{k} n_i^{(h)} \sum_{i=1}^{g} n_i^{(i)})}{\log(k \cdot g)} \tag{5.18}$$

Maximization of mutual information is desired.

(c) *Relative criteria* measure the cluster quality by comparing the actual cluster structure with the generated one based on their relative merit. Dunn-index [27] is a relative criterion that measures both compactness and separateness of clusters. It is defined as:

$$\text{Dunn-index} = \min_{C_k \in C} \left\{ \min_{C_l \in C, k \neq l} (\frac{\delta(C_k, C_l)}{\max_{C_m \in C}\{\Delta(C_m)\}}) \right\} \tag{5.19}$$

where $\delta(C_k, C_l)$ is the distance between clusters and defined as:

$$\delta(C_k, C_l) = \min_{x \in C_k, y \in C_l} \{d(x,y)\} \tag{5.20}$$

and $\Delta(C_m)$ is the diameter of cluster and defined as:

$$\Delta(C_m) = \max_{x,y \in C_m} \{d(x,y)\} \tag{5.21}$$

If the diameters of the clusters are small and the distances between the clusters are large, then a compact and well-separated clustering solution can be obtained. Thus, maximization of Dunn-index is desired.

As some benchmark datasets have class imbalance problem with skewed class distribution, accuracy alone is not sufficient for classification evaluation. The Precision, Recall and F-measure, which are commonly used to evaluate rare class prediction, are employed. For intrusion detection problem, the Attack Detection Rate (ADR) represents the recall rate of detecting all the attack classes, while False Alarm Rate (FAR) represents the False Positive Rate (FPR) of recognizing normal network traffic. These criteria are calculated from the confusion matrix in Table 5.2, and defined as follows.

Table 5.2. Confusion matrix

Actual Class		Predicted Class	
		Positive Class	Negative Class
	Positive Class	True Positive (TP)	False Negative (FN)
	Negative Class	False Positive (FP)	True Negative (TN)

$$\text{Recall} = \frac{TP}{TP+FN} \quad (5.22)$$

$$\text{Precision} = \frac{TP}{TP+FP} \quad (5.23)$$

$$\text{F-measure} = \frac{(\beta^2+1)(\text{Precision} \cdot \text{Recall})}{\beta^2 \cdot \text{Precision} + \text{Recall}} \text{ where } \beta = 1 \quad (5.24)$$

$$\text{Accuracy} = \frac{TP+TN}{TP+TN+FN+FP} \quad (5.25)$$

$$\text{False Positive Rate} = \frac{FP}{FP+TN} \quad (5.26)$$

5.4.3 Cluster Analysis on Benchmark Datasets

Table 5.3 shows the comparative results of the clustering solutions obtained by K-Means, E-M, ant-based clustering and ACCM, which are trained and evaluated using the training datasets under 10-fold cross-validation.

Comparing ACCM with the ant-based clustering algorithm, the degree of cluster compactness and separateness in ACCM are obviously improved. In particular, it can significantly reduce the number of statistically equivalent clusters and search near-optimal number of clusters for KDD-Cup99 data. In general, the comparative results indicate that ACCM generates more compact clusters as well as lower error rates than the other clustering algorithms.

Considering the clustering results on the "Wisconsin breast cancer" and "Pima Indians diabetes" datasets, although E-M does not achieve the minimum error rates, it can generate compact clusters for minor "malign" class in cancer dataset and minor "positive diabetes" class in diabetes dataset such that it obtains good cluster entropy

Table 5.3. The cluster validity, error rate and number of clusters found on benchmark datasets. Average results of 10-fold cross-validation are reported. *Note that in K-means and E-M algorithms the pre-defined input number of clusters k is set to the known number of classes for comparison. The bold face indicates the best result and underlined face indicates the second best

	K-Means *	E-M *	Ant-based Clustering [5]	ACCM
Iris (class distribution: 50, 50, 50)				
Intra-cluster distance	0.94	0.90	0.88	**0.85**
Dunn-index	2.53	2.57	2.89	**2.97**
Cluster entropy	0.30	0.26	0.23	**0.21**
Mutual information	0.67	0.71	0.77	**0.80**
Error rate (%)	11.3	9.33	2.97	**2.67**
Number of clusters	3.00	3.00	3.00	3.00
Wine (class distribution: 59, 71, 48)				
Intra-cluster distance	2.59	2.56	2.56	**2.55**
Dunn-index	3.79	4.08	4.14	**4.27**
Cluster entropy	0.17	0.14	0.14	**0.12**
Mutual information	0.86	0.90	0.88	**0.92**
Error rate (%)	5.62	2.81	2.88	**2.20**
Number of clusters	3.00	3.00	3.00	3.00
Wisconsin Breast Cancer (class distribution: 444, 239)				
Intra-cluster distance	1.64	1.64	1.67	**1.63**
Dunn-index	**5.46**	5.44	5.44	5.45
Cluster entropy	0.24	**0.13**	0.31	0.23
Mutual information	0.75	**0.92**	0.71	0.79
Error rate (%)	4.39	5.86	5.93	**3.81**
Number of clusters	2.00	2.00	2.00	2.00
Pima Indians Diabetes (class distribution: 500, 268)				
Intra-cluster distance	**2.31**	2.36	2.35	2.35
Dunn-index	**4.19**	4.14	4.18	4.17
Cluster entropy	0.44	**0.36**	0.45	0.44
Mutual information	0.58	**0.62**	0.51	0.54
Error rate (%)	25.10	27.98	25.62	**24.32**
Number of clusters	2.00	2.00	2.00	2.00
Yeast (class distribution: 463, 429, 244, 163, 51, 44, 37, 30, 20, 5)				
Intra-cluster distance	1.54	1.49	1.82	**1.42**
Dunn-index	1.68	1.91	1.78	**1.97**
Cluster entropy	0.71	**0.68**	0.79	0.70
Mutual information	0.25	0.29	0.20	**0.31**
Error rate (%)	58.56	57.53	62.02	**57.14**
Number of clusters	10.00	10.00	6.50	8.70
KDDCup-99 Intrusion Detection (class distribution: refer to [24])				
Intra-cluster distance	7.13	7.51	7.60	**7.12**
Dunn-index	92.11	87.58	85.75	**94.34**
Cluster entropy	0.26	0.29	**0.21**	0.23
Mutual information	0.69	0.67	0.65	**0.73**
Error rate (%)	6.83	6.98	8.72	**5.37**
Number of clusters	23.00	23.00	34.60	24.70

and mutual information. For the "Yeast" and "KDD-Cup99" complex datasets, it is found that ACCM is always superior to E-M and K-Means in terms of intra-cluster distance, Dunn-index, mutual information as well as error rates since both the E-M and K-Means do not handle well on clustering high-dimensional data with noisy patterns.

An example of ACCM clustering on Wisconsin breast cancer data is given in Fig. 5.3, in which two different colors represent the actual class labels of major class "benign" and minor class "malign". Note that the clustering space is formulated as a toroidal space, thus the surrounding edges are interconnected. The data objects are randomly distributed at the initial stage. After ACCM starts up, small clusters of similar objects grow by attracting more object-carrying ants to deposit more similar objects therein, while the incompact clusters are split and eliminated by attracting more unladen ants to pick up the objects therein. Fig. 5.3 clearly shows that two spatially separated clusters can be found in the final solution.

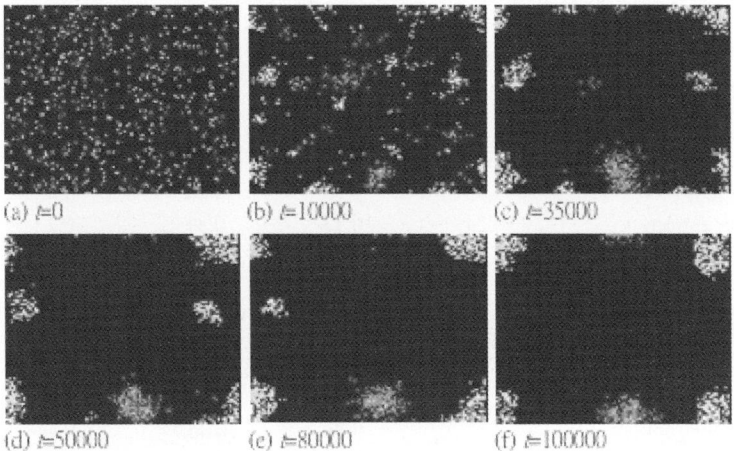

Fig. 5.3. Example of the clustering process on Wisconsin breast cancer dataset. Note that t represents the number of iterations

5.4.4 ACCM with Feature Extraction for Intrusion Detection

To further improve the clustering solution and alleviate the "curse of dimensionality" problem in clustering network connection data, four unsupervised feature extraction algorithms are evaluated and compared. Principle Component Analysis (PCA) applies second-order statistics to extract principle components (PCs) as mutually orthogonal and linear combinations of original features for dimensionality reduction. In this work, PCA is applied to remove outliers that may corrupt the cluster structure, and fasten the clustering process by dimensionality reduction. The K-Means and E-M clustering algorithms as well as ACCM are trained with different number of PCs extracted from training data and evaluated using the test data. The achieved results in terms of average ADR and FPR are reported in Fig. 5.4.

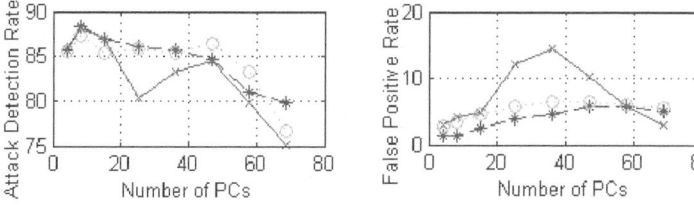

Fig. 5.4. Average attack detection rate and false positive rate on KDDCup-99 test data using different number of PCs (Symbols: 'o'-K-Means; 'x'-E-M and '*'-ACCM)

It is found that using 8 PCs with significantly large eigvenvalues yields good performance in terms of both measures. In addition, it further reveals that E-M suffers from an unstable computation for probabilities when outliers increase with dimensionality. On the contrary, the ACCM performs more stably than K-Means and E-M when number of PCs increases, implying that ACCM is insensitive to the noisy data, which usually exists in network traffic.

The statistical independence, which can be measured by higher-order statistics, is generally a stronger property than the uncorrelatedness offered by PCA to extract latent features from unlabeled data. Independent Component Analysis (ICA) has recently received much theoretical and practical interest in the areas of Blind Source Separation (BSS) as well as biomedical signal and image processing. It decomposes the observed data or mixture of signals into source components that are as statistically independent from each other as possible. ICA has been proved to be effective to extract independent features of non-Gaussian distribution. Meanwhile, the recent study has shown the network traffic is generally self-similar with non-Gaussian and Poisson-like distributions. To this end, we advocate the evaluation of applying ICA to extract latent features from network data, in order to enhance the clustering results.

Three well-known ICA algorithms: InfomaxICA [28], Extended Infomax ICA [29] (hereafter Ext-InfomaxICA) and FastICA [30] are evaluated in our experiments. The PCA is applied as preprocessing step for different ICA algorithms to reduce noisy data and avoid over-learning in ICA. Since determining number of independent components (ICs) to be estimated in ICA is a difficult task and requires human intervention, the 8 PCs found previously that cover enough variance are then used as the input to ICA. The correlation matrices in Fig. 5.5 reveal that the originally correlated and dependent features are reduced by PCA and further diminished by ICA algorithms. The scatter plots of the first three PCs/ICs with large eigenvalues extracted from PCA and ICAs are depicted in Fig. 5.6. It appears that Ext-InfomaxICA and FastICA usually transform network connection data into the subspace where the groupings of similar data are cigar-shaped and elongate.

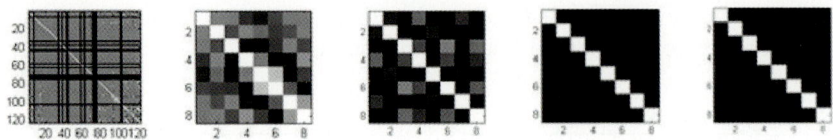

Fig. 5.5. From left to right: Correlation matrix of original re-sampled data (123 features), PCA (8PC), InfomaxICA (8IC), Ext. InfomaxICA (8IC) and FastICA transformed data (8IC)

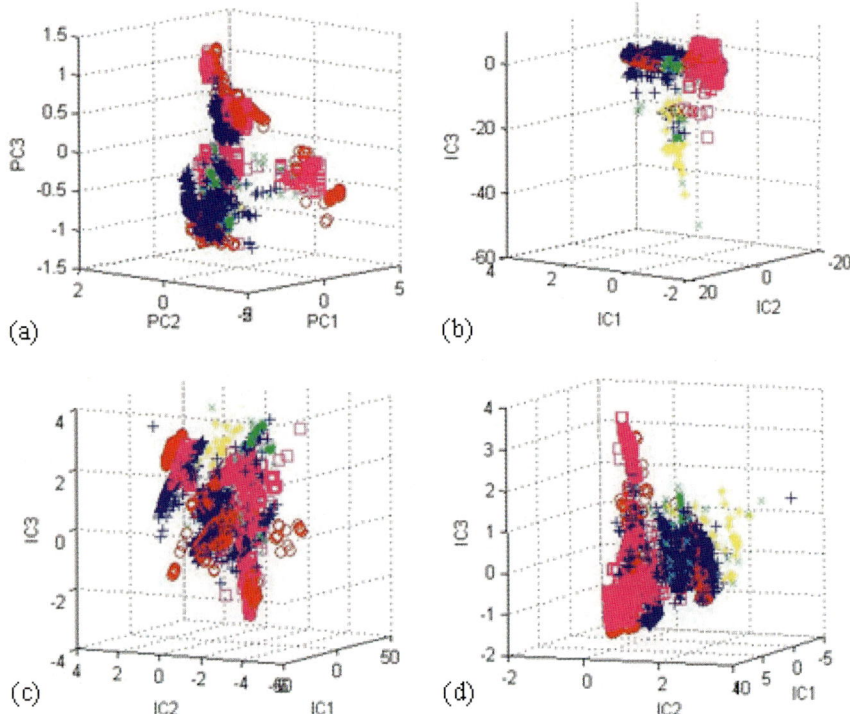

Fig. 5.6. 3-D scatter plots of the subspace of the first three 3 PCs/ICs. (a) PCA; (b) InfomaxICA; (c) Ext-InfomaxICA; and (d) FastICA. (Symbols: Blue '+'-Normal traffic; Red 'o'-DOS attacks; Yellow '*'-U2R attacks; Green 'x'-U2R attacks; and Magenta ''-Probe attacks)

After the PCA/ICA preprocessing, ACCM applies cosine similarity measure instead of Euclidean distance in training, as the ICA basis vectors are not mutually orthogonal. During evaluation on test data, the Mahalanobis distances between the PCA-ICA transformed instances and the retrieved clusters are measured. Given the actual class labels for cluster validation, ACCM labels the test instances with the major class of objects in the closest cluster.

In order to determine which feature extraction method can improve clustering on network connection data, the classification performance obtained with different combinations of clustering and feature extractions algorithms are evaluated and shown in Table 5.4.

Table 5.4. The ADR(%) and FPR(%) of different clustering algorithms on KDD-Cup99 test data. Note that the bold faces indicate the best ADR and FPR for each algorithm

Using 8 PCs/ICs	K-Means		E-M		ACCM	
	ADR	FPR	ADR	FPR	ADR	FPR
PCA	87.27	**3.32**	88.14	**4.07**	88.39	1.35
InfomaxICA	88.95	4.64	89.08	5.81	91.68	2.79
Ext-InfomaxICA	88.51	3.76	89.12	4.16	92.07	**1.17**
FastICA	**89.17**	4.29	**90.94**	4.24	**92.23**	1.53

The results show that FastICA improves ADR, but suffers from higher FPR as compared with both PCA and Ext-InfomaxICA. In terms of ADR, there is no significant difference between Ext-InfomaxICA and FastICA. As the negative impact of high FPR is more critical when amount of daily normal traffic is always larger than intrusion attempts, Ext-InfomaxICA is chosen as the feature extraction method to obtain lower FPR. In addition, it is interesting to note that although PCA gives low ADR, it offers relatively low FPR as compared with the ICA methods since it does not make any assumption on the distributions of the latent features of normal traffic records. InfomaxICA incurs a relatively high FPR due to its assumption on super-Gaussian distribution of the latent features of normal traffic. The confusion matrix in Table 5.5 shows that ACCM is capable of grouping intrusion attacks into different clusters in particular for DOS, R2L and Probe attack classes. In addition, a large amount of Normal connections can be grouped into highly pure Normal clusters.

Table 5.5. Confusion matrix: classification results on training data

Class (# of pattern)	DOS	U2R	R2L	Probe	Normal	Recall (%)
DOS (5467)	5393	0	10	24	40	98.65
U2R (52)	0	28	17	0	7	53.85
R2L (1126)	25	13	958	29	101	85.08
Probe (4107)	57	0	8	3952	90	96.23
Normal (10000)	74	0	52	43	9831	98.31
Precision (%)	97.19	68.29	91.67	97.63	97.64	
F-measure (%)	97.91	60.22	88.25	96.92	97.97	

The trained ACCM is evaluated using independent KDD-Cup99 test data with both known and unseen attacks. The confusion matrix in Table 5.6 shows that high recall rates of DOS and Probe can be obtained. As compared with Table 5.5, it reveals that ACCM is particularly effective to detect the unseen attacks belong to these categories.

The performance of ACCM is further compared with other approaches in Table 5.7. The results indicate that ACCM outperforms the other approaches in detecting

Table 5.6. Confusion matrix: classification results on test data using trained ACCM

Class(# of pattern)	DOS	U2R	R2L	Probe	Normal	Recall (%)
DOS (229853)	223693	26	94	637	5403	97.32
U2R (228)	0	70	19	37	102	30.74
R2L (16189)	0	13	2047	108	14021	12.64
Probe (4166)	187	0	0	3645	334	87.51
Normal (60593)	502	0	174	32	59885	98.83
Precision (%)	99.69	64.22	87.70	81.74	75.10	
F-measure (%)	98.49	41.58	22.10	84.53	85.35	

DOS, U2R and R2L attacks meanwhile offers a low false positive rate for Normal traffic. Based on the classification cost matrix used in KDD-Cup99 contest [31], the classification cost per example of ACCM (0.2267) is lower than that of the KDD-Cup99 winner (0.2331).

Table 5.7. Recall and classification cost [31] of different approaches on test data. Note that the bold face indicates the best and underlined face indicates the second best

	K-Means	E-M	[5]	[32]	KDD-Cup-99 Winner	ACCM
DOS	94.2	94.7	94.5	**97.3**	97.1	**97.3**
U2R	27.4	28.2	25.9	<u>29.8</u>	13.2	**30.7**
R2L	6.5	7.1	8.3	<u>9.6</u>	8.4	**12.6**
Probe	86.9	87.2	72.4	**88.7**	83.3	<u>87.5</u>
Normal	96.2	95.8	89.1	N/A	**99.5**	<u>98.8</u>
Classification cost:	0.2750	0.2668	0.2794	N/A	0.2331	0.2267

5.5 Conclusions

As intrusion attacks become more sophisticated and polymorphous nowadays, there is a growing demand of reliable and intelligent IDS. Drawing inspiration from ant colony optimization, we have presented an unsupervised learning approach – ACCM for anomaly intrusion detection. The ACCM improves existing ant-based clustering algorithms by searching near-optimal clustering heuristically. The proposed improvements such as local regional entropy, pheromone infrastructure, modified short-term memory and tournament selection scheme aim to balance the exploration and exploitation forces for cluster merging and splitting optimization. The ACCM automatically determines the number of clusters that is critically required to be input in other clustering algorithms such as K-Means, E-M and Fuzzy C-Means clustering. In addition, ACCM is insensitive to the outliers, which usually exist in the network traffic and corrupt the cluster structure.

The experimental results demonstrate that, comparing with the existing ant-based clustering algorithm, ACCM significantly reduces the number of statistically equivalent clusters and generates near-optimal clustering solution. In general, the cluster quality and classification performance of ACCM are, respectively, more stable and accurate than K-Means and E-M clustering, as shown in the comparative results on the benchmark datasets. The PCA and recent advanced ICA algorithms are studied and evaluated for dimensionality reduction and feature extraction on network connection data. The empirical results indicate that applying ACCM with the Extended Infomax ICA algorithm is effective to detect known or unseen intrusion attacks with high detection rate and recognize normal network traffic with low false positive rate.

5.6 Future Works

In this work, the cluster quality and classification accuracy of our proposed ACCM has been demonstrated by some real-world benchmark datasets and KDD-Cup99 IDS benchmark data. In future work, we plan to improve the time performance of ACCM for clustering more larger and higher dimensional data. In addition, we shall investigate the hybridization of ACCM and other clustering algorithms. This proposal mainly explores whether ACCM can cooperate with local-search clustering algorithm to extract more globally optimal cluster structures using different objective functions at different clustering levels. In order to develop an intelligent and scalable IDS for the large switched network nowadays, the multiple and heterogeneous ant colonies clustering approach will be designed to integrate ACCM into our developing multi-agent IDS architecture.

References

1. MacQueen J. B. (1967) Some methods for classification and analysis of multivariate observations, In: Fifth Berkeley Symposium on Mathematical Statistics and Probability, University of California Press, Berkeley, CA, Vol. 1, pp. 281-297
2. Deneubourg J. L., Goss S., Franks N., Sendova-Franks A., Detrain C., Chretien L. (1991) The dynamics of collective sorting: robot-like ant and ant-like robot. In: First Conf. on Simulation of Adaptive Behavior: from animals to animats, Cambridge, MA: MIT Press, pp. 356-365
3. Lumer E., Faieta B. (1994) Diversity and adaptation in populations of clustering ants. In: Third Int. Conf. on Simulation of Adaptive Behavior: from animals to animats 3, Cambridge, MA: MIT Press, pp. 499-508
4. Bonabeau E., Dorigo M., Theraulaz G. (1999) Swarm intelligence: From natural to artificial system. Oxford University Press, New York
5. Handl J., Knowles J., Dorigo M. (2004) Strategies for the increased robustness of ant-based clustering. In: Engineering Self-Organising Systems, LNCS 2977, Springer-Verlag, Heidelberg, pp. 90-104
6. Dempster, A. P., Laird, N. M., Rubin, D. B. (1977) Maximum likelihood from incomplete data via the EM algorithm. J. Royal Statistical Society, Series B, 39(1):1-38

7. Handl J., Meyer B. (2002) Improved ant-based clustering and sorting in a document retrieval interface. In: Seventh International Conference on Parallel Problem Solving from Nature, LNCS 2439, Springer-Verlag, Berlin, Germany, pp. 913-923
8. Yang Y., Kamel M. (2003) Clustering ensemble using swarm intelligence. In: IEEE Swarm Intelligence Symposium 2003, pp. 65-71
9. Hoe K. M., Lai W. K., Tai S. Y. (2002) Homogeneous ants for web document similarity modeling and categorization. In: Third Int. Workshop on Ant Algorithms (ANTS2002), Brussels, Belgium, LNCS 2463, Springer-Verlag, Berlin, Heidelberg, Germany, pp. 256-261
10. Ramos V., Abraham A. (2003) Swarms on continuous data. In: Fifth Congress on Evolutionary Computation (CEC2003), Canberra, Australia, IEEE Press, pp. 1370-1375
11. Chialvo D. R., Millonas M. M. (1995) How swarms build cognitive maps. In: Luc Steels (ed), The Biology and Technology of Intelligent Autonomous Agents, 144, NATO ASI Series, pp. 439-450
12. Abraham A., Ramos V. (2003) Web using mining using artificial ant colony clustering and linear genetic programming, In: Fifth Congress on Evolutionary Computation (CEC2003), Canberra, Australia, IEEE Press, pp. 1384-1391
13. Tsang W., Kwong S. (2005) Unsupervised anomaly intrusion detection using ant colony clustering model. In: Fourth IEEE International Workshop on Soft Computing as Transdisciplinary Science and Technology (WSTST2005), Muroran, Japan, pp. 223-232
14. Ramos V., Abraham A. (2005) ANTIDS: Self-organized ant-based clustering model for intrusion detection system. In: Fourth IEEE International Workshop on Soft Computing as Transdisciplinary Science and Technology (WSTST2005), Muroran, Japan, pp. 977-986
15. Banerjee S., Grosan C., Abraham A., Mahanti P. K. (2005) Intrusion detection on sensor networks using emotional ants. Int. J. of Applied Science and Computations, USA
16. Albuquerque P., Dupuis A. (2002) A parallel cellular ant colony algorithm for clustering and sorting. In: S. Bandini, B. Chopard and M. Tomassini (eds.), In: Fifth Int. Conf. on Cellular Automata for Research and Industry, (ACRI2002), LNCS 2493. Springer, pp. 220-230
17. Monmarché N., Slimane M., Venturini G. (1999) AntClass: discovery of clusters in numeric data by an hybridization of an ant colony with the Kmeans algorithm. Technical Report 213, Laboratoire d'Informatique, E3i, University of Tours
18. Mikami T., Wada M. (2001) Data visualization method for growing self-organizing networks with ant clustering algorithm. In: Sixth European Conference on Artificial Life (ECAL2001), Prague, Czech Republic, LNAI 2159, pp. 623-626
19. Kanade P. M., Hall L. O. (2003) Fuzzy Ants as a Clustering Concept. In: 22nd Int. Conf. of the North American Fuzzy Information Processing Society, pp. 227-232
20. Schockaert S., De Cock M., Cornelis C., Kerre E. E. (2004) Efficient clustering with fuzzy ants. In: Applied Computational Intelligence, World Scientific Press, pp. 195-200
21. Shannon C. E., Weaver W. (1949) The Mathematical Theory of Communication. The University of Illinois Press, Urbana, IL
22. Theraulaz G., Gautrais J., Camazine S., Deneubourg J.-L. (2003) The formation of spatial patterns in social insects: from simple behaviours to complex structures. Philosophical Transactions of the Royal Society of London A, 361, pp. 1263-1282
23. Parunak H. V. D., Brueckner S., Sauter J. A., Posdamer J. (2001) Mechanics and military applications for synthetic pheromones. In: 2001 Workshop on Autonomy Oriented Computation, Montreal, Canada

24. UCI Machine Learning Repository [Online]. Available: http://www.ics.uci.edu/~mlearn/MLRepository.html
25. Boley D. (1998) Principle direction divisive partitioning. J. Data Mining and Knowledge Discovery, Vol. 2, Dec 1998, pp. 325-344
26. Strehl A., Ghosh J., Mooney R. (2000) Impact of similarity measures on web-page clustering. In: 7th National Conf. on Artificial Intelligence: Workshop of Artificial Intelligence for Web Search, AAAI, Austin, Texas, USA, pp. 58-64
27. Dunn J. C. (1973) A fuzzy relative of the ISODATA process and its use in detecting compact well-separated clusters. J. Cybernetics, Vol. 3, 1973, pp. 32-57
28. Bell A. J., Sejnowski T. J. (1995) An information-maximization approach to blind separation and blind deconvolution. J. Neural Computation, Vol. 7, Nov. 1995, pp.1129-1159
29. Lee T.-W., Girolami M., Sejnowski T. J. (1999) Independent component analysis using an extended informax algorithm for mixed sub-gaussian and super-gaussian sources. J. Neural Computation, Vol. 11, Feb. 1999, pp. 417-441
30. Hyvärinen A. (1999) Fast and robust fixed-point algorithms for independent component analysis. IEEE Trans. Neural Networks, Vol. 10, May. 1999, pp. 626-634
31. Elkan C. (2000) Results of the KDD'99 classifier learning. In: ACM SIGKDD Int. Conf. on Knowledge Discovery and Data Mining, Boston, MA, 1(2): 63-64
32. Maheshkumar S., Gursel S. (2003) Application of machine learning algorithms to KDD intrusion detection dataset within misuse detection context. In: Int. Conf. on Machine Learning, Models, Technologies and Applications, Las Vegas, Nevadat, USA, CSREA Press, pp. 209-215

6

Particle Swarm Optimization for Pattern Recognition and Image Processing

Mahamed G.H. Omran[1], Andries P. Engelbrecht[2], and Ayed Salman[3]

[1] Faculty of Computing & IT, Arab Open University, Kuwait
   mjomran@gmail.com
[2] Department of Computer Science, University of Pretoria, South Africa
   engel@cs.up.ac.za
[3] Department of Computer Engineering, Kuwait University, Kuwait
   ayed@eng.kuniv.edu.kw

Summary. Pattern recognition has as its objective to classify objects into different categories and classes. It is a fundamental component of artificial intelligence and computer vision. This chapter investigates the application of an efficient optimization method, known as Particle Swarm Optimization (PSO), to the field of pattern recognition and image processing. First a clustering method that is based on PSO is discussed. The application of the proposed clustering algorithm to the problem of unsupervised classification and segmentation of images is investigated. Then PSO-based approaches that tackle the color image quantization and spectral unmixing problems are discussed.

6.1 Introduction

As humans, it is easy (even for a child) to recognize letters, objects, numbers, voices of friends, etc. However, making a computer solve these types of problems is a very difficult task. Pattern recognition is the science with the objective to classify objects into different categories and classes. It is a fundamental component of artificial intelligence and computer vision. Pattern recognition methods are used in various areas such as science, engineering, business, medicine, etc. Interest in pattern recognition is fast growing in order to deal with the prohibitive amount of information we encounter in our daily life. Automation is desperately needed to handle this information explosion. This chapter presents the application of an efficient optimization method, known as Particle Swarm Optimization (PSO), to the field of pattern recognition and image processing. PSO is based on the social behavior of bird flocks.

There are many difficult problems in the field of pattern recognition and image processing. These problems are the focus of much active research in order to find efficient approaches to address them. However, the outcome of the research is still unsatisfactory.

Local search approaches were generally used to solve difficult problems in the field of pattern recognition and image processing. However, the selected set of problems in this chapter are NP-hard and combinatorial. Hence, evolutionary algorithms are generally more suitable to solve these difficult problems because they are population-based stochastic approaches. Thus, evolutionary algorithms can avoid being trapped in a local optimum and can often find a global optimal solution. A PSO is a population-based stochastic optimization approach modeled after the simulation of the social behavior of bird flocks. PSO is easy to implement and has been successfully applied to solve a wide range of optimization problems [56]. Thus, due to its simplicity and efficiency in navigating large search spaces for optimal solutions, PSOs are used in the presented research to develop efficient, robust and flexible algorithms to solve a selective set of difficult problems in the field of pattern recognition and image processing. Out of these problems, data clustering clustering is elaborately tackled in this chapter, specifically clustering of image data. The motivation for the focus on data clustering is the fact that data clustering is an important process in pattern recognition and machine learning. Actually, clustering is a primary goal of pattern recognition. Furthermore, it is a central process in Artificial Intelligence. In addition, clustering algorithms are used in many applications, such as image segmentation, vector and color image quantization, spectral unmixing, data mining, compression, etc. Therefore, finding an efficient clustering algorithm is very important for researchers in many different disciplines.

The remainder of the chapter is organized as follows: Sect. 6.2 provides a brief introduction to the three problems addressed in this chapter (i.e. clustering, color image quantization and spectral unmixing). Sect. 6.3 discusses particle swarm optimization. Sect. 6.4 presents a PSO-based clustering algorithm and compares it with other popular clustering algorithms. A PSO-based color image quantization algorithm is given in Sect. 6.5. An end-member selection method for spectral unmixing that is based on PSO is shown in Sect. 6.6. Finally, Sect. 6.7 concludes the chapter.

6.2 Background

This section provides the reader with a brief introduction to the three problems tackled in this chapter. First the clustering problem is defined and representative clustering methods are presented. Then a brief overview of color image quantization is given. Finally, the spectral unmixing problem is discussed.

6.2.1 The clustering problem

Data is the process of identifying natural groupings or clusters within multidimensional data based on some similarity measure (e.g. Euclidean distance) [2, 4]. It is an important process in pattern recognition and machine learning [27].

Furthermore, data clustering is a central process in Artificial Intelligence (AI) [26]. Clustering algorithms are used in many applications, such as image segmentation [25, 3, 50], vector and color image quantization [54, 51, 64], data mining [16], compression [28], machine learning [11], etc. A cluster is usually identified by a cluster center (or *centroid*) [14]. Data clustering is a difficult problem in unsupervised pattern recognition as the clusters in data may have different shapes and sizes [4].

The following terms are used in this chapter:

- A *pattern* (or *feature vector*), $\mathbf{z}$, is a single object or data point used by the clustering algorithm [2].
- A *feature* (or *attribute*) is an individual component of a pattern [2].
- A *cluster* is a set of similar patterns, and patterns from different clusters are not similar [7].
- *Hard* (or *Crisp*) clustering algorithms assign each pattern to one and only one cluster.
- *Fuzzy* clustering algorithms assign each pattern to each cluster with some degree of membership.
- A *distance measure* is a metric used to evaluate the similarity of patterns [2].

The clustering problem can be formally defined as follows [15]: Given a data set $\mathbf{Z} = \{\mathbf{z}_1, \mathbf{z}_2, \ldots, \mathbf{z}_p, \ldots, \mathbf{z}_{N_p}\}$ where $\mathbf{z}_p$ is a pattern in the N_d-dimensional feature space, and N_p is the number of patterns in $\mathbf{Z}$, then the clustering of $\mathbf{Z}$ is the partitioning of $\mathbf{Z}$ into K clusters $\{\mathbf{C}_1, \mathbf{C}_2, \ldots, \mathbf{C}_K\}$ satisfying the following conditions:

- Each pattern should be assigned to a cluster, i.e.

$$\cup_{k=1}^{K} \mathbf{C}_k = \mathbf{Z}$$

- Each cluster has at least one pattern assigned to it, i.e.

$$\mathbf{C}_k \neq \emptyset, \quad k = 1, \ldots, K$$

- Each pattern is assigned to one and only one cluster (in case of hard clustering only), i.e.

$$\mathbf{C}_k \cap \mathbf{C}_{kk} = \emptyset \quad \text{where} \quad k \neq kk.$$

As previously mentioned, clustering is the process of identifying natural groupings or clusters within multidimensional data based on some similarity measure. Hence, similarity measures are fundamental components in most clustering algorithms [2].

The most popular way to evaluate a similarity measure is the use of distance measures. The most widely used distance measure is the Euclidean distance, defined as

$$d(\mathbf{z}_u, \mathbf{z}_w) = \sqrt{\sum_{j=1}^{N_d}(z_{u,j} - z_{w,j})^2} = ||\mathbf{z}_u - \mathbf{z}_w||_2 \qquad (6.1)$$

Most clustering algorithms are based on two popular techniques known as *hierarchical* and *partitional* clustering [29, 61]. In this chapter, the focus will be on the partitional clustering algorithms because they are more popular than the hierarchical clustering algorithms [4].

Partitional clustering algorithms divide the data set into a specified number of clusters. These algorithms try to minimize certain criteria (e.g. a square error function) and can therefore be treated as optimization problems. However, these optimization problems are generally NP-hard and combinatorial [61].

Partitional clustering algorithms are generally iterative algorithms that converge to local optima [27]. Employing the general form of iterative clustering used by [27], the steps of an iterative clustering algorithm are:

1. Randomly initialize the K cluster centroids
2. Repeat
   a) **For** each pattern, $\mathbf{z}_p$, in the data set **do**
      Compute its membership $u(\mathbf{m}_k|\mathbf{z}_p)$ to each centroid $\mathbf{m}_k$ and its weight $w(\mathbf{z}_p)$
      endloop
   b) Recalculate the K cluster centroids, using

$$\mathbf{m}_k = \frac{\sum_{\forall \mathbf{z}_p} u(\mathbf{m}_k|\mathbf{z}_p) w(\mathbf{z}_p) \mathbf{z}_p}{\sum_{\forall \mathbf{z}_p} u(\mathbf{m}_k|\mathbf{z}_p) w(\mathbf{z}_p)} \quad (6.2)$$

until a stopping criterion is satisfied.

In the above algorithm, $u(\mathbf{m}_k|\mathbf{z}_p)$ is the membership function which quantifies the membership of pattern $\mathbf{z}_p$ to cluster C_k. The membership function, $u(\mathbf{m}_k|\mathbf{z}_p)$, must satisfy the following constraints:

1. $u(\mathbf{m}_k|\mathbf{z}_p) \geq 0$, $p = 1, \ldots, N_p$ and $k = 1, \ldots, K$
2. $\sum_{k=1}^{K} u(\mathbf{m}_k|\mathbf{z}_p) = 1$, $p = 1, \ldots, N_p$

Crisp clustering algorithms use a *hard* membership function (i.e. $u(\mathbf{m}_k|\mathbf{z}_p) \in \{0, 1\}$), while fuzzy clustering algorithms use a *soft* member function (i.e. $u(\mathbf{m}_k|\mathbf{z}_p) \in [0, 1]$) [27].

The weight function, $w(\mathbf{z}_p)$, in Eq. 6.2 defines how much influence pattern $\mathbf{z}_p$ has in recomputing the centroids in the next iteration, where $w(\mathbf{z}_p) > 0$ [27]. The weight function was proposed by Zhang [10].

Different stopping criteria can be used in an iterative clustering algorithm, for example:

- stop when the change in centroid values are smaller than a user-specified value,
- stop when the quantization error is small enough, or
- stop when a maximum number of iterations has been exceeded.

In the following, popular iterative clustering algorithms are described by defining the membership and weight functions in Eq. 6.2.

The K-means Algorithm

The most widely used partitional algorithm is the iterative K-means approach [20]. The objective function that the K-means optimizes is

$$J_{K-means} = \sum_{k=1}^{K} \sum_{\forall \mathbf{z}_p \in C_k} d^2(\mathbf{z}_p, \mathbf{m}_k) \quad (6.3)$$

Hence, the *K*-means algorithm minimizes the intra-cluster distance [27]. The *K*-means algorithm starts with *K* centroids (initial values for the centroids are randomly selected or derived from *a priori* information). Then, each pattern in the data set is assigned to the closest cluster (i.e. closest centroid). Finally, the centroids are recalculated according to the associated patterns. This process is repeated until convergence is achieved.

The membership and weight functions for *K*-means are defined as

$$u(\mathbf{m}_k|\mathbf{z}_p) = \begin{cases} 1 & \text{if } d^2(\mathbf{z}_p, \mathbf{m}_k) = \arg\min_k\{d^2(\mathbf{z}_p, \mathbf{m}_k)\} \\ 0 & \text{otherwise} \end{cases} \quad (6.4)$$

and

$$w(\mathbf{z}_p) = 1 \quad (6.5)$$

Hence, *K*-means has a hard membership function. Furthermore, *K*-means has a constant weight function, thus, all patterns have equal importance [27].

The *K*-means algorithm has the following main advantages [50]:

- it is very easy to implement, and
- its time complexity is $O(N_p)$ making it suitable for very large data sets.

However, the *K*-means algorithm has the following drawbacks [19]:

- the algorithm is data-dependent,
- it is a greedy algorithm that depends on the initial conditions, which may cause the algorithm to converge to suboptimal solutions, and
- the user needs to specify the number of clusters in advance.

The Fuzzy C-means Algorithm

A fuzzy version of *K*-means, called Fuzzy C-means (FCM) (sometimes called fuzzy *K*-means), was proposed by Bezdek [30, 31]. FCM is based on a fuzzy extension of the least-square error criterion. The advantage of FCM over *K*-means is that FCM assigns each pattern to each cluster with some degree of membership (i.e. fuzzy clustering). This is more suitable for real applications where there are some overlap between the clusters in the data set. The objective function that the FCM optimizes is

$$J_{FCM} = \sum_{k=1}^{K} \sum_{p=1}^{N_p} u_{k,p}^q d^2(\mathbf{z}_p, \mathbf{m}_k) \quad (6.6)$$

where q is the fuzziness exponent, with $q \geq 1$. Increasing the value of q will make the algorithm more fuzzy; $u_{k,p}$ is the membership value for the *p*-th pattern in the *k*-th cluster satisfying the following constraints:

1. $u_{k,p} \geq 0$, $p = 1, \ldots, N_p$ and $k = 1, \ldots, K$
2. $\sum_{k=1}^{K} u_{k,p} = 1$, $p = 1, \ldots, N_p$

The membership and weight functions for FCM are defined as [27]:

$$u(\mathbf{m}_k|\mathbf{z}_p) = \frac{||\mathbf{z}_p - \mathbf{m}_k||^{-2/(q-1)}}{\sum_{k=1}^{K} ||\mathbf{z}_p - \mathbf{m}_k||^{-2/(q-1)}} \quad (6.7)$$

and
$$w(\mathbf{z}_p) = 1 \qquad (6.8)$$

Hence, FCM has a soft membership function and a constant weight function. In general, FCM performs better than K-means [26] and it is less affected by the presence of uncertainty in the data [5]. However, as in K-means it requires the user to specify the number of clusters in the data set. In addition, it may converge to local optima [2].

Krishnapuram and Keller [47, 48] proposed a possibilistic clustering algorithm, called *possibilistic* C-means. Possibilistic clustering is similar to fuzzy clustering; the main difference is that in possibilistic clustering the membership values may not sum to one [50]. Possibilistic C-means works well in the presence of noise in the data set. However, it has several drawbacks, namely [50],

- it is likely to generate coincident clusters,
- it requires the user to specify the number of clusters in advance,
- it converges to local optima, and
- it depends on initial conditions.

Swarm Intelligence Approaches

Clustering approaches inspired by the collective behaviors of ants have been proposed by Lumer and Faieta [21], Wu and Shi [9], Labroche *et al.* [44]. The main idea of these approaches is that artificial ants are used to pick up items and drop them near similar items resulting in the formation of clusters.

Omran *et al.* [40] proposed the first PSO-based clustering algorithm. This algorithm is discussed in Sect. 6.4. The results of Omran *et al.* [40, 43] show that PSO outperformed K-means, FCM and other *state-of-the-art* clustering algorithms. The same algorithm of Omran *et al.* [40] was used by Van der Merwe and Engelbrecht [18] to cluster general data sets. It was applied on a set of multi-dimensional data (e.g. the Iris plant data base). In general, the results show that the PSO-based clustering algorithm performs better than the K-means algorithm, which verify the results of Omran *et al.* [40]. Furthermore, Xiao *et al.* [60] used PSO for gene clustering.

More recently, Paterlini and Krink [52] compared the performance of K-means, GAs, PSO and Differential Evolution (DE) [49] for a representative point evaluation approach to partitional clustering. The results show that GAs, PSO and DE outperformed the K-means algorithm. The results also show that DE performed better than GAs and PSO. However, the performance of DE is significantly degraded if noise exists in the data set.

6.2.2 Color Image Quantization

Color image quantization is the process of reducing the number of colors presented in a digital color image [32]. Color image quantization can be formally defined as follows [38]: Given a set of $N_{S'}$ colors, $\mathbf{S}' \subset \mathbb{R}^{N_d}$. The color quantization is a map, $f_q : \mathbf{S}' \to \mathbf{S}''$, where $\mathbf{S}''$ is a set of $N_{S''}$ colors such that $\mathbf{S}'' \subset \mathbf{S}'$ and $N_{S''} < N_{S'}$. The objective is to minimize the quantization error resulting from replacing a color $\mathbf{c} \in \mathbf{S}'$ with its quantized value $f_q(\mathbf{c}) \in \mathbf{S}''$.

Color image quantization is an important problem in the fields of image processing and computer graphics [38]:

- It can be used in lossy compression techniques [38];
- It is suitable for mobile and hand-held devices where memory is usually small [58];
- It is suitable for low-cost color display and printing devices where only a small number of colors can be displayed or printed simultaneously [45].
- Most graphics hardware use color lookup tables with a limited number of colors [8].

Color image quantization consists of two major steps:

1. Creating a colormap (or palette) where a small set of colors (typically 8–256 [45] is chosen from the (2^{24}) possible combinations of red, green and blue (RGB).
2. Mapping each color pixel in the color image to one of the colors in the colormap.

Therefore, the main objective of color image quantization is to map the set of colors in the original color image to a much smaller set of colors in the quantized image [65]. Furthermore, this mapping, as already mentioned, should minimize the difference between the original and the quantized images [8]. The color quantization problem is known to be NP-complete [59]. This means that it is not feasible to find the global optimal solution because this will require a prohibitive amount of time. To address this problem, several approximation techniques have been used. One popular approximation method is to use a standard local search strategy such as *K*-means. *K*-means has already been applied to the color image quantization problem [53, 39]. However, as previously mentioned, *K*-means is a greedy algorithm which depends on the initial conditions, which may cause the algorithm to converge to suboptimal solutions. This drawback is magnified by the fact that the distribution of local optima is expected to be broad in the color image quantization problem due to the three dimensional color space. In addition, this local optimality is expected to affect the visual image quality. The local optimality issue can be addressed by using stochastic optimization schemes.

Self-Organizing Maps (SOMs) [55] is a single-layered unsupervised artificial neural network where input patterns are associated with output nodes via weights that are iteratively modified until a stopping criterion is met. SOM combines competitive learning (in which different nodes in the Kohonen network compete to be the winner when an input pattern is presented) with a topological structuring of nodes, such that adjacent nodes tend to have similar weight vectors.

SOMs were used by Dekker [1] to quantize color images. The approach selects an initial colormap, and then modifies the colors in the colormap by moving them in the direction of the image color pixels. However, to reduce the execution time, only samples of the colors in the image are used.

6.2.3 Spectral Unmixing

In remote sensing, classification is the main tool for extracting information about the surface cover type. Conventional classification methods assign each pixel to one class (or species). This class can represent water, vegetation, soil, etc. The classification methods generate a map showing the species with highest concentration. This map is known as the *thematic map*. A thematic map is useful when the pixels in the image represent pure species (i.e. each pixel represents the spectral signature of one species). Hence, thematic maps are suitable for imagery data with a small ground sampling distance (GSD) such as LANDSAT Thematic Mapper (GSD = 30 m). However, thematic maps are not as useful for large GSD imagery such as NOAA's AVHRR (GSD = 1.1 km) because in this type of imagery pixels are usually not pure. Therefore, pixels need to be assigned to several classes along with their respective concentrations in that pixel's footprint. Spectral unmixing (or *mixture modeling*) is used to assign these classes and concentrations. Spectral unmixing generates a set of maps showing the proportions of all species present in each pixel footprint. These maps are called the *abundance images*. Hence, each abundance image shows the concentration of one species in a scene. Therefore, spectral unmixing provides a more complete and accurate classification than a thematic map generated by conventional classification methods.

Spectral unmixing can be used for the compression of multispectral imagery. Using spectral unmixing, the user can prioritize the species of interest in the compression process. This is done by first applying the spectral unmixing on the original images to generate the abundance images. The abundance images representing the species of interest are then prioritized by coding them with a relatively high bit rate. Other abundance images are coded using a relatively low bit rate. At the decoder, the species-prioritized reconstructed multispectral imagery is generated via a re-mixing process on the decoded abundance images [37]. This approach is feasible if the spectral unmixing algorithm results in a small (negligible) residual error.

Linear Pixel Unmixing (or Linear Mixture Modeling)

Spectral unmixing is generally performed using a linear mixture modeling approach. In linear mixture modeling the spectral signature of each pixel vector is assumed to be a linear combination of a limited set of fundamental spectral components known as *end-members*. Hence, spectral unmixing can be formally defined as follows:

$$\mathbf{z}_p = \mathbf{X}.\mathbf{f} + \mathbf{e} = f_1\chi_1 + f_2\chi_2 + \ldots + f_i\chi_i + \ldots + f_{N_e}\chi_{N_e} + \mathbf{e} \qquad (6.9)$$

where $\mathbf{z}_p$ is a pixel signature of N_b components, $\mathbf{X}$ is an $N_b \times N_e$ matrix of end-members $\chi_{1,\ldots,N_e}$, f_i is the fractional component of end-member i (i.e. proportion of footprint covered by species i), $\mathbf{f}$ is the vector of fractional components $(f_1, f_2, \cdots, f_{N_e})^T$, χ_i is the end-member i of N_b components, $\mathbf{e}$ is the residual error vector of N_b components, N_b is the number of spectral bands and N_e is the number of components, $N_e \leq N_b$.

Provided that the number of end-members is less than or equal to the true spectral dimensionality of the scene, the solution via classical least-squares estimation is,

$$\mathbf{f} = (\mathbf{X}^T\mathbf{X})^{-1}\mathbf{X}^T\mathbf{z}_p \qquad (6.10)$$

Therefore, there are two requirements for linear spectral unmixing:

- The spectral signature of the end-members needs to be known.
- The number of end-members is less than or equal to the true spectral dimensionality of the scene (i.e. dimension of the feature space). This is known as the *condition of identifiability*.

The condition of identifiability restricts the application of linear spectral unmixing when applied to multispectral imagery because the end-members may not correspond to physically identifiable species on the ground. Moreover, the number of distinct species in the scene may be more than the true spectral dimensionality of the scene. For example: for Landsat TM with seven spectral bands ($N_b = 7$), the true spectral dimension is at most five ($N_e = 5$) based on principal component analysis.

Selection of the End-Members

To overcome the condition of identifiability, Maselli [23] proposed a method of dynamic selection of an optimum end-member subset. In this technique, an optimum subset of all available end-members is selected for spectral unmixing of each pixel vector in the scene. Thus, although every pixel vector will not have a fractional component for each end-member, the ensemble of all pixel vectors in the scene will collectively have fractional contributions for each end-member.

For each pixel vector, a unique subset of the available end-members is selected which minimizes the residual error after decomposition of that pixel vector. To determine the optimum end-members for pixel vector $\mathbf{z}_p$, the pixel vector is projected onto all available normalized end-members. The most efficient projection, which corresponds to the highest dot product value c_{max}, indicates the first selected end-member χ_{max}. It can be shown that this procedure is equivalent to finding the end-member with the smallest spectral angle with respect to $\mathbf{z}_p$. The residual pixel signature, $\mathbf{r}_{\mathbf{z}_p} = \mathbf{z}_p - c_{max}\chi_{max}$ is then used to identify the second end-member by repeating the projection onto all remaining end-members. The process continues until a prefixed maximum (N_e) number of end-members from the total of N_m available end-members has been identified.

More recently, Saghri *et al.* [36] proposed a method to obtain end-members from the scene with relatively small residual errors. In this method, the set of end-members are chosen from a thematic map resulting from a modified ISODATA [24] (ISODATA is an enhancement of the *K*-means algorithm with the ability to split and merge clusters). The modified ISODATA uses the spectral angle measure instead of the Euclidean distance measure to reduce the effect of shadows and sun angle effects. The end-members are then set as the centroids of the compact and well-populated clusters. Maselli's approach discussed above is then used to find the optimum end-member subset from the set of available end-members for each pixel in the scene. Linear spectral unmixing is then applied to generate the abundance images.

According to [36], the proposed approach has several advantages: the resulting end-members correspond to physically identifiable (and likely pure) species on the ground, the residual error is relatively small and minimal human interaction time is required. However, this approach has a drawback in that it uses ISODATA which depends on the initial conditions.

6.3 Particle Swarm Optimization

Particle swarm optimizers are population-based optimization algorithms modeled after the simulation of social behavior of bird flocks [34, 35]. In a PSO system, a swarm of individuals (called *particles*) fly through the search space. Each particle represents a candidate solution to the optimization problem. The position of a particle is influenced by the best position visited by itself (i.e. its own experience) and the position of the best particle in its neighborhood (i.e. the experience of neighboring particles). When the neighborhood of a particle is the entire swarm, the best position in the neighborhood is referred to as the global best particle, and the resulting algorithm is referred to as the *gbest* PSO. When smaller neighborhoods are used, the algorithm is generally referred to as the *lbest* PSO [63]. The performance of each particle (i.e. how close the particle is to the global optimum) is measured using a fitness function that varies depending on the optimization problem.

Each particle in the swarm is represented by the following characteristics:

- $\mathbf{x}_i$: The *current position* of the particle;
- $\mathbf{v}_i$: The *current velocity* of the particle;
- $\mathbf{y}_i$: The *personal best position* of the particle.

The personal best position of particle i is the best position (i.e. one resulting in the best fitness value) visited by particle i so far. Let f denote the objective function. Then the personal best of a particle at time step t is updated as

$$\mathbf{y}_i(t+1) = \begin{cases} \mathbf{y}_i(t) & \text{if } f(\mathbf{x}_i(t+1)) \geq f(\mathbf{y}_i(t)) \\ \mathbf{x}_i(t+1) & \text{if } f(\mathbf{x}_i(t+1)) < f(\mathbf{y}_i(t)) \end{cases} \quad (6.11)$$

If the position of the global best particle is denoted by the vector $\hat{\mathbf{y}}$, then

$$\hat{\mathbf{y}}(t) \in \{\mathbf{y}_0, \mathbf{y}_1, \ldots, \mathbf{y}_s\} = \min\{f(\mathbf{y}_0(t)), f(\mathbf{y}_1(t)), \ldots, f(\mathbf{y}_s(t))\} \quad (6.12)$$

where s denotes the size of the swarm. For the *lbest* model, a swarm is divided into overlapping neighborhoods of particles. For each neighborhood N_j, a best particle is determined with position $\hat{\mathbf{y}}_j$. This particle is referred to as the *neighborhood best particle*, defined as

$$\hat{\mathbf{y}}_j(t+1) \in \{N_j | f(\hat{\mathbf{y}}_j(t+1)) = \min\{f(\mathbf{y}_i(t))\}, \forall \mathbf{y}_i \in N_j\} \quad (6.13)$$

where

$$N_j = \{\mathbf{y}_{i-l}(t), \mathbf{y}_{i-l+1}(t), \ldots, \mathbf{y}_{i-1}(t), \mathbf{y}_i(t), \mathbf{y}_{i+1}(t), \ldots, \mathbf{y}_{i+l-1}(t), \mathbf{y}_{i+l}(t)\} \quad (6.14)$$

Neighborhoods are usually determined using particle indices [46], however, topological neighborhoods can also be used [62].

For each iteration of a PSO algorithm, the velocity $\mathbf{v}_i$ update step is specified for each dimension $j \in 1,\ldots,N_d$, where N_d is the dimension of the problem. Hence, $v_{i,j}$ represents the j-th element of the velocity vector of the i-th particle. Thus the velocity of particle i is updated using the following equation:

$$v_{i,j}(t+1) = wv_{i,j}(t) + c_1 r_{1,j}(t)(y_{i,j}(t) - x_{i,j}(t)) + c_2 r_{2,j}(t)(\hat{y}_j(t) - x_{i,j}(t)) \quad (6.15)$$

where w is the inertia weight [62], c_1 and c_2 are the acceleration constants and $r_{1,j}, r_{2,j} \sim U(0,1)$.

The position of particle i, $\mathbf{x}_i$, is then updated using the following equation:

$$\mathbf{x}_i(t+1) = \mathbf{x}_i(t) + \mathbf{v}_i(t+1) \quad (6.16)$$

Velocity updates can be clamped through a user defined maximum velocity, V_{max}, which would prevent them from exploding, thereby causing premature convergence [17].

The PSO algorithm performs the update equations above, repeatedly, until a specified number of iterations have been exceeded, or velocity updates are close to zero. The quality of particles is measured using a fitness function which reflects the optimality of a particular solution.

6.4 A PSO-based Clustering Algorithm with Application to Unsupervised Image Classification

A clustering method that is based on PSO is presented in this section. The algorithm [40] finds the centroids of a user specified number of clusters, where each cluster groups together similar patterns. The application of the proposed clustering algorithm to the problem of unsupervised classification and segmentation of images is investigated. To illustrate its wide applicability, the proposed algorithm is then applied to synthetic, MRI and satellite images. Experimental results show that, in general, the PSO clustering algorithm performs better than other well-known clustering algorithms (namely, K-means and Fuzzy C-means) in all measured criteria.

Different measures can be used to express the quality of a clustering algorithm. The most general measure of performance is the quantization error, defined as

$$J_e = \frac{\sum_{k=1}^{K} \sum_{\forall \mathbf{z}_p \in \mathbf{C}_k} d(\mathbf{z}_p, \mathbf{m}_k)/n_k}{K} \quad (6.17)$$

where $\mathbf{C}_k$ is the k-th cluster, and n_k is the number of pixels in $\mathbf{C}_k$.

In the context of data clustering, a single particle represents the K cluster centroids. That is $\mathbf{x}_i = (\mathbf{m}_{i,1},\ldots,\mathbf{m}_{i,k},\ldots,\mathbf{m}_{i,K})$ where $\mathbf{m}_{i,k}$ refers to the k-th cluster centroid vector of the i-th particle. Therefore, a swarm represents a number of candidate data clusterings. The quality of each particle is measured using

$$f(\mathbf{x}_i, \mathbf{Z}_i) = w_1 \overline{d}_{max}(\mathbf{Z}_i, \mathbf{x}_i) + w_2(z_{max} - d_{min}(\mathbf{x}_i)) + w_3 J_e \qquad (6.18)$$

where z_{max} is the maximum value in the data set (i.e. in the context of digital images, $z_{max} = 2^s - 1$ for an s-bit image); $\mathbf{Z}_i$ is a matrix representing the assignment of patterns to the clusters of particle i. Each element $z_{i,k,p}$ indicates if pattern $\mathbf{z}_p$ belongs to cluster $\mathbf{C}_k$ of particle i. The constants w_1, w_2 and w_3 are user-defined constants used to weigh the contribution of each of the sub-objectives. Also,

$$\overline{d}_{max}(\mathbf{Z}_i, \mathbf{x}_i) = \max_{k=1,\ldots,K} \left\{ \sum_{\forall \mathbf{z}_p \in \mathbf{C}_{i,k}} d(\mathbf{z}_p, \mathbf{m}_{i,k}) / n_{i,k} \right\} \qquad (6.19)$$

is the maximum average Euclidean distance of particles to their associated clusters, and

$$d_{min}(\mathbf{x}_i) = \min_{\forall k, kk, k \neq kk} \{d(\mathbf{m}_{i,k}, \mathbf{m}_{i,kk})\} \qquad (6.20)$$

is the minimum Euclidean distance between any pair of clusters. In the above, $n_{i,k}$ is the number of patterns that belong to cluster $\mathbf{C}_{i,k}$ of particle i.

The fitness function in Eq. 6.18 has as objective to simultaneously minimize the quantization error, as quantified by J_e, and the intra-cluster distance between patterns and their cluster centroids, as quantified by $\overline{d}_{max}(\mathbf{Z}_i, \mathbf{x}_i)$, and to maximize the inter-cluster distance between any pair of clusters, as quantified by, $d_{min}(\mathbf{x}_i)$.

According to the definition of the fitness function, a small value of $f(\mathbf{x}_i, \mathbf{Z}_i)$ suggests compact and well-separated clusters (i.e. *good* clustering).

The fitness function is thus a multi-objective problem. Approaches to solve multi-objective problems have been developed mostly for evolutionary computation approaches [12]. Recently, approaches to multi-objective optimization using PSO have been developed by [57], [33] and [13]. Since our scope is to illustrate the applicability of PSO to data clustering, and not on multi-objective optimization, a simple weighted approach is used to cope with multiple objectives. Different priorities are assigned to the subobjectives via appropriate initialization of the values of w_1, w_2 and w_3.

The PSO clustering algorithm is summarized below:

1. Initialize each particle to contain K randomly selected cluster centroids
2. For $t = 1$ to t_{max}
    a) For each particle i
        i. For each pattern $\mathbf{z}_p$
            - calculate $d(\mathbf{z}_p, \mathbf{m}_{i,k})$ for all clusters $\mathbf{C}_{i,k}$ using Eq. 6.1
            - assign $\mathbf{z}_p$ to $\mathbf{C}_{i,k}$ where

$$d(\mathbf{z}_p, \mathbf{m}_{i,k}) = \min_{\forall k=1,\ldots,K} \{d(\mathbf{z}_p, \mathbf{m}_{i,k})\}$$

        ii. Calculate the fitness, $f(\mathbf{x}_i, \mathbf{Z}_i)$
    b) Find the personal best position for each particle and the global best solution, $\hat{\mathbf{y}}(t)$
    c) Update the cluster centroids using Eqs. 6.15 and 6.16

In general, the complexity of the above algorithm is $O(sKt_{max}N_p)$. The parameters s, K and t_{max} can be fixed in advance. Typically s, K and $t_{max} \ll N_p$. Therefore, the time complexity of the algorithm is $O(N_p)$. Hence, in general the algorithm has linear time complexity in the size of a data set.

As previously mentioned, an advantage of using PSO is that a parallel search for an optimal clustering is performed. This population-based search approach reduces the effect of the initial conditions, compared to K-means, especially for relatively large swarm sizes.

6.4.1 Experimental Results

The PSO-based clustering algorithm has been applied to three types of imagery data, namely synthetic, MRI and LANDSAT 5 MSS (79 m GSD) images. These data sets have been selected to test the algorithms, and to compare them with other algorithms, on a range of problem types, as listed below:

Synthetic Image: Figure 6.1(a) shows a 100×100 8-bit gray scale image created to specifically show that the PSO algorithm does not get trapped in the local minimum. The image was created using two types of brushes, one brighter than the other.

MRI Image: Figure 6.1(b) shows a 300×300 8-bit gray scale image of a human brain, intentionally chosen for its importance in medical image processing.

Remotely Sensed Imagery Data: Figure 6.1(c) shows band 4 of the four-channel multispectral test image set of the Lake Tahoe region in the US. Each channel is comprised of a 300×300, 8-bit per pixel (remapped from the original 6 bit) image. The test data are one of the North American Landscape Characterization (NALC) Landsat multispectral scanner data sets obtained from the U.S. Geological Survey (USGS).

The results reported in this section are averages and standard deviations over 20 simulations. All comparisons are made with reference to $J_e, \overline{d}_{max}$ and d_{min}. Furthermore, a total number of clusters of 3, 8 and 4 were used respectively for the synthetic, MRI and Tahoe images. In all cases, for PSO, 50 particles were trained for 100 iterations; for K-means and FCM, 5000 iterations were used (that is, all algorithms have performed 5000 function evaluations). $V_{max} = 255, w = 0.72$ and $c_1 = c_2 = 1.49$. The chosen values of w, c_1 and c_2 are popular in the literature and ensure convergence [17]. For the fitness function in Eq. 6.18, $w_1 = w_2 = 0.3$, $w_3 = 0.4$ were used for the synthetic image, $w_1 = 0.2, w_2 = 0.5, w_3 = 0.3$ were used for the MRI image, and $w_1 = w_2 = w_3 = 0.333333$ were used for the Tahoe image. These values were set empirically. For FCM, q was set to 2 since it is the commonly used value [22].

Table 6.1 shows that PSO generally outperformed K-means and FCM in d_{min} and $\overline{d}_{max}$, while performing comparably with respect to J_e (for the synthetic image, PSO performs significantly better than K-means and FCM with respect to J_e). These results show that the PSO-based clustering algorithm is a viable alternative that merits further investigation. For a more thorough study, the reader is referred to Omran et al. [43].

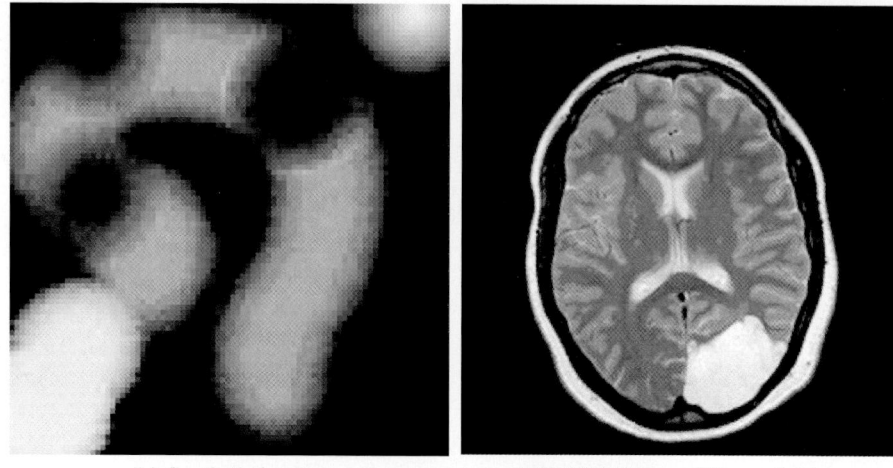

(a) Synthetic image (b) MRI Image of Human brain

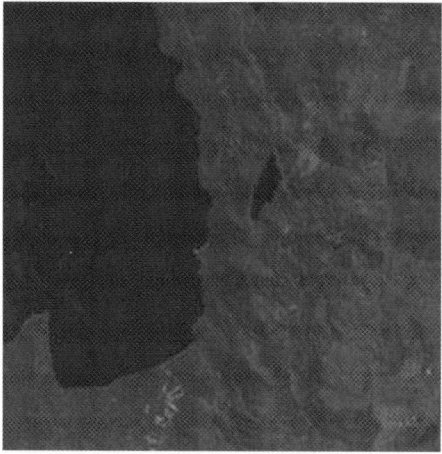

(c) Band 4 of the Landsat MSS test image of Lake Tahoe

Fig. 6.1. Data set consisting of synthetic, MRI and LANDSAT images

6.5 A PSO-based Color Image Quantization (PSO-CIQ) Algorithm

A PSO-based color image quantization algorithm [41], referred to as PSO-CIQ, is presented in this section. The algorithm randomly initializes each particle in the swarm to contain K centroids (i.e. color triplets). The K-means clustering algorithm is then applied to each particle at a user-specified probability to refine the chosen centroids. Each pixel is then assigned to the cluster with the closest centroid. The PSO is then applied to refine the centroids obtained from the K-means algorithm. The

Table 6.1. Comparison between *K*-means, FCM and PSO

Image		J_e	d_{max}	d_{min}
Synthetic	K-means	20.21225 ± 0.937836	28.0405 ± 2.7779388	78.4975 ± 7.0628718
	FCM	20.73192 ± 0.650023	28.55921 ± 2.221067	82.43411 ± 4.404686
	PSO	16.98891 ± 0.023937	24.69605 ± 0.130334	93.6322 ± 0.248234
MRI	K-means	7.3703 ± 0.042809	13.21437 ± 0.761599	9.93435 ± 7.308529
	FCM	7.20598 ± 0.166418	10.85174 ± 0.960273	19.51775 ± 2.014138
	PSO	7.594520 ± 0.449454	10.18609 ± 1.237529	26.70592 ± 3.008073
Tahoe	K-means	3.280730 ± 0.095188	5.234911 ± 0.312988	9.40262 ± 2.823284
	FCM	3.164670 ± 0.000004	4.999294 ± 0.000009	10.97061 ± 0.000015
	PSO	3.523967 ± 0.172424	4.681492 ± 0.110739	14.66486 ± 1.177861

proposed algorithm is then applied to commonly used images. It is shown from the conducted experiments that the proposed algorithm generally results in a significant improvement of image quality compared to other well-known approaches.

Define the following symbols:

- N_p denotes the number of image pixels,
- K denotes the number of clusters (i.e. colors in the colormap),
- $\mathbf{z}_p$ denotes the coordinates of pixel p, and
- $\mathbf{m}_k$ denotes the centroid of cluster k (representing one color triple in the colormap).

In this section, the terms centroid and color triple are used interchangeably.

The most general measure of performance is the mean square error (MSE) of the quantized image using a specific colormap. The MSE is defined as follows,

$$MSE = \frac{\sum_{k=1}^{K} \sum_{\forall \mathbf{z}_p \in C_k} (\mathbf{z}_p - \mathbf{m}_k)^2}{N_p} \quad (6.21)$$

where $\mathbf{C}_k$ is the k-th cluster.

In the context of color image quantization, a single particle represents a colormap (i.e. a particle consists of K cluster centroids representing RGB color triples). The RGB coordinates in each color triple are floating-point numbers. Each particle is constructed as $\mathbf{x}_i = (\mathbf{m}_{i,1}, \ldots, \mathbf{m}_{i,k}, \ldots, \mathbf{m}_{i,K})$ where $\mathbf{m}_{i,k}$ refers to the k-th cluster centroid vector of the i-th particle. Therefore, a swarm represents a number of candidate colormaps. The quality of each particle is measured using the MSE (defined in Eq. 6.21) as follows:

$$f(\mathbf{x}_i) = MSE(\mathbf{x}_i) \quad (6.22)$$

The algorithm initializes each particle randomly from the color image to contain K centroids (i.e. color triplets). The set of K color triplets represents the colormap. The K-means clustering algorithm is then applied to each particle at a user-specified probability, p_{kmeans}. The K-means algorithm is used in order to refine the chosen colors and to reduce the search space. Each pixel is then assigned to the cluster with the closest centroid. The fitness function of each particle is calculated using Eq. 6.22. The PSO velocity and update Eqs. 6.15 and 6.16 are then applied. The procedure is repeated until a stopping criterion is satisfied. The colormap of the global best particle after t_{max} iterations is chosen as the optimal result. The PSO-CIQ algorithm is summarized below:

1. Initialize each particle by randomly choosing K color triplets from the image
2. For $t = 1$ to t_{max}
   a) For each particle i
      i. Apply K-means for a few iterations with a probability p_{kmeans}
      ii. For each pixel z_p
         - Calculate $d^2(z_p - m_{i,k})$ for all clusters $C_{i,k}$
         - Assign z_p to $C_{i,kk}$ where
           $d^2(z_p - m_{i,kk}) = \min_{\forall k=1,\ldots,k}\{d^2(z_p - m_{i,k})\}$
      iii. Calculate the fitness, $f(x_i)$
   b) Find the personal best position for each particle and the global best solution, $\hat{y}(t)$
   c) Update the centroids using Eqs. 6.15 and 6.16

In general, the complexity of the PSO-CIQ algorithm is $O(sKt_{max}N_p)$. The parameters s, K and t_{max} can be fixed in advance. Typically s, K and $t_{max} \ll N_p$. Therefore, the time complexity of PSO-CIQ is $O(N_p)$. Hence, in general the algorithm has linear time complexity in the size of the data set.

6.5.1 Experimental Results

The PSO-CIQ algorithm was applied to a set of four commonly used color images namely: *Lenna, mandrill, jet* and *peppers* (shown in Figure 6.2(a)). The size of each image is 512×512 pixels. All images are quantized to 16, 32 and 64 colors.

The results reported in this section are averages and standard deviations over 10 simulations. The PSO-CIQ parameters were initially set as follows: $p_{kmeans} = 0.1, s = 20, t_{max} = 50$, number of K-means iterations is 10, $w = 0.72, c_1 = c_2 = 1.49$ and $V_{max} = 255$ for all the test images. These parameters were used in this section unless otherwise specified. For the GCMA (a GA-based color image quantization algorithm) [45] a population of 20 chromosomes was used, and evolution continued for 50 generations. For the SOM, a Kohonen network of 4×4 nodes was used when quantizing an image to 16 colors, a Kohonen network of 8×4 nodes was used when quantizing an image to 32 colors, and a Kohonen network of 8×8 nodes was used when quantizing an image to 64 colors. All SOM parameters were set as in Pandya and Macy [6]: the learning rate $\eta(t)$ was initially set to 0.9 then decreased by 0.005 until it reached 0.005, the neighborhood function $\Delta_w(t)$ was initially set to (4+4)/4 for 16 colors, (8+4)/4 for 32 colors, and (8+8)/4 for 64 colors. The neighborhood function is then decreased by 1 until it reached zero.

Table 6.2 summarizes the results for the four images. The results of the GCMA represent the best case over several runs and are copied from Scheunders [45]. The results are compared based on the MSE measure (defined in Eq. 6.21). The results showed that, in general, PSO-CIQ outperformed GCMA in all the test images except for the mandrill image and the case of quantizing the Jet image to 64 colors. Furthermore, PSO-CIQ generally performed better than SOM for both Lenna and peppers images. SOM and PSO-CIQ performed comparably well when applied to the mandrill image. SOM generally performed better than PSO-CIQ when applied to the Jet image. Figure 6.2 shows the visual quality of the quantized images generated by PSO-CIQ when applied to peppers. For more thorough study, the reader is referred to Omran *et. al.* [41].

Table 6.2. Comparison between SOM, GCMA and PSO-CIQ

Image	K	SOM	GCMA	PSO-CIQ
Lena	16	235.6 ± 0.490	332	209.841 ± 0.951
	32	126.400 ± 1.200	179	119.167 ± 0.449
	64	74.700 ± 0.458	113	77.846 ± 16.132
Peppers	16	425.600 ± 13.162	471	399.63 ± 2.636
	32	244.500 ± 3.854	263	232.046 ± 2.295
	64	141.600 ± 0.917	148	137.322 ± 3.376
Jet	16	121.700 ± 0.458	199	122.867 ± 2.0837
	32	65.000 ± 0.000	96	71.564 ± 6.089
	64	38.100 ± 0.539	54	56.339 ± 11.15
Mandrill	16	629.000 ± 0.775	606	630.975 ± 2.059
	32	373.600 ± 0.490	348	375.933 ± 3.42
	64	234.000 ± 0.000	213	237.331 ± 2.015

6.6 The PSO-based End-Member Selection (PSO-EMS) Algorithm

This section presents the PSO-EMS algorithm [42] by first presenting a measure to quantify the quality of a spectral unmixing algorithm, after which the PSO-EMS algorithm is shown.

To measure the quality of a spectral unmixing algorithm, the root mean square (RMS) residual error can be used, defined as follows:

$$E = \sum_{j=1}^{N_b} \sqrt{\hat{M}_j} \qquad (6.23)$$

where

$$\hat{M} = \frac{\sum_{p=1}^{N_p}(\mathbf{z}_p - \mathbf{X}.\mathbf{f})^2}{N_p} \qquad (6.24)$$

with N_p the number of pixels in the image.

In the context of spectral unmixing, a single particle represents N_m end-members. That is $\mathbf{x}_i = (\chi_{i,1}, \ldots, \chi_{i,k}, \ldots, \chi_{i,N_m})$ where $\chi_{i,k}$ refers to the k-th end-member vector

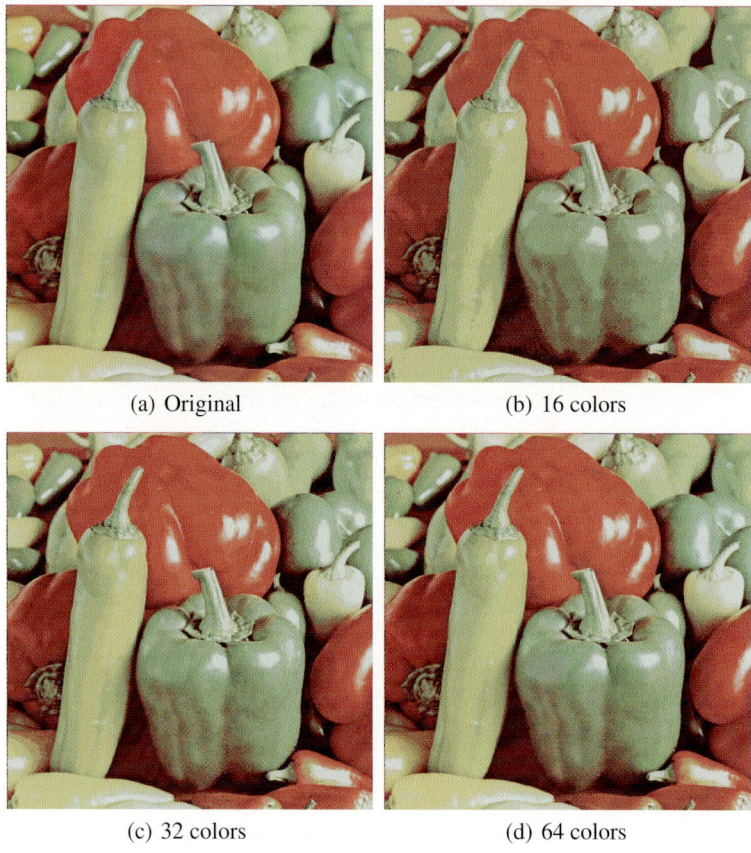

Fig. 6.2. Quantization results for the peppers image using PSO-CIQ

of the i-th particle. Therefore, a swarm represents a number of candidate end-members. The quality of each particle is measured using the RMS residual error (defined in Eq. 6.23) as follows:

$$f(\mathbf{x}_i) = E \qquad (6.25)$$

The algorithm works as follows: each particle is randomly initialized from the multispectral image set to contain N_m end-members. The K-means clustering algorithm (using a few iterations) is then applied to a random set of particles with a user-specified probability, ρ_{kmeans}. The K-means algorithm is used in order to refine the chosen end-members, reduce the search space. Then for each particle i, the N_m end-members of the particle form the pool of available candidate end-members for the subsequent spectral unmixing procedure. Maselli's approach [23] is used to dynamically select the N_e optimum end-member subsets from the pool of N_m end-members. Each pixel vector is then spectrally decomposed as a linear combination of its optimum subset of end-members. The RMS residual error for particle i is then calculated. The PSO velocity and update equations (Eqs. 6.15 and 6.16) are then applied. The procedure is repeated until a stopping criterion is satisfied. The N_m end-members of the best particle are used to generate the abundance images.

6.6.1 The Generation of Abundance Images

For each species represented by an end-member, the ensemble of all fractional components forms a concentration map (i.e. abundance map). The fractional concentration maps are then optimally mapped to an eight-bit integer format for display and storage purposes. This is done using the following non-linear mapping function [36]:

$$\Omega = \frac{255(f^{exp} - f_{min}^{exp})}{f_{max}^{exp} - f_{min}^{exp}} + 0.5 \qquad (6.26)$$

where Ω is the mapped integer fractional component in the range of $0 \leq \Omega \leq 255$, f is the fractional component, f_{min} is the minimum fractional component, f_{max} is the maximum fractional component and exp is the floating-point exponent parameter in the range of $0 \leq exp \leq 1.0$. In this chapter, exp is set to 0.6 for the abundance images as suggested by [36].

The PSO-EMS algorithm is summarized below:

1. Initialize each particle to contain N_m randomly selected end-members
2. For $t = 1$ to t_{max}
   a) For each particle i
      i. Apply K-means for a few iterations with a probability ρ_{kmeans}
      ii. For each pixel $\mathbf{z}_p$
         - Find the N_e optimum end-member subset
         - Apply linear spectral unmixing using Eq. 6.9
      iii. Calculate the fitness, $f(\mathbf{x}_i)$
   b) Find the personal best position for each particle and the global best solution, $\hat{\mathbf{y}}(t)$
   c) Update the end-members using Eqs. 6.15 and 6.16
3. Generate the abundance images using the N_m end-members of particle $\hat{\mathbf{y}}(t)$

In general, the complexity of the PSO-EMS algorithm is $O(st_{max}N_p)$. The parameters s and t_{max} can be fixed in advance. Typically s and $t_{max} \ll N_p$. Therefore, the time complexity of PSO-EMS is $O(N_p)$. Hence, in general the algorithm has linear time complexity in the size of the data set.

6.6.2 Experimental results

The PSO-EMS algorithm has been applied to two types of imagery data, namely LANDSAT 5 MSS (79 m GSD) and NOAA's AVHRR (1.1 km GSD) images. These image sets have been selected to test the algorithms on a variety of platforms with a relatively large GSD which represent good candidates for spectral unmixing in order to get sub-pixel resolution. The two image sets are described below:

LANDSAT 5 MSS: Fig. 6.1(c) shows band 4 of the four-channel multispectral test image set of the Lake Tahoe region in the US. Each channel is comprised of a 300×300, 8-bit per pixel (remapped from the original 6 bit) image and corresponds to a GSD of 79 m. The test image set is one of the North American Landscape Characterization (NALC) Landsat multispectral scanner data sets obtained from the U.S. Geological Survey (USGS). The result of a preliminary principal component study of this data set indicates that its intrinsic true spectral dimension N_e is 3. As in [36], a total of six end-members were obtained from the data set (i.e. $N_m = 6$).

NOAA's AVHRR: Fig. 6.3 shows the five-channel multispectral test image set of an almost cloud-free territory of the entire United Kingdom (UK). This image set was obtained from the University of Dundee Satellite Receiving Station. Each channel (one visible, one near-infra red and three in the thermal range) is comprised of a 847×1009, 10-bit per pixel (1024 gray levels) image and corresponds to a GSD of 1.1 km. The result of a preliminary principal component study of this data set indicates that its intrinsic true spectral dimension N_e is 3. As in [36], a total of eight end-members were obtained from the data set (i.e. $N_m = 8$).

The PSO-EMS algorithm is compared with the end-member selection method proposed by [36] (discussed in Sect. 6.2.3), which is referred to in this chapter as ISO-UNMIX. The time complexity of ISO-UNMIX is $O(N_p)$. Saghri *et al.* [36] showed that ISO-UNMIX performed very well compared to other popular spectral unmixing methods.

The results reported in this section are averages and standard deviations over 10 simulations. The PSO-EMS parameters were initially set as follows: $\rho_{kmeans} = 0.1, s = 20, t_{max} = 100$, number of K-means iterations is 10, $w = 0.72, c_1 = c_2 = 1.49$ and $V_{max} = 255$ for the Lake Tahoe image set and $V_{max} = 1023$ for the UK image set. No attempt was made to tune the PSO parameters (i.e. w, c_1 and c_2) to each data set. The rationale behind this decision is the fact that in real-world applications the evaluation time is significant and such parameter tuning is usually a time consuming process. These parameters are used in this section unless otherwise specified.

Table 6.3 summarizes the results for the two image sets. The results are compared based on the RMS residual error (defined in Eq. 6.23). The results showed that, for both image sets, PSO-EMS performed better than the ISO-UNMIX in terms of the RMS residual error. Figs. 6.4 and 6.5 show the abundance images generated from ISO-UNMIX and PSO-EMS, respectively, when applied to the UK image set. For display purposes the fractional species concentrations were mapped to 8 bits per pixels abundance images. For a more thorough study, the reader is referred to [42].

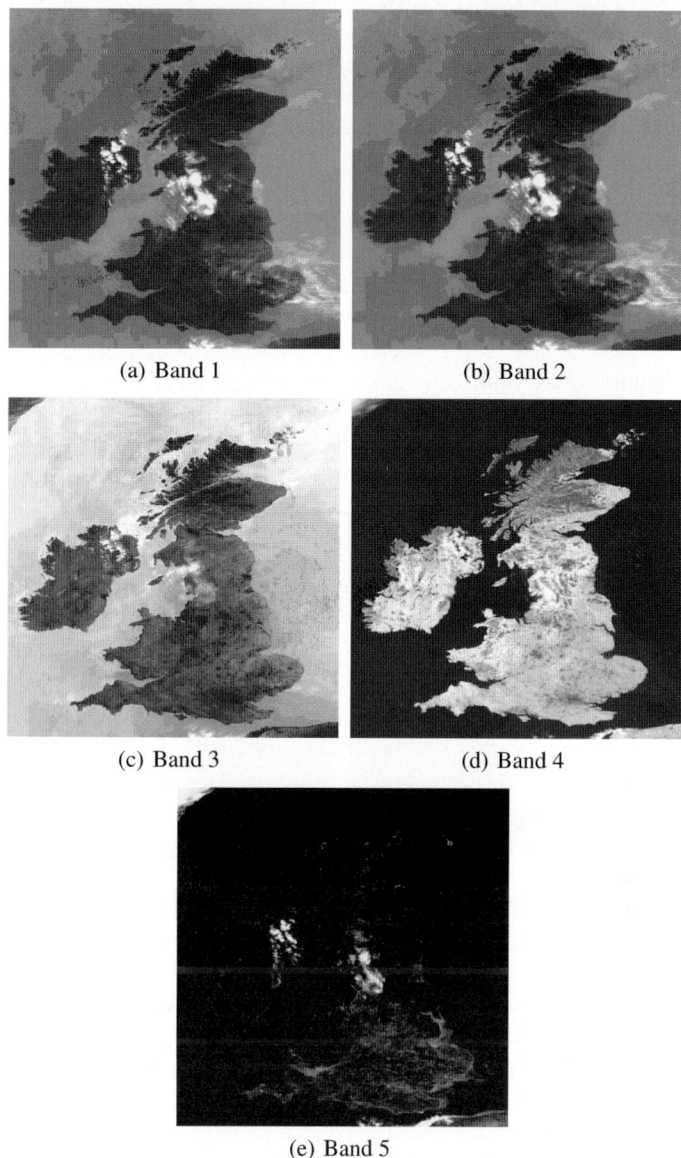

(a) Band 1   (b) Band 2

(c) Band 3   (d) Band 4

(e) Band 5

Fig. 6.3. AVHRR Image of UK, Size: 847x1009, 5 bands, 10-bits per pixel

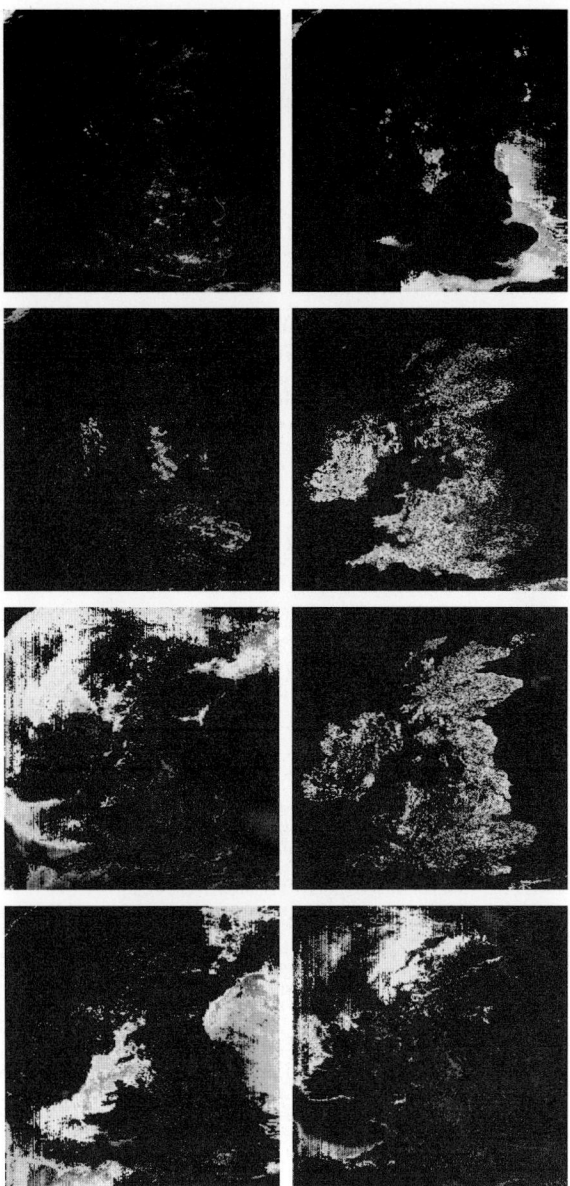

Fig. 6.4. Species concentration maps resulting from the application of ISO-UNMIX to unmix the UK test image set

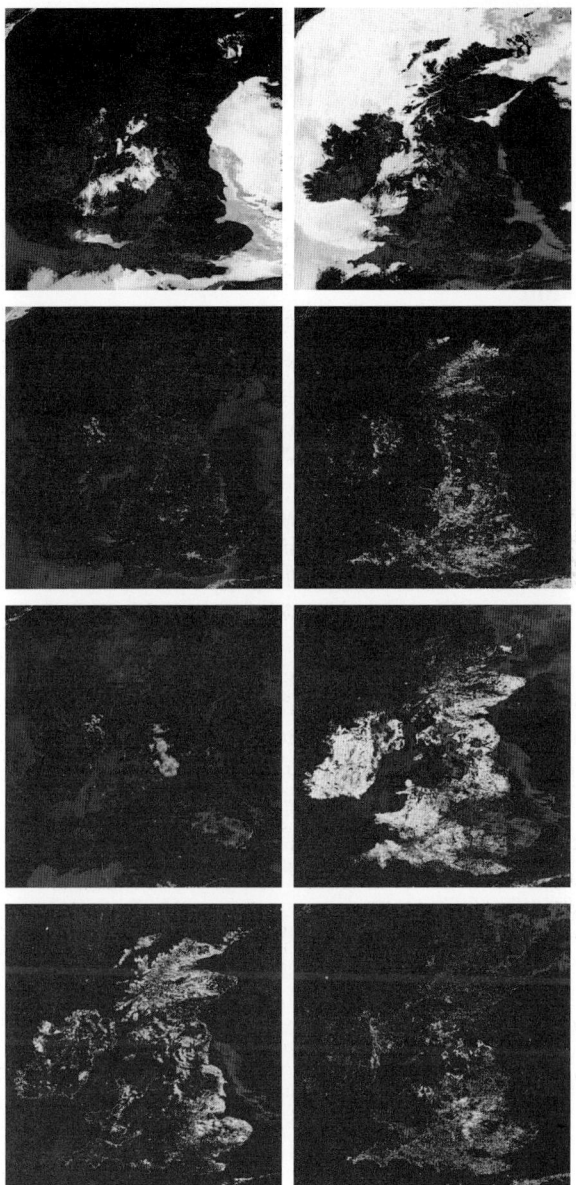

Fig. 6.5. Species concentration maps resulting from the application of PSO-EMS to unmix the UK test image set

Table 6.3. Comparison between ISO-UNMIX and PSO-EMS

Image	Algorithm	RMS
LANDSAT 5 MSS	ISO-UNMIX	0.491837
	PSO-EMS	0.462197 ± 0.012074
NOAA's AVHRR	ISO-UNMIX	3.725979
	PSO-EMS	3.510287 ± 0.045442

6.7 Conclusion

This chapter investigated the application of an efficient optimization approach known as Particle Swarm Optimization to the field of pattern recognition and image processing.

First a clustering approach using PSO was discussed. The objective of the proposed algorithm is to simultaneously minimize the quantization error and intra-cluster distances, and to maximize the inter-cluster distances. The application of the proposed clustering algorithm to the problem of unsupervised classification and segmentation of images was investigated. The proposed algorithm was compared against well-known clustering algorithms. In general, the PSO algorithms produced better results with reference to inter- and intra-cluster distances, while having quantization errors comparable to the other algorithms.

Then the chapter addressed two difficult problems in the field of pattern recognition and image processing. The two problems are: color image quantization and spectral unmixing. First, the chapter discussed a PSO-based color image quantization algorithm (PSO-CIQ). The PSO-CIQ algorithm was compared against other well-known color image quantization techniques. In general, the PSO-CIQ performed better than other techniques when applied to a set of commonly used images.

Finally, a new spectral unmixing approach using PSO (PSO-EMS) was shown. The objective of the PSO-EMS algorithm is to determine the appropriate set of end-members for a given multispectral image set. The PSO-EMS algorithm performed well when applied to test image sets from various platforms such as LANDSAT 5 MSS and NOAA's AVHRR.

From the results presented in this chapter, it can be concluded that the PSO is an efficient optimization algorithm for difficult pattern recognition and image processing problems. These problems are considered difficult because they are NP-hard and combinatorial problems.

References

1. Dekker A. Kohonen Neural Networks for Optimal Colour Quantization. *Network: Computation in Neural Systems*, 5:351–367, 1994.
2. Jain A, Murty M, and Flynn P. Data Clustering: A Review. *ACM Computing Surveys*, 31(3):264–323, 1999.

3. Jain A and Dubes R. *Algorithms for Clustering Data*. Prentice Hall, 1988.
4. Jain A, Duin R, and Mao J. Statistical Pattern Recognition: A Review. *IEEE Transactions on Pattern Analysis and Machine Intelligence*, 22(1):4–37, 2000.
5. Liew A, Leung S, and Lau W. Fuzzy Image Clustering Incorporating Spatial Continuity. In *IEE Proceedings Vision, Image and Signal Processing*, volume 147, 2000.
6. Pandya A and Macy R. *Pattern Recognition with Neural Networks in C++*. CRC Press, 1996.
7. Everitt B. *Cluster Analysis*. Heinemann Books, 1974.
8. Freisleben B and Schrader A. An Evolutionary Approach to Color Image Quantization. In *Proceedings of IEEE International Conference on Evolutionary Computation*, pages 459–464, 1997.
9. Wu B and Shi Z. A Clustering Algorithm based on Swarm Intelligence. In *Proceedings of the International Conference on Info-tech and Info-net*, pages 58–66, 2001.
10. Zhang B. Generalized K-Harmonic Means - Boosting in Unsupervised Learning. Technical Report HPL-2000-137, Hewlett-Packard Labs, 2000.
11. Carpineto C and Romano G. A Lattice Conceptual Clustering System and Its Application to Browsing Retrieval. *Machine Learning*, 24(2):95–122, 1996.
12. Coello Coello C. *An Empirical Study of Evolutionary Techniques for Multiobjective Optimization in Engineering Design*. PhD thesis, Tulane University, 1996.
13. Coello Coello C and Lechuga M. MOPSO: A Proposal for Multiple Objective Particle Swarm Optimization. In *Congress on Evolutionary Computation*, volume 2, pages 1051–1056, 2002.
14. Lee C and Antonsson E. Dynamic Partitional Clustering Using Evolution Strategies. In *The Third Asia-Pacific Conference on Simulated Evolution and Learning*, 2000.
15. Veenman C, Reinders M, and Backer E. A Cellular Coevolutionary Algorithm for Image Segmentation. *IEEE Transactions on Image Processing*, 12(3):304–316, 2003.
16. Judd D, Mckinley P, and Jain A. Large-scale Parallel Data Clustering. *IEEE Transactions on Pattern Analysis and Machine Intelligence*, 20(8):871–876, 1998.
17. Van den Bergh F. *An Analysis of Particle Swarm Optimizers*. PhD thesis, Department of Computer Science, University of Pretoria, 2002.
18. Van der Merwe D and Engelbrecht A. Data Clustering using Particle Swarm Optimization. In *IEEE Congress on Evolutionary Computation*, pages 215–220, 2003.
19. Davies E. *Machine Vision: Theory, Algorithms, Practicalities*. Academic Press, 2nd edition, 1997.
20. Forgy E. Cluster Analysis of Multivariate Data: Efficiency versus Interpretability of Classification. *Biometrics*, 21:768–769, 1965.
21. Lumer E and Faieta B. Diversity and Adaptation in Populations of Clustering Ants. In *Proceedings of the Third International Conference on Simulation and Adaptive Behavior*, pages 501–508, 1994.
22. Hoppner F, Klawonn F, Kruse R, and Runkler T. *Fuzzy Cluster Analysis, Methods for Classification, Data Analysis and Image Recognition*. John Wiley & Sons Ltd, 1999.
23. Maselli F. Multiclass Spectral Decomposition of Remotely Sensed Scenes by Selective Pixel Unmixing. *IEEE Transactions on Geoscience and Remote Sensing*, 36(5):1809–1819, 1998.
24. Ball G and Hall D. A Clustering Technique for Summarizing Multivariate Data. *Behavioral Science*, 12:153–155, 1967.
25. Coleman G and Andrews H. Image Segmentation by Clustering. In *Proceedings of IEEE*, volume 67, pages 773–785, 1979.

26. Hamerly G. *Learning Structure and Concepts in Data using Data Clustering*. PhD thesis, University of California, San Diego, USA, 2003.
27. Hamerly G and Elkan C. Alternatives to the *K*-means Algorithm that Find Better Clusterings. In *Proceedings of the ACM Conference on Information and Knowledge Management (CIKM-2002)*, pages 600–607, 2002.
28. Abbas H and Fahmy M. Neural Networks for Maximum Likelihood Clustering. *Signal Processing*, 36(1):111–126, 1994.
29. Frigui H and Krishnapuram R. A Robust Competitive Clustering Algorithm with Applications in Computer Vision. *IEEE Transactions on Pattern Analysis and Machine Intelligence*, 21(5):450–465, 1999.
30. Bezdek J. A Convergence Theorem for the Fuzzy ISODATA Clustering Algorithms. *IEEE Transactions on Pattern Analysis and Machine Intelligence*, 2:1–8, 1980.
31. Bezdek J. *Pattern Recognition with Fuzzy Objective Function Algorithms*. Plenum Press, 1981.
32. Braquelaire J and Brun L. Comparison and Optimization of Methods of Color Image Quantization. *IEEE Transactions on Image Processing*, 6(7):1048–1052, 1997.
33. Fieldsend J and Singh S. A Multi-objective Algorithm based upon Particle Swarm Optimization, an Efficient Data Structure and Turbulence. In *The 2002 UK Workshop on Computational Intelligence*, pages 34–44, 2002.
34. Kennedy J and Eberhart R. Particle Swarm Optimization. In *Proceedings of IEEE International Conference on Neural Networks*, volume 4, pages 1942–1948, 1995.
35. Kennedy J and Eberhart R. *Swarm Intelligence*. Morgan Kaufmann, 2001.
36. Saghri J, Tescher A, Jaradi F, and Omran M. A Viable End-Member Selection Scheme for Spectral Unmixing of Multispectral Satellite Imagery Data. *Journal of Imaging Science and Technology*, 44(3):196–203, 2000.
37. Saghri J, Tescher A, and Omran M. Class-Prioritized Compression of Multispectral Imagery Data. *Journal of Electronic Imaging*, 11(2):246–256, 2002.
38. Velho L, Gomes J, and Sobreiro M. Color Image Quantization by Pairwise Clustering. In *Proceedings of the Tenth Brazilian Symposium on Computer Graphics and Image Processing*, pages 203–207, 1997.
39. Celenk M. A Color Clustering Technique for Image Segmentation. *Computer Vision, Graphics and Image Processing*, 52:145–170, 1990.
40. Omran M, Salman A, and Engelbrecht AP. Image Classification using Particle Swarm Optimization. In *Conference on Simulated Evolution and Learning*, volume 1, pages 370–374, 2002.
41. Omran M, Engelbrecht AP, and Salman A. A PSO-based Color Image Quantizer. *Special issue of Informatica Journal in multimedia mining in Soft Computing*, 29(3):263–271, 2005.
42. Omran M, Engelbrecht AP, and Salman A. A PSO-based End-Member Selection Method for Spectral Unmixing of Multispectral Satellite Images. *International Journal of Computational Intelligence*, 2(2):124–132, 2005.
43. Omran M, Engelbrecht AP, and Salman A. Particle Swarm Optimization Method for Image Clustering. *International Journal of Pattern Recognition and Artificial Intelligence*, 19(3):297–322, 2005.
44. Labroche N, Monmarche N, and Venturini G. Visual Clustering based on the Chemical Recognition System on Ants. In *Proceedings of the European Conference on Artificial Intelligence*, 2002.
45. Scheunders P. A Genetic C-means Clustering Algorithm Applied to Image Quantization. *Pattern Recognition*, 30(6), 1997.

46. Suganthan P. Particle Swarm Optimizer with Neighborhood Optimizer. In *Proceedings of the Congress on Evolutionary Computation*, pages 1958–1962, 1999.
47. Krishnapuram R and Keller J. A Possibilistic Approach to Clustering. *IEEE Transactions on Fuzzy Systems*, 1(2):98–110, 1993.
48. Krishnapuram R and Keller J. The Possibilistic C-Means algorithm: Insights and Recommendations. *IEEE Transactions on Fuzzy Systems*, 4(3):385–393, 1996.
49. Storn R and Price K. Differential Evolution - A Simple and Efficient Adaptive Scheme for Global Optimization over Continuous Spaces. Technical Report TR-95-012, International Computer Science Institute, 1995.
50. Turi R. *Clustering-Based Colour Image Segmentation*. PhD thesis, Monash University, 2001.
51. Baek S, Jeon B, Lee D, and Sung K. Fast Clustering Algorithm for Vector Quantization. *Electronics Letters*, 34(2):151–152, 1998.
52. Paterlini S and Krink T. Differential Evolution and Particle Swarm Optimization in Partitional Clustering. *Computational Statistics and Data Analysis*, 50(2006):1220–1247, 2005.
53. Shafer S and Kanade T. *Color Vision*, pages 124–131. Wiley, 1987.
54. Kaukoranta T, FrÃ?nti P, and Nevalainen O. A New Iterative Algorithm for VQ Codebook Generation. In *International Conference on Image Processing*, pages 589–593, 1998.
55. Kohonen T. *Self-Organization and Associative Memory*. Springer-Verlag, 3rd edition, 1989.
56. Hu X. Particle Swarm Optimization: Bibliography. 2004.
57. Hu X and Eberhart R. Adaptive Particle Swarm Optimization: Detection and Response to Dynamic Systems. In *Proceedings of Congress on Evolutionary Computation*, pages 1666–1670, 2002.
58. Rui X, Chang C, and Srikanthan T. On the initialization and Training Methods for Kohonen Self- Organizing Feature Maps in Color Image Quantization. In *Proceedings of the First IEEE International Workshop on Electronic Design, Test and Applications*, 2002.
59. Wu X and Zhang K. A Better Tree-Structured Vector Quantizer. In *Proceedings IEEE Data Compression Conference*, pages 392–401, 1991.
60. Xiao X, Dow E, Eberhart R, Ben Miled Z, and Oppelt R. Gene Clustering using Self-Organizing Maps and Particle Swarm Optimization. In *Proceeding of Second IEEE International Workshop on High Performance Computational Biology*, 2003.
61. Leung Y, Zhang J, and Xu Z. Clustering by Space-Space Filtering. *IEEE Transactions on Pattern Analysis and Machine Intelligence*, 22(12):1396–1410, 2000.
62. Shi Y and Eberhart R. A Modified Particle Swarm Optimizer. In *Proceedings of the IEEE International Conference on Evolutionary Computation*, pages 69–73, 1998.
63. Shi Y and Eberhart R. Parameter Selection in Particle Swarm Optimization. In *Evolutionary Programming VII: Proceedings of EP'98*, pages 591–600, 1998.
64. Xiang Z. Color Image Quantization by Minimizing the Maximum Inter-cluster Distance. *ACM Transactions on Graphics*, 16(3):260–276, 1997.
65. Xiang Z and Joy G. Color Image Quantization by Agglomerative Clustering. *IEEE Computer Graphics and Applications*, 14(3):44–48, 1994.

7

Data and Text Mining with Hierarchical Clustering Ants

Hanene Azzag[1], Christiane Guinot[1,2], and Gilles Venturini[1]

[1] Laboratoire d'Informatique,
École Polytechnique de l'Université de Tours - Département Informatique
64, Avenue Jean Portalis, 37200 Tours, France.
Phone: +33 2 47 36 14 14, Fax: +33 2 47 36 14 22
http://www.antsearch.univ-tours.fr/WebRTIC/
{hanene.azzag, venturini}@univ-tours.fr

[2] C.E.R.I.E.S.,
20, rue Victor Noir, 92521 Neuilly sur Seine Cedex.
christiane.guinot@ceries-lab.com

Summary. In this paper is presented a new model for data clustering, which is inspired from the self-assembly behavior of real ants. Real ants can build complex structures by connecting themselves to each others. It is shown is this paper that this behavior can be used to build a hierarchical tree-structured partitioning of the data according to the similarities between those data. Several algorithms have been detailed using this model : deterministic or stochastic algorithms that may use or not global or local thresholds. We have also introduce an incremental version of our artificial ants algorithm.

Those algorithms have been evaluated using artificial and real databases. Our algorithms obtain competitive results when compared to the Kmeans, Ascending Hierarchical Clustering, AntClass and AntClust (two biomimetic methods). Our methods have been applied to three real world applications: the analysis of human healthy skin, the on-line mining of web sites usage, and the automatic construction of portal sites.

Finally, we have developed two possibilities to explore the portal site. The first possibility consists in representing the tree in HTML pages in order to explore the portal site with a conventional browser. The second possibility we have studied is to integrate the results of our algorithms in our virtual reality data mining tool VRMiner [39].

7.1 Introduction

The initial goal of this work was to build a new hierarchical clustering algorithm and to apply it to several domains including the automatic construction of portal sites for the Web. A portal site can be viewed as a hierarchical partitioning of a set of documents which aims at recursively reproducing the following property: at each node (or "mother" category), the sub-categories are as similar as possible to their mother, but they are as much dissimilar as possible to their "sister" categories. One of the major problems to solve in this area is the definition of this hierarchy of documents. In actual systems, this hierarchy must be given by a human expert [1][2][3]. So there is a need for automating the hierarchy construction, especially when one deals with an important number of documents.

We propose a new approach which builds a tree-structured partitioning of the data that checks the recursive property mentioned above which can be applied to many other domains. This method simulates a new biological model: the way ants build structures by assembling their bodies together. Ants start from a point and progressively become connected to this point, and recursively to these firstly connected ants. Ants move in a distributed way over the living structure in order to find the best place where to connect. This behavior can be adapted to build a tree from the data to be clustered.

The rest of this chapter is organized as follows: section 7.2 describes the ants biological behavior, how it can be modeled for building trees, and the related approaches found in the literature. Section 7.3 presents the different algorithmic choices which result in several algorithms. In section 7.4, we have presented a comparative study between these algorithms and the K-means, Ascending Hierarchical Clustering (AHC), AntClass and AntClust on several databases. Section 7.5 presents three real world applications and section 7.6 introduce an incremental version of our artificial ants algorithm. Section 7.7 concludes on the perspectives which can be derived from this work.

7.2 Biological and computer models

7.2.1 Ants based algorithms for clustering

The numerous abilities of ants have inspired researchers for more than ten years regarding designing new clustering algorithms [4][5]. The initial and pioneering work in this area is due to Deneubourg and his colleagues [4]. These authors have been interested in the way real ants sort objects in their nest by carrying them and dropping them in appropriate locations without a centralized control policy. They have proposed the following principles [6]: artificial ants move in a 2D grid where all objects to be sorted are initially scattered. An ant has a local perception of the objects around its location and does not communicate directly with other ants. Instead, ants influence themselves using the configurations of objects on the grid. An ant has a given and variable probability of picking an encountered object. This probability is high when the frequency of this type of object was low in the recent history of the ant. Similarly, an ant will drop a carried object with a high probability when similar objects have been perceived in the ants surrounding. Using this simple and distributed principles, ants are able to collectively sort the objects.

The next step toward data clustering has been done in [5]. These authors have adapted the previous algorithm by considering that an object is a datum and by tuning the picking/dropping probabilities according to the similarities between data. These ants based algorithms inherit from real ants interesting properties, such as the local/global optimization of the partitioning, the absence of need of a priori information on an initial partitioning or number of classes, or parallelism. Furthermore, the results are presented as a visualization, a property which is coherent with an important actual trend in data mining called "visual data mining" where results are presented in a visual and interactive way to the domain expert.

As a consequence, several other authors have used and extended this model, especially toward real and more difficult data with very interesting results, such as in graph partitionning for VSLI technology [7, 8], in Web usage mining [9] or in document clustering [10].

Other models of "clustering ants" have been studied (see for instance [11], but the model which has been studied the most is the way ants sort objects in their nest.

In this work, we propose a new model of clustering ants which is based on the self-assembly behavior observed in these insects and which is described in the next section.

7.2.2 Self-assembly in real ants

The self-assembly behavior of individuals can be observed in several insects like bees or ants for instance (see a survey with spectacular photographs in [12]). We are interested here in the complex structures which are build by ants. We have specifically studied how two species were building such structures, namely the Argentina ants *Linepithema humiles* and ants of gender *Oecophylla* and that we briefly describe here [13][14]: these insects may become fixed to one another to build live structures with different functions. Ants may thus build "chains of ants" in order to fill a gap between two points. These structures disaggregate after a given time.

The general principles that rule these behaviors are the followings: ants start from an initial point (called the support). They begin to connect themselves to this support and then progressively to previously connected ants. When an ant is connected, it becomes a part of the structure and other ants may move over this ant or connect themselves to it. The structure grows over time according to the local actions performed by the ants. Moving ants are influenced by the local shape of the structure and by a visual attractor (for instance, the point to reach). Ants which are in the middle of the structure cannot easily disconnect themselves.

7.2.3 A computer model of ants self-assembly for hierarchical clustering

From those elements, we define the outlines of our computer model which simulates this behavior for tree building (see Figure 7.1). The n ants $a_1, ..., a_n$ represent each of the n data $d_1, ..., d_n$ of the database. These ants are initially placed on the support which is denoted by a_0. Then, we successively simulate one action for each ant. An ant can be in two states: it can be free (i.e. disconnected) and may thus move over the structure in order to find a place where it can connect, or it can be connected to the

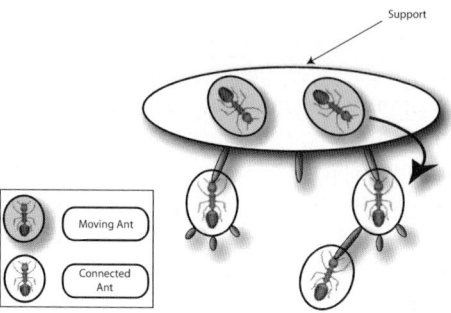

Fig. 7.1. General principles of tree building with artificial ants.

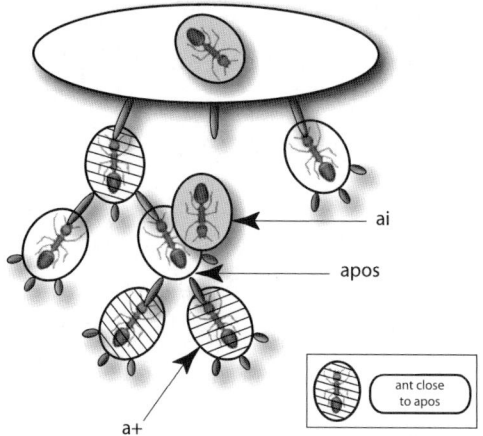

Fig. 7.2. The computation of an ant's neighborhood.

structure without any possibility of moving apart from to disconnect. In this work, an ant may become connected to only one other ant, which ensures that ants will build a tree.

Ants locally perceive the structure: a moving ant a_i located over a connected ant a_{pos} perceives a neighborhood N_{pos} (see the striped ants in Figure 7.2), which is limited (1) to the (mother) ant to which a_{pos} is connected, and (2) to the (daughter) ants which are connected to a_{pos}. a_i can perceive the similarities between the data d_i it is representing and the data represented by ants of N_{pos}. According to these similarities, a_i may either get connected to a_{pos}, or move to one of the ants in N_{pos}.

We also include in this study the use of "ants disconnections" in order to change choices made previously in the growth of the structure. Once all ants are connected to the structure, then our algorithms stop. The resulting tree can be interpreted as a partitioning of the data 7.3.4. The properties that we wish to obtain for data clustering in the context of portal site building are the following: each sub-tree T represents one category compound of all ants in this sub-tree. Let a be the ant which is at the root of T. We would like that (1) a is as the best representative of this category (ants placed below a are as much similar to a as possible), (2) the "daughters" ants of a which represent sub-categories of a are as dissimilar to each others as possible (well separated sub-categories). This property should be recursive and possibly checked anywhere in the hierarchical structure.

This model is different from previously studied ant-based algorithms which make use of pheromones (like the Ant Colony Optimization (ACO) approach [15]). In ACO, ants may deposit a pheromone trail to attract other ants. But here ants do not communicate via an external support but are themselves the support. One must notice however that tree building with ants has already been tackled in the literature. On the one hand, we mention the work done on genetic programming with artificial ants [16] and the ACO approach [17]. Ants build trees by assigning probabilities (i.e. pheromones) to symbols. On the other hand, the ACO approach has also been used in the specific context of phylogenetic tree building [18]. Ants use pheromones to determine the order of the symbols that will be used to build an unrooted tree. However, these two models are not centered on the data clustering problem and have no common points with the self-assembly behavior that we are simulating.

7.2.4 Self-assembly and robotics

We briefly outline in this section the pioneering work which has been done in robotics in the context of artificial self-assembly behavior. Since about 15 years, researchers in robotics are trying to build modular, or reconfigurable or metamorphic robots. These robots are in fact an assembly of simple modules. Each module has its own processor and local communication unit, and has the ability to connect in several ways to other modules. In general, a module can change the properties of the connections (like their orientation) which can lead the formation of complex self-assembling structures. Modules may thus move on the structure and connect to a given location.

The aim of such robotic systems is to adapt the robot configuration to the task at hand, and to increase the robustness of the robot (one module can be replaced by another one). One of the first work in this area has been presented in [19] where a robot is compound of several modules which can autonomously connect to each others in order to change the overall 2D configuration of the modular robot. Several other examples of such 2D reconfigurable robots have been initially studied (see also [20]).

More recently, complex modules that may be connected in 3D have appeared. The M-TRAN series of robots [21] can change their configuration in 3D (i.e. from a four legged robot to a snake-like robot). Another example is the Atron robot [22]. A very interesting combination of physical robotic system with a behavioral distributed strategy has been performed in the Swarm-bot project [23]: the basic module is a mobile robot with 6 wheels. This robot may become connected to another robot or to object to be carried to a given location. The perceptual abilities of each module is more evolved than in previous reconfigurable robots (with the use of a camera). Such robots are able to create a structure in order to drag an object to a given location or to cross an empty space (these tasks are impossible to solve with a single robot).

These studies have introduced similar concepts as those used in our work, but are mainly intended to solve robotics problems and not data mining problems unlike the algorithms we will present in the next section.

7.3 Two stochastic and deterministic algorithms

7.3.1 Common principles

We have studied several algorithms in order to explore the different possibilities and to experimentally validate/invalidate them. Each algorithm uses the same global framework:

1. Initialize: ants are all placed on the support a_0. Possibly compute the thresholds, possibly sort the data (see explanations in the following),
2. Build the structure: choose an action for each ant a_i according to behavioral rules. This step is repeated until a stopping criterion is checked (all ants are connected, or a given number of iterations have been performed, etc),
3. Terminate: assign each non connected ant to the structure.

We describe now the ants behavioral rules, which will result in the building of a tree . We have studied several versions of our algorithms. We present only the two best algorithms. The first one use stochastic moves for the ants and it performs self-adaptation of the local thresholds used by ants (AntTree$_{STOCH}$). The other one is deterministic and uses no parameters at all : only the local shape of the structure is used to determine what the ants will do (AntTree$_{NO-THRESHOLDS}$).

7.3.2 Stochastic algorithm: AntTree$_{STOCH}$

Non determinism is often used in biological simulations not only because the animal behavior can be stochastic but also because complete determinism would require a very fine level of model analysis which may not be necessary (or possible). In our case, it has been motivated by the fact that random behavior can change the hierarchical structure and has improved the results over similar but deterministic versions [24].

We consider that each ant a_i has two thresholds $T_{Sim}(a_i)$ and $T_{Dissim}(a_i)$ which will be used to determine its behavior (move or connect). Each ant can be connected to exactly one ant, and we limit the number of ants which can be connected to an other ant to a value denoted by L_{max}. This will be the maximum branching factor of the tree. We consider the two following decisions rules, which correspond to the cases where a_i is located on the support or on an other ant:

R_1 ($a_{pos} = a_0$ (the support))
    1. If no ant is connected yet to the support a_0 Then connect a_i to a_0
    2. Else

a) If $\text{Sim}(a_i, a^+) \geq T_{Sim}(a_i)$ Then move a_i toward a^+
b) Else
   i. If $\text{Sim}(a_i, a^+) < T_{Dissim}(a_i)$ Then connect a_i to the support a_0 (or, if no incoming link available on a_0, decrease $T_{Sim}(a_i)$ and move a_i toward a^+)
   ii. Else a_i stays on its position, $T_{Sim}(a_i)$ is decreased and $T_{Dissim}(a_i)$ is increased

R_2 (a_{pos} = another ant)
1. Let a_k denote a randomly selected neighbor of a_{pos}
2. If $\text{Sim}(a_i, a_{pos}) \geq T_{Sim}(a_i)$ Then
   a) If $\text{Sim}(a_i, a^+) < T_{Dissim}(a_i)$ Then connect a_i to a_{pos} (or, if no incoming link is available on a_{pos}, move a_i toward a_k)
   b) Else decrease $T_{Sim}(a_i)$, increase $T_{Dissim}(a_i)$, and move a_i toward a_k
3. Else move a_i toward a_k (possibly modify the thresholds to avoid loops)

where a_{pos} denotes the ant (or the support) over which a_i is located, and where a^+ denotes the ant connected to a_{pos} which is the most similar to a_i.

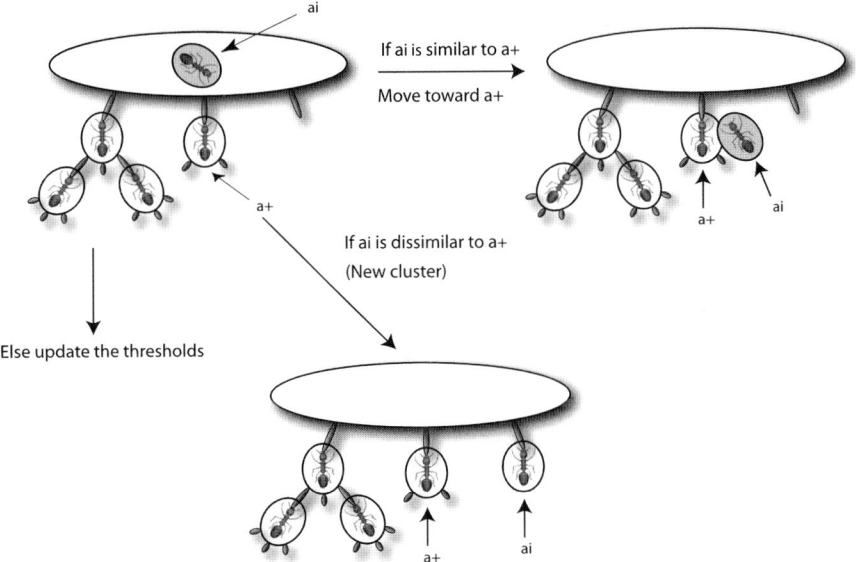

Fig. 7.3. First behavioral rule of AntTree$_{STOCH}$: ant a_i is located on the support a_0.

The first rule can be explained as follows (see Figure 7.3): when the tree is a root only, then the first ant connects to the support. Then, the second case of this first rule is when an ant located on the support must decide either to move to a connected ant or to connect to the support. The decision is taken according to the ant's thresholds: if a_i is similar enough to a connected ant, then it will move toward this ant, and will probably try to connect in this subtree. Else, it will try to connect to the support (and

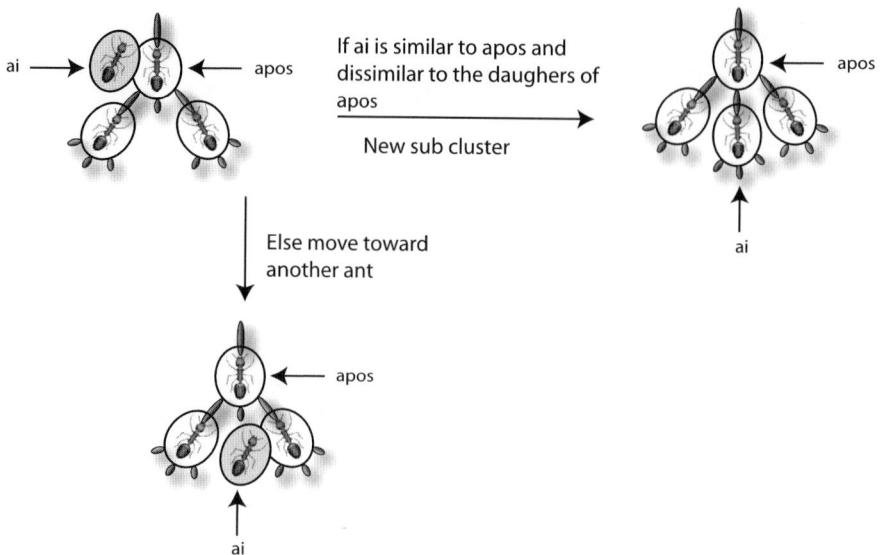

Fig. 7.4. Second behavioral rule of AntTree$_{STOCH}$: ant a_i is located on an other ant.

thus create a new subtree), but provided that it is dissimilar enough the ants already connected to the support (this subtree will represent a well separated class). If none of these cases applies, then the ant remains on the support and its thresholds will be updated in such a way that a_i will become more tolerant.

The second rule considers the same problem, but related to a subtree rather than the root (see Figure 7.4). a_i is located on an ant a_{pos}. If it is similar enough to a_{pos} and dissimilar enough to the daughters of a_{pos}, then it will connect to a_{pos} (and it thus creates a new subtree, i.e. a subclass closely related to a_{pos} but well separated from the other subclasses of a_{pos}). Else, if it cannot create a subtree, then the ant will randomly move in the tree. The stochastic aspect of this algorithm is due to the randomness of moves in this rule.

The thresholds are updated in the following way. The similarity threshold $T_{Sim}(a_i)$ can be initially equal to 1, $Sim_{max}(a_i)$, $Sim_{mean}(a_i)$ or $\frac{Sim_{mean}(a_i)+Sim_{max}(a_i)}{2}$ where $Sim_{mean}(a_i)$ and $Sim_{max}(a_i)$ respectively denote the mean and the maximum similarities between a_i and the other data. Then, this threshold is updated in the following way: $T_{Sim}(a_i) \leftarrow T_{Dissim}(a_i) \times \alpha_1$. Similarly, the dissimilarity threshold $T_{Dissim}(a_i)$ can be initialized to 0, $Sim_{min}(a_i)$, $Sim_{mean}(a_i)$ or $\frac{Sim_{mean}(a_i)+Sim_{min}(a_i)}{2}$ where $Sim_{min}(a_i)$ denotes the minimum of similarities between a_i and the other data. The update of this threshold is performed as follows: $T_{Dissim}(a_i) \leftarrow T_{Dissim}(a_i) + \alpha_2$. α_1 and α_2 are two constants that will be experimentally determined.

The fact that these thresholds are local to each ant rather than global has proven to be important because they are better adapted to the local distribution of the data (outliers for instance may need different values).

7.3.3 Deterministic algorithm with no thresholds and no parameters : AntTree$_{NO-THRESHOLDS}$

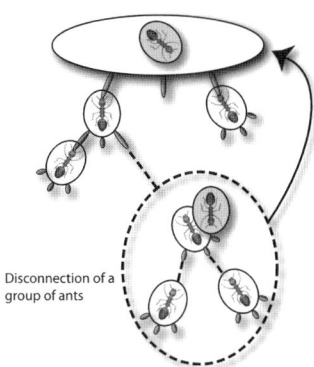

Fig. 7.5. Disconnection of ants.

Adjusting thresholds or parameters is a well known and difficult task for any algorithm and especially the algorithms inspired from biological models where many options must be selected (see [25] for evolutionary algorithms). We propose in this section an AntTree algorithm which does not use any thresholds and more generally any parameters.

A second motivation is also that, in real world applications of clustering, the non determinism of results may be confusing for the user. For example, the applications described in section 7.5 are such that users prefer to obtain always the same results for two different runs.

This deterministic algorithm uses the possibility to disconnect ants (like drops of ants for the *Linepithema humiles* and the disaggregate of chains for *Oecophylla longinoda*). The basic idea is simple: an ant a_i will become connected to the ant a_{pos} provided that it is dissimilar enough to a^+, but this time the threshold will only depend on ants already connected to a_{pos}, i.e. the threshold is determined by the structure itself.

Let us denote by $T_{Dissim}(a_{pos})$ the lowest similarity value which can be observed among the daughters of a_{pos}. a_i is connected to a_{pos} if and only if the connection of a_i decreases further this value. Since this minimum value can only be computed with at least two ants, then the two first ants are automatically connected without any tests. This may result in "abusive" connections for the second ant. Therefore the second ant is removed and disconnected as soon as a third ant is connected (for this latter ant, we are certain that the dissimilarity test has been successful). When this second ant is removed, all ants that were connected to it are also dropped, and all these ants are placed back onto the support (see Figure 7.5). This algorithm can thus be stated as follows:

R_1 (no ant or only one ant connected to a_{pos}): a_i connects to a_{pos}

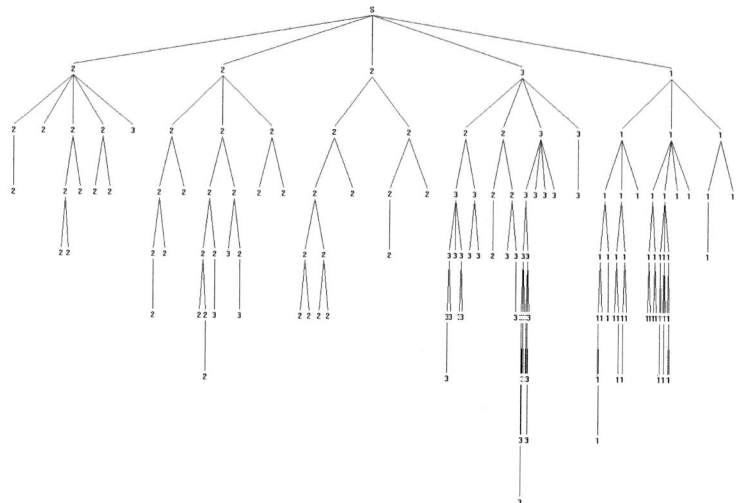

Fig. 7.6. Hierarchical structure obtained by AntTree$_{NO-THRESHOLDS}$ with Iris database. Number indicates the real classes.

R'_1 (2 ants connected to a_{pos}, and for the first time):
  1. Disconnect the second ant (the most dissimilar to a_i) from a_{pos} (and recursively all ants connected to it)
  2. Place all these ants back onto the support a_0
  3. Connect a_i to a_{pos}

R_2 (more than 2 ants connected to a_{pos}, or 2 ants connected to a_{pos} but for the second time):
  1. Let $T_{Dissim}(a_{pos})$ be the lowest dissimilarity value between daughters of a_{pos} (i.e. $T_{Dissim}(a_{pos}) = Min\ Sim(a_j, a_k)$ where a_j and $a_k \in$ {ants connected to a_{pos}}),
  2. If a_i is dissimilar enough to a^+ ($Sim(a_i, a^+) < T_{Dissim}(a_{pos})$), then a_i connects to a_{pos}
  3. Else a_i moves toward a^+

When ants are placed back on the support, they may find an other place where to connect using the same behavioral rules. It can be observe that, for any node of the tree, the value $T_{Dissim}(a_{pos})$ is only decreasing, which ensures the termination and convergence of the algorithm.

7.3.4 Properties

We have shown in Figures 7.6 and 7.7 typical learned trees. We have evaluated for all databases several statistical measures for the trees: the mean branching factor, the ratio between the similarity of one node/data and its children, and the mean similarity between the children. This last measure is intended to check that the initial property we want to obtain for portal sites is verified. The results of these evaluations are not

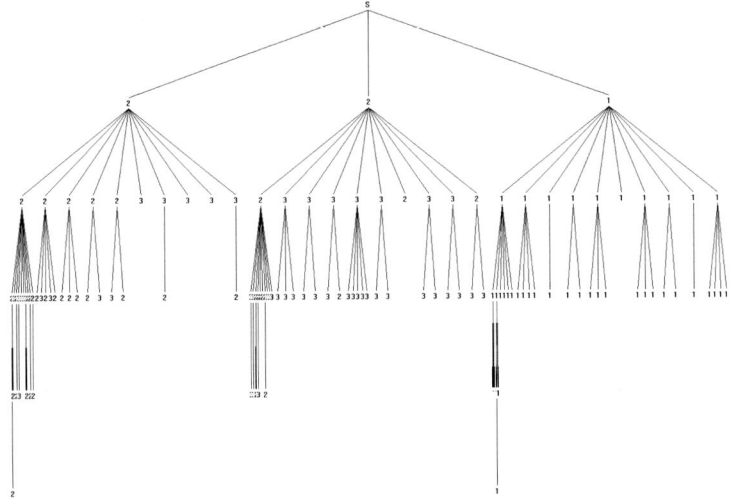

Fig. 7.7. Hierarchical structure obtained by AntTree$_{STOCH}$ with Iris database. Number indicates the real classes.

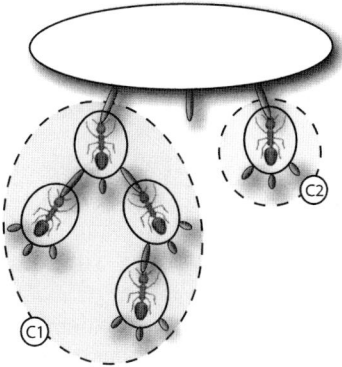

Fig. 7.8. Interpreting the tree as clusters.

presented here due to limited space, but the conclusions where positive: the desired property is checked. The branching factor is close to 2 for AntTree$_{NO-THRESHOLDS}$ and close to 3 for AntTree$_{STOCH}$. This can be visually checked in Figures 7.6 and 7.7. AntTree$_{NO-THRESHOLDS}$ connects easily two ants, but the local thresholds favor less subcategories than in AntTree$_{STOCH}$.

The complexity of these algorithms is difficult to compute. We have tried to do so but worst cases are not realistic, and average cases depends on the data. The complexity is theoretically within $O(n \ln n)$ and $O(n^2)$. However, according to the results, the use of a tree is one of the best way to reduce the complexity in a clustering process (see section 7.6).

We can use the hierarchical structure as the results, but one may also interpret the tree as a flat partitioning (see Figure 7.8): each sub-tree which is directly connected to the support a_0 is considered as a cluster. This step is necessary for comparison purposes (see next section).

In general, the first ants (i.e. data) of the database will become connected first in the tree. Therefore the order of the data is important for our algorithms. Since the data in the benchmarks are often sorted according to their class attribute, using the original order could greatly bias the results. We have thus studied different ways to sort the data according to their mean similarity to each others: random, increasing and decreasing orders. This sort is performed in $O(n \ln n)$ and does not increase the overall complexity of the algorithms (see next section).

Finally, we mention to the interested reader that a limited Java demonstration is available on our Web site (see the Demo page).

7.4 Experimental results with numeric, symbolic and textual databases

7.4.1 Testing methodology

We have evaluated and compared our algorithms on a set of 22 databases which have from $N = 47$ to 5000 examples (see Table 7.1). The databases ART1 to ART6 are artificial ones and have been generated with gaussian and uniform laws. Other databases have been extracted from the Machine Learning Repository [26], and one database comes from the CE.R.I.E.S. [27] and deals with the healthy human skin analysis (see section 7.5.1). The real classes of data are of course not given to the algorithms but instead are used in the final evaluation of the obtained partitioning.

The evaluation of the results is performed with the number of found clusters C_f, with the purity P of clusters (percentage of correctly clustered data in a given cluster), and with a classification error Ec (proportion of data couples which are not correctly clustered, i.e. in the same real cluster but not in the same found cluster, and vice versa) defined as follows [28]:

$$Ec = \frac{2}{N(N-1)} \sum_{(i,j) \in \{1,\ldots,N\}^2, i<j} \varepsilon_{ij} \qquad (7.1)$$

where :

$$\varepsilon_{ij} = \begin{cases} 0 & \text{if } (Cr_i = Cr_j \wedge Cf_i = Cf_j) \vee \\ & (Cr_i \neq Cr_j \wedge Cf_i \neq Cf_j) \\ 1 & \text{else} \end{cases} \qquad (7.2)$$

and where C_r denotes the number of "real classes" and C_f the number of "found classes". The results are averaged over 15 trials. We have also presented standard deviations which are respectively denoted by σ_{Cf}, σ_P and σ_{Ec}.

Table 7.1. The 22 databases used in our tests (C_r is the number of classes, M is the number of attributes and N the number of examples).

Databases	C_r	M	N	Type
ART1	4	2	400	Numeric
ART2	2	2	1000	Numeric
ART3	4	2	1100	Numeric
ART4	2	2	200	Numeric
ART5	9	2	900	Numeric
ART6	4	8	400	Numeric
GLASS	7	9	214	Numeric
HAYES-ROTH	3	5	132	Mixed
HEART	2	13	270	Mixed
HOUSE-VOTES-84	2	16	435	Symbolic
IRIS	3	4	150	Numeric
LYMPHOGRAPHY	4	18	148	Mixed
PIMA	2	8	768	Numeric
SEGMENT	7	19	2310	Numeric
SOYBEAN	4	35	47	Numeric
THYROID	3	5	215	Numeric
TIC-TAC-TOE	2	9	958	Symbolic
VEHICLE	4	18	846	Numeric
WAVEFORM	3	21	5000	Numeric
WINE	3	13	178	Numeric
ZOO	7	17	101	Mixed
CERIES	6	4	259	Numeric

For all the algorithms that we propose, we used the same similarity measure $Sim(d_i, d_j)$. This measure is a simple a 2-norm when the data is numeric only (Euclidean distance, see equation 7.3). When the data is both numeric and symbolic, we have used a 1-norm for numeric values (based on absolute values) and the Hamming distance for symbolic values (see equations 7.4 and 7.5). The result of this measure belongs to $[0, 1]$. All numeric values are previously normalized in $[0, 1]$.

$$Sim(i,j) = 1 - \sqrt{\frac{1}{M} \sum_{k=1}^{M} (v_{i_k} - v_{j_k})^2} \quad (7.3)$$

v_{i_k} is the kth numeric value of the data v_i.

$$Sim(i,j) = 1 - \frac{1}{M} \sum_{k=1}^{M} (v_{i_k} \oplus v_{j_k}) \quad (7.4)$$

v_{i_k} is the kth symbolic attribute of the data v_i.

$$Sim(i,j) = 1 - \frac{1}{M}\left(\overset{M-Numeric}{\underset{k=1}{\sum}}|v_{i_k} - v_{j_k}| + \overset{M-Symbolic}{\underset{k=M-Numeric+1}{\sum}}(v_{i_k} \oplus v_{j_k})\right) \quad (7.5)$$

where M-Numeric is the number of numeric attributes and M-Symbolic the number of symbolic attributes.

7.4.2 Parameters study

As mentioned before, AntTree$_{STOCH}$ has several parameters that we would like to determine once for all: α_1 and α_2, which rule the increasing/decreasing of the $T_{Sim}(i)$ and $T_{Dissim}(i)$ thresholds for each ant a_i, the initial values to which these thresholds must be set, the initial sorting of the data (random, increasing, decreasing). We have decided to test the following values: three possible sorting of the data (random, increasing, decreasing), four initial values for $T_{Sim}(i)$ and $T_{Dissim}(i)$ (respectively (1;0), ($Sim_{max}(a_i);Sim_{min}(a_i)$), ($Sim_{mean}(a_i);Sim_{mean}(a_i)$), ($\frac{Sim_{mean}(a_i)+Sim_{max}(a_i)}{2}; \frac{Sim_{mean}(a_i)+Sim_{min}(a_i)}{2}$)), height values for α_1 (0,7; 0,8; 0,90; 0,92; 0,95; 0,97; 0,98; 0,99), and ten values for α_2 (0,005; 0,01; 0,02; 0,03; 0,05; 0,07; 0,1; 0,2; 0,3; 0,4). This yields $3 \times 4 \times 8 \times 10 \times 15$ runs for each of the 22 databases (a total of 316 800 runs). Results are presented in Table 7.2. Apparently, the sorting of the data is not necessary for this algorithm. Initializing the thresholds to ($\frac{Sim_{mean}(a_i)+Sim_{max}(a_i)}{2}; \frac{Sim_{mean}(a_i)+Sim_{min}(a_i)}{2}$) seems to be a good compromise between (1;0) which results in a longer adaptation, and ($Sim_{mean}(a_i);Sim_{mean}(a_i)$) where ants get quickly and abusively connected to the tree. As a consequence, we will use the following parameters for AntTree$_{STOCH}$: random sorting, thresholds initialized to ($\frac{Sim_{mean}(a_i)+Sim_{max}(a_i)}{2}; \frac{Sim_{mean}(a_i)+Sim_{min}(a_i)}{2}$), $\alpha_1 = 0,95$ and $\alpha_2 = 0,2$. We have also initialize L_{max} (maximum branching factor of the tree) to $2 \times Max(C_r)$ (= 20).

For AntTree$_{NO-THRESHOLDS}$, the only parameter to be tested is the initial sorting of the data. Table 7.3 shows that the increasing order leads to the worst results: the most dissimilar data are not adapted here for starting the classes. The value of $T_{Dissim}(a_{pos})$ decreases too quickly, especially for the support, which leads to a very small number of classes (and thus poor results). So for this algorithm, we have used the decreasing order which gives the best results.

The other algorithms which have been used in the next section (i.e. K-means, AntClass, AntClust, AHC) are not sensitive to the ordering of the data because they select the data in a specific way, either randomly for K-means, AntClass and AntClust, or according to a heuristic for AHC.

7.4.3 Tested algorithms

We decided to compare our algorithms with two standard and well known methods in data clustering, namely the AHC and the K-means [29], as well as with two of our previous ants based approaches named AntClass [30] and AnClust [31].

AHC is a standard, very well known and competitive method. It also uses hierarchical principles and can be applied to any type of data. The K-means is well

Table 7.2. Results obtained by AntTree$_{STOCH}$ for our test parameters. Respectively $\frac{Sim_{mean}(a_i)+Sim_{max}(a_i)}{2}$ and $\frac{Sim_{mean}(a_i)+Sim_{min}(a_i)}{2}$ are represented by $Sim_{mean-max}(a_i)$ and $Sim_{mean-min}(a_i)$. E_C is the average classification error computed by the 22 numerical and symbolic databases, σ_{Ec} is the corresponding standard deviation.

Rank	Sorting	Initialization	α_1	α_2	$E_C\ [\sigma_{Ec}]$
1	random	$Sim_{mean-max}(a_i)$ $Sim_{mean-min}(a_i)$	0,95	0,2	0,2683 [0.0256]
2	random	$Sim_{max}(a_i)$ $Sim_{min}(a_i)$	0,9	0,2	0,2726 [0.0273]
3	decreasing	$Sim_{mean-max}(a_i)$ $Sim_{mean-min}(a_i)$	0,97	0,3	0,2744 [0.0035]
4	increasing	$Sim_{max}(a_i)$ $Sim_{min}(a_i)$	0,8	0,4	0,2805 [0.00201]
5	random	1 0	0,92	0,3	0,2830 [0.024610409]

Table 7.3. Results obtained by AntTree$_{NO-THRESHOLDS}$ on numerical databases using increasing, decreasing and random order

Sorting	Increasing	Decreasing	Random
$E_C\ [\sigma_{Ec}]$	0,4029	0,2782	0,3718 [0,0699]

adapted to numeric data because it directly makes use of the notion of center of mass. This method is efficient but suffers from an initialization problem. AntClass is an hybridization between an ant-based algorithm and the K-means method, and is supposed to solve the initialization of the clusters more efficiently. AntClust is an ant-based clustering algorithm inspired from the chemical recognition system of ants. The clustering algorithm associates an object to the odor of an ant and at the end artificial ants that share a similar odor are grouped in the same cluster.

In the K-means, we decided to initialize the initial partition to 10 or 20 randomly generated clusters. We have denoted each method respectively by 10-means and 20-means. With 10 clusters, the initial number of classes is close to the real one ($Max(C_r) = 10$). With 20 clusters, we can compare the results to AntTree$_{STOCH}$ where L_{max} is set to $2 \times Max(C_r)$.

In the AHC algorithm, we have used the Ward criterion to automatically cut the dendrogram just before the largest jump, in order to obtain a flat clustering of the data. This method has no problems in initialization but its temporal and spatial complexities are known to be important. In our implementation, we have favored the temporal complexity by using fast matrix operations (which consume much memory).

7.4.4 Results with numeric databases

We can compare all methods with the three parameters which are measured at each run: the number of found clusters C_f, the purity of clusters P_R, the classification error E_c.

AHC is the method which approximates the best the number of classes. One must notice however that the real number of clusters is not necessarily representative of the real number of groups (or point clouds) in the data. For the Iris or Wine databases, basic visualizations show that the real number of classes corresponds to the number of observed groups in the data. This is not the case for other databases like the Pima database which contains two real classes but many more groups. AHC found three classes for the Pima database, and thus seems to better approximate the real number of classes, but in fact we know that there are many groups in the data. This is the reason why the other algorithms found many more clusters (13,47 for $AntTree_{STOCH}$, 19,27 for 20-means). So AHC obtains better performances because it systematically builds less classes than the other methods, which is a favorable behavior for the data at hand. Our method obtains better results than the K-means.

The best purity values are obtained by 20-means which gets the best results for 11 databases over 16. This is explained by the fact that the purity measure is sensitive to the number of found clusters. If many clusters are found, then these clusters are more likely to be homogeneous and thus the purity values will be higher (the extreme case is obtained when one data = one cluster, where the purity equals 1). So purity must be related to the number of found clusters, which are much higher for 20-means than for the other methods. One may notice that some databases are more difficult than the others (Art5, Glass, Ceries, Pima, Segment, Vehicle or Waveform). In these bases, the clusters are not well separated from each others, and this is why $AntTree_{STOCH}$ outperforms AHC.

With the classification error, the best results are obtained by AHC (9 databases over 16) because AHC better approximates the number of real classes. The worst results are obtained by 20-means. For our algorithms $AntTree_{STOCH}$ and 10-means obtain the same performance and are better than $AntTree_{NO-THRESHOLDS}$.

7.4.5 Results with symbolic databases

The symbolic or mixed databases are more difficult than the others. $AntTree_{STOCH}$ better approximates the real number of classes than the other methods (AHC performs well also). The purity values are globally lower than with numeric databases.

The lowest classification error is obtained by AHC (5 databases over 6). However, AHC performs poorly on the Hayes Roth database because $AntTree_{STOCH}$ finds more clusters than AHC and thus lowers the classification error.

Table 7.4. Results obtained by AntTree$_{STOCH}$ on numeric and symbolic databases

Databases	C_r	$E_C[\sigma_{Ec}]$	$C_f[\sigma_{Cf}]$	$P_R[\sigma_{P_R}]$
ART1	4	0,19 [0,03]	5,2 [0,75]	0,77 [0,07]
ART2	2	0,24 [0,04]	4,93 [0,77]	0,95 [0,03]
ART3	4	0,25 [0,06]	5,53 [0,50]	0,86 [0,04]
ART4	2	0,28 [0,03]	5,53 [0,96]	0,99 [0,01]
ART5	9	0,17 [0,01]	6 [0,82]	0,5 [0,02]
ART6	4	0,07 [0,03]	6,4 [1,02]	0,97 [0,01]
CERIES	6	0,15 [0,01]	9,53 [1,31]	0,75 [0,04]
GLASS	7	0,34 [0,03]	8,07 [1,12]	0,55 [0,05]
IRIS	3	0,18 [0,04]	3,93 [0,77]	0,84 [0,07]
PIMA	2	0,48 [0,01]	13,47 [1,26]	0,72 [0,01]
SEGMENT	7	0,14 [0,01]	14,6 [1,62]	0,67 [0,04]
SOYBEAN	4	0,1 [0,02]	6,67 [0,87]	0,99 [0,02]
THYROID	3	0,27 [0,07]	5,93 [1,00]	0,87 [0,03]
VEHICULE	4	0,31 [0,01]	8,47 [1,02]	0,45 [0,02]
WAVEFORM	3	0,31 [0,02]	6,6 [0,95]	0,66 [0,05]
WINE	3	0,21 [0,03]	8,2 [1,64]	0,88 [0,03]
HAYES ROTH	3	0,37 [0,01]	7,8 [0,83]	0,55 [0,03]
HEART	2	0,45 [0,01]	11,13 [1,36]	0,79 [0,01]
HOUSE VOTES	2	0,36 [0,03]	12,13 [1,45]	0,91 [0,02]
LYMPHOGRAPHY	4	0,44 [0,01]	10,73 [1,34]	0,75 [0,03]
TIC-TAC-TOE	2	0,54 [0,00]	18,33 [0,79]	0,68 [0,01]
ZOO	7	0,06 [0,03]	6,73 [1,12]	0,87 [0,04]

7.4.6 Processing times

We present in this section a comparative study based on processing times for the five algorithms (AntTree$_{STOCH}$, AntTree$_{NO-THRESHOLDS}$, AHC, 10-means, 20-means) on a similar basis (C++ implementation running on a Pentium4, 3.2GHz, 512Mb). For each database, we have separated the processing times needed to extract/compute the similarity values from the time needed by each algorithm to output the found clusters. The processing times common to all methods are presented in Table 7.9. The results for each method are presented in Table 7.10. This second table takes into account the time needed for sorting the data (AntTree$_{NO-THRESHOLDS}$) and the time needed for computing the initial values of the thresholds like the mean similarity (AntTree$_{STOCH}$).

The best results are obtained by the K-means, but we have shown in the previous section that these methods get the worst results in terms of quality. Compared to AHC, our algorithms obtain significantly better results (up to 1000 times faster for large databases like Waveform). Since the results are comparable in terms of quality (see previous section), this is a very positive advantage for our methods, and it opens the possibility to apply our methods to real world problems where on-line processing

Table 7.5. Results obtained by AntTree$_{NO-THRESHOLDS}$ on numeric and symbolic databases.

Databases	C_r	E_C	C_f	P_R
ART1	4	0,19	8	0,77
ART2	2	0,24	6	0,98
ART3	4	0,17	7	0,88
ART4	2	0,14	4	0,98
ART5	9	0,16	8	0,52
ART6	4	0,07	4	0,93
CERIES	6	0,21	6	0,60
GLASS	7	0,3	9	0,45
IRIS	3	0,12	5	0,94
PIMA	2	0,51	8	0,67
SEGMENT	7	0,17	9	0,56
SOYBEAN	4	0,14	6	0,91
THYROID	3	0,45	9	0,88
VEHICLE	4	0,35	8	0,38
WAVEFORM	3	0,32	7	0,60
WINE	3	0,23	8	0,87
HAYES ROTH	3	0,38	6	0,52
HEART	2	0,42	4	0,72
HOUSE VOTES	2	0,37	4	0,80
LYMPHOGRAPHY	4	0,45	5	0,58
TIC-TAC-TOE	2	0,53	7	0,65
ZOO	7	0,17	5	0,66

is necessary (see section 7.5). More precisely, AntTree$_{NO-THRESHOLDS}$ is the fastest for all databases. The initial sorting of the data (performed in $o(n \ln n)$) has not much influence on the final execution time.

7.4.7 Comparison with biomimetic methods

We compared in Tables 7.11 and 7.12 the results obtained with our methods to those previously obtained by two ants-based clustering methods, namely AntClust [31] AntClass [30]. For these two algorithms, we do not have the purity values, so the comparison is based on the classification error. AntTree and AntClust obtain similar results, and the worst results are obtained by AntClass. A similar comment can be made for the number of classes. The good results obtained by AntClust are explained by the fact that the algorithm ends with a heuristic : clusters with small size are discarded, and the corresponding data reassigned to larger clusters with a nearest neighbor method. The results obtained by AntClust before this heuristic are not as good as the final results. We could have used this heuristic for AntTree also. We mention also the fact that AntClass is a hybrid method (ants with K-means), and can only be applied to numeric data.

Table 7.6. Results obtained by AHC on numeric and symbolic databases.

Databases	C_r	E_C	C_f	P_R	Tps
ART1	4	0,15	5	0,84	1,31
ART2	2	0,14	3	0,98	26,31
ART3	4	0,16	3	0,89	35,48
ART4	2	0,13	3	1,00	0,20
ART5	9	0,08	10	0,78	16,48
ART6	4	0,03	5	1,00	1,31
CERIES	6	0,24	3	0,56	0,50
GLASS	7	0,43	3	0,49	0,25
IRIS	3	0,13	3	0,88	0,11
PIMA	2	0,48	3	0,65	11,03
SEGMENT	7	0,33	3	0,41	355,70
SOYBEAN	4	0,08	6	1,00	0,03
THYROID	3	0,35	5	0,84	0,44
VEHICLE	4	0,39	3	0,35	14,80
WAVEFORM	3	0,32	3	0,58	4163,55
WINE	3	0,20	6	0,84	0,25
HAYES ROTH	3	0,43	4	0,42	0,08
HEART	2	0,42	3	0,76	0,45
HOUSE VOTES	2	0,21	3	0,93	1,56
LYMPHGRAPHY	4	0,43	3	0,69	0,09
TIC-TAC-TOE	2	0,52	4	0,65	17,05
ZOO	7	0,03	5	0,86	0,03

Table 7.7. Results obtained by 10-means on numeric databases.

Databases	C_r	E_C	C_f	P_R	Tps
ART1	4	0,18 [0,01]	8,93 [0,85]	0,84 [0,02]	0,01 [0,01]
ART2	2	0,39 [0,01]	8,87 [0,96]	0,97 [0,01]	0,02 [0,00]
ART3	4	0,32 [0,01]	8,73 [0,93]	0,91 [0,01]	0,02 [0,01]
ART4	2	0,32 [0,03]	6,33 [0,94]	1,00 [0,00]	0,00 [0,01]
ART5	9	0,08 [0,01]	9,00 [0,82]	0,78 [0,05]	0,02 [0,00]
ART6	4	0,11 [0,02]	9,07 [0,93]	1,00 [0,00]	0,01 [0,01]
CERIES	6	0,11 [0,01]	9,40 [0,71]	0,84 [0,03]	0,01 [0,01]
GLASS	7	0,30 [0,01]	9,13 [0,62]	0,57 [0,03]	0,01 [0,01]
IRIS	3	0,18 [0,03]	6,60 [0,88]	0,93 [0,03]	0,00 [0,00]
PIMA	2	0,50 [0,00]	9,93 [0,25]	0,71 [0,01]	0,02 [0,01]
SEGMENT	7	0,11 [0,00]	9,00 [0,73]	0,72 [0,03]	0,15 [0,02]
SOYBEAN	4	0,14[0,01]	9,07 [0,77]	0,98 [0,02]	0,00 [0,00]
THYROID	3	0,42 [0,02]	9,67 [0,47]	0,93 [0,01]	0,00 [0,01]
VEHICLE	4	0,28 [0,00]	9,67 [0,60]	0,50 [0,03]	0,04 [0,01]
WAVEFORM	3	0,29 [0,00]	10,00 [0,00]	0,80 [0,01]	0,41 [0,06]
WINE	3	0,23 [0,02]	9,60 [0,49]	0,91 [0,02]	0,01 [0,01]

Table 7.8. Results obtained by 20-means on numeric databases.

Databases	C_r	E_C	C_f	P_R	Tps
ART1	4	0,21 [0,00]	15,47 [1,31]	0,86 [0,01]	0,01 [0,01]
ART2	2	0,43 [0,00]	15,40 [1,25]	0,98 [0,00]	0,02 [0,01]
ART3	4	0,36 [0,00]	15,87 [1,20]	0,93 [0,01]	0,03 [0,01]
ART4	2	0,39 [0,02]	10,53 [1,41]	1,00 [0,00]	0,01 [0,01]
ART5	9	0,09 [0,00]	16,47 [1,45]	0,74 [0,01]	0,03 [0,01]
ART6	4	0,16 [0,01]	13,40 [1,02]	1,00 [0,00]	0,01 [0,01]
CERIES	6	0,14 [0,01]	16,80 [1,80]	0,88 [0,02]	0,01 [0,01]
GLASS	7	0,26 [0,01]	16,20 [0,98]	0,65 [0,05]	0,01 [0,01]
IRIS	3	0,23 [0,02]	12,13 [1,96]	0,96 [0,01]	0,00 [0,01]
PIMA	2	0,52 [0,00]	19,27 [0,85]	0,74 [0,01]	0,03 [0,01]
SEGMENT	7	0,11 [0,00]	16,13 [1,45]	0,79 [0,033]	0,18 [0,02]
SOYBEAN	4	0,18 [0,01]	13,73 [1,48]	0,99 [0,02]	0,00 [0,00]
THYROID	3	0,47 [0,01]	17,53 [1,26]	0,94 [0,01]	0,00 [0,00]
VEHICLE	4	0,25 [0,00]	17,73 [1,18]	0,56 [0,02]	0,05 [0,01]
WAVEFORM	3	0,31 [0,00]	19,67 [0,47]	0,81 [0,01]	0,71 [0,16]
WINE	3	0,27 [0,01]	16,87 [1,36]	0,93 [0,02]	0,01 [0,01]

7.4.8 Comparative study on textual databases

As far as portal site construction is concerned, we must apply our algorithms to textual databases. We have selected 6 databases where the number of texts varies from 259 to 4504 (see Table 7.13). The CE.R.I.E.S. database contains typical texts about healthy human skin. The AntSearch database deals with scientific documents extracted from the web using a search engine (73 texts on scheduling problems, 84 on pattern recognition, 81 on networks and Tcp-Ip, 94 on 3D and VRML courses). The WebAce1 and WebAce2 databases have been built by the authors of the WebAce project [32] and contain web pages from Yahoo! [33]. Finally, the WebKb database [34] contains 8282 documents that represent pages of people working in several american universities. These pages are classified into 7 classes ("student", "staff", "project", "faculty", "department", "course", "others"). For our tests, we have ignored the last category because its pages did not seem to be really similar, and this finally amounts to 4504 documents.

We now need to compute a similarity measure between the texts. For this purpose, we have represented the documents with the standard vector model [35]. We have use "Zipf's law" [36] to remove words which occurrence in a document collection is too high or too low. We have used the tf*idf method described in equation 7.6 to give a weight to each word that appears in the texts. Then, we compute the cosine measure [37] between two documents in order to obtain a similarity measure that belongs to [0, 1].

$$tf * idf = log\left(f(t_i, d) + 1\right) \cdot log(1/n) \qquad (7.6)$$

Table 7.9. Processing times (in seconds) needed to compute the similarity values for each database (and which are common to all tested algorithms).

Databases	T_{Exe}	N
ART1	0,03	400
ART2	0,16	1000
ART3	0,20	1100
ART4	0,02	200
ART5	0,13	900
ART6	0,05	400
CERIES	0,05	259
GLASS	0,02	214
IRIS	0,03	150
PIMA	0,16	768
SEGMENT	2,72	2310
SOYBEAN	0,03	47
THYROID	0,03	215
VEHICLE	0,38	846
WAVEFORM	13,84	5000
WINE	0,03	178
HAYES ROTH	0,02	132
HEART	0,05	270
HOUSE VOTES	0,44	435
LYMPHOGRAPHY	0,06	148
TIC-TAC-TOE	1,28	958
ZOO	0,05	101

where $f(t_i, d)$ denotes the word count of the term t_i in document d and n is the number of all documents.

We have used the same methodology as in the previous databases. First, we have scanned the parameters space. The best results of AntTree$_{STOCH}$ are obtained with the parameters mentioned in Table 7.14. The performances obtained by AntTree$_{NO-THRESHOLDS}$ are presented in Table 7.15.

We have compared our algorithms to AHC which results are presented in Table 7.16. AntTree$_{STOCH}$ obtains the best average classification error, and AntTree$_{NO-THRESHOLDS}$ and AHC seem to be equivalent: AHC obtains a better purity but AntTree$_{NO-THRESHOLDS}$ obtains a better number of found classes. When the similarity measure is not as efficient as in the numeric case for instance, then AHC does not perform so well. The advantage of AntTree$_{STOCH}$ in this context is the fact that it locally adapts the thresholds for each data.

Table 7.10. Processing times for each algorithm.

Databases	STOCH T_{Exe}	NO-TH T_{Exe}	AHC T_{Exe}	10-means T_{Exe}	20-means T_{Exe}
ART1	0,02 [0,01]	0,02	0,92	0,00 [0,00]	0,00 [0,01]
ART2	0,16 [0,01]	0,08	23,73	0,01 [0,01]	0,02 [0,01]
ART3	0,21 [0,01]	0,09	32,34	0,01 [0,01]	0,02 [0,01]
ART4	0,01 [0,01]	0,00	0,11	0,00 [0,01]	0,00 [0,01]
ART5	0,17 [0,01]	0,06	14,25	0,01 [0,01]	0,02 [0,01]
ART6	0,04 [0,01]	0,02	0,91	0,00 [0,01]	0,00 [0,01]
CERIES	0,01 [0,00]	0,00	0,25	0,00 [0,00]	0,00 [0,01]
GLASS	0,01 [0,01]	0,00	0,14	0,00 [0,01]	0,00 [0,01]
IRIS	0,01 [0,01]	0,00	0,05	0,00 [0,00]	0,00 [0,00]
PIMA	0,13 [0,01]	0,05	9,52	0,01 [0,01]	0,01 [0,01]
SEGMENT	1,16 [0,06]	0,55	343,76	0,05 [0,02]	0,08 [0,03]
SOYBEAN	0,00 [0,00]	0,00	0,00	0,00 [0,00]	0,00 [0,00]
THYROID	0,01 [0,01]	0,00	0,14	0,00 [0,01]	0,00 [0,01]
VEHICLE	0,17 [0,01]	0,08	13,09	0,02 [0,00]	0,04 [0,01]
WAVEFORM	8,39 [1,11]	3,77	4061,36	0,25 [0,11]	0,56 [0,17]
WINE	0,00 [0,00]	0,02	0,06	0,00 [0,00]	0,00 [0,01]
HAYES	0,00 [0,01]	0,00	0,03	-	-
HEART	0,01 [0,01]	0,00	0,30	-	-
HOUSE	0,04 [0,01]	0,00	1,16	-	-
LYMPH	0,00 [0,01]	0,00	0,05	-	-
TIC	0,16 [0,01]	0,06	14,98	-	-
ZOO	0,00 [0,01]	0,00	0,02	-	-

Table 7.11. Summary of the results obtained by AntTree$_{NO-THRESHOLDS}$ and AntTree$_{STOCH}$ on numeric databases.

Database	AntTree$_{NO-THRESHOLDS}$			AntTree$_{STOCH}$	
	C_R	E_C	C_f	$E_C[\sigma_{E_C}]$	$C_f[\sigma_{C_f}]$
ART1	4	0,19	8	0,19 [0,03]	5,20 [0,75]
ART2	2	0,24	6	0,24 [0,04]	4,93 [0,77]
ART3	4	0,17	7	0,25 [0,06]	5,53 [0,50]
ART4	2	0,14	4	0,28 [0,03]	5,53 [0,96]
ART5	9	0,16	8	0,17 [0,01]	6,00 [0,82]
ART6	4	0,07	4	0,07 [0,03]	6,40 [1,02]
IRIS	3	0,12	5	0,18 [0,04]	3,93 [0,77]
GLASS	7	0,30	9	0,34 [0,03]	8,07 [1,12]
PIMA	2	0,51	8	0,48 [0,01]	13,47 [1,26]
SOYBEAN	4	0,14	6	0,10 [0,02]	6,67 [0,87]
THYROID	3	0,45	9	0,27 [0,07]	5,93 [1,00]

Table 7.12. Comparative results obtained with AntClust and AntClass on numeric databases.

Base de données	C_R	AntClust $E_C[\sigma_{E_c}]$	$C_f[\sigma_{C_f}]$	AntClass $E_C[\sigma_{E_c}]$	$C_f[\sigma_{C_f}]$
ART1	4	0,22 [0,04]	4,72 [0,83]	0,15 [0,05]	4,22 [1,15]
ART2	2	0,07[0,02]	2,16 [0,37]	0,41 [0,01]	12,32 [2,01]
ART3	4	0,14[0,01]	2,82 [0,96]	0,35 [0,0]	14,66 [2,68]
ART4	2	0,22[0,05]	3,88 [0,82]	0,29 [0,23]	1,68 [0,84]
ART5	9	0,27 [0,02]	6,42 [1,86]	0,08 [0,01]	11,36 [1,94]
ART6	4	0,05 [0,01]	4,14 [0,40]	0,11 [0,13]	3,74 [1,38]
IRIS	3	0,22 [0,01]	2,92 [0,85]	0,19 [0,08]	3,52 [1,39]
GLASS	7	0,36 [0,02]	5,58 [1,36]	0,40 [0,06]	5,60 [2,01]
PIMA	2	0,46 [0,01]	11,08 [2,03]	0,47 [0,02]	6,10 [1,84]
SOYBEAN	4	0,08 [0,04]	4,00 [0,40]	0,54 [0,17]	1,60 [0,49]
THYROID	3	0,16 [0,04]	4,74 [0,92]	0,22 [0,09]	5,84 [1,33]

Table 7.13. The textual databases used in our tests.

Bases	Size (# of documents)	Size(Mb)	C_R
CE.R.I.E.S.	259	3.65	17
AntSearch	332	13.2	4
WebAce1	185	3.89	10
WebAce2	2340	19	6
WebKb	4504	28.9	6

Table 7.14. Results obtained by AntTree$_{STOCH}$ on textual databases with the following parameters: increasing order, $(Sim_{mean-Max}(a_i)$; $Sim_{mean-Min}(a_i))$, $\alpha_1 = 0,70$ and $\alpha_2 = 0,01$.

Base	$E_C[\sigma_{E_c}]$	C_R	$C_f[\sigma_{C_f}]$	$P_R[\sigma_{P_R}]$
CE.R.I.E.S.	0,24 [0,00]	17,00	7,00 [0,00]	0,28 [0,00]
Antsearch	0,22 [0,00]	4,00	20,00 [0,00]	0,75[0,00]
WebAce1	0,16 [0,00]	10,00	12,00 [0,00]	0,44 [0,00]
WebAce2	0,40 [0,00]	6,00	4,00 [0,00]	0,66 [0,00]
WebKb	0,36 [0,00]	6,00	8,00 [0,00]	0,38 [0,00]
Mean	0,28	-	-	0,50

7.5 Real world applications

7.5.1 Human skin analysis

AntTree has been applied to a real world database in collaboration with the biometry unit of the CE.R.I.E.S., a research center on human skin funded by CHANEL. Several databases have been tested. As an example, one of them consists in 259 examples where each example represents the description of the visual and tactile

Table 7.15. Results obtained by AntTree$_{NO-THRESHOLDS}$ on textual databases with decreasing order.

Base	E_C	C_R	C_f	P_R
CE.R.I.E.S.	0,31	17,00	6,00	0,23
AntSearch	0,24	4,00	6,00	0,72
WebAce1	0,29	10,00	6,00	0,32
WebAce2	0,32	6,00	7,00	0,77
WebKb	0,32	6,00	9,00	0,40
Mean	0,30	-	-	0,49

Table 7.16. Results obtained by AHC on textual databases.

Base	E_C	C_R	C_f	P_R
CE.R.I.E.S.	0,36	17,00	3,00	0,29
Antsearch	0,17	4,00	6,00	0,79
WebAce1	0,28	10,00	4,00	0,27
WebAce2	0,29	6,00	3,00	0,79
WebKb	0,42	6,00	3,00	0,39
Moy	0,30	-	-	0,51

facial skin characteristics of a Caucasian woman between 20 and 50 years old. Each example is described with thirty skin characteristics, i.e. attributes. These characteristics were evaluated on the right cheek or on the forehead. The initial aim of the study was to establish a classification of healthy facial skin, based on defined skin characteristics, using multivariate data analysis approach with SAS software release 6.12. : ascending hierarchical clustering method applied on principal components resulting of multiple correspondence analysis. Finally, a strengthened proposal of a six clusters-classification was obtained, which was mainly characterized by skin color, oiliness, roughness and features associated with the vascular system of the skin. The interested reader may refer for instance to [27] for more details.

As shown in Figure 7.9, we have used AntTree to confirm the relevance of the found clusters. The expert runs AntTree and may then explore and visually analyze the resulting clusters (see section 7.5.3). Our tool has been useful to confirm the clustering found by other traditional methods. In the example shown in Figure 7.9, there are 8 sub-trees on the root (the support), which corresponds to 8 classes. The hierarchical organization of the data is quite interesting and useful : for instance, in cluster 6 (6th sub-tree) which is compound of individuals from real classes 2 and 3, the distribution of individuals is further separated into two sub-trees, one that corresponds to class 2 and the other one to class 3. This confirms the ability of AntTree to organize a data set in a hierarchical way.

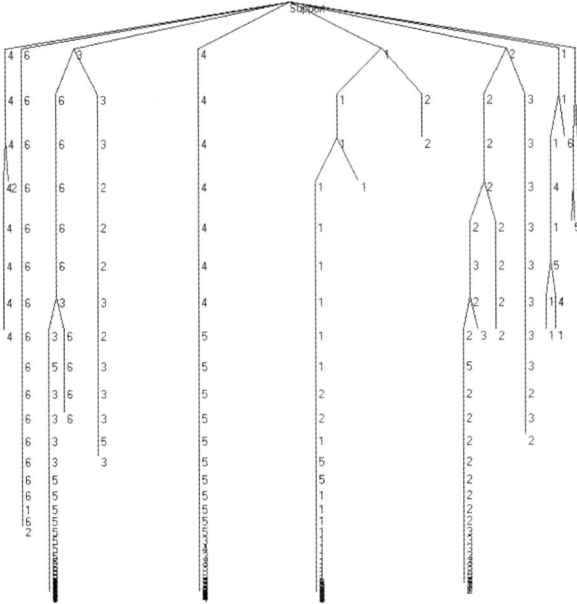

Fig. 7.9. Example of a learned tree for the CE.R.I.E.S. databases, with a hyperbolic display. Numbers represent the classes found by a data analysis technique (see text for more explanation).

7.5.2 Web usage mining

Web usage mining consists in analyzing and understanding how users behave when visiting a web site [38]. Basic statistics can be easily derived like for instance the most visited pages, the origin of users, the referent (i.e. the page from which they have accessed the site etc). If one wants to analyze more deeply the user behavior, or when too many data have been generated by numerous site visitors, then clustering algorithms are necessary to sum up the information. For instance, one may want to know what typical behaviors can be observed in terms of visited pages. The clustering algorithm should thus create clusters of user sessions (a session is the list of pages visited by a given user). This problem is difficult because thousands of sessions may have to be clustered, and also because each session is described for instance as a vector (each dimension is the time spend on each page) with high dimensionality (number of pages in the site). To these difficulties we add the following one: data must be analyzed in real time (on-line) rather than off line as in many web usage mining software. The reason is the following one: AntTree has been integrated in a web site generator called Webxygen (please consult www.webxygen.org for more details), a system which is intended to host many sites and to concentrate in a single software many advanced features for these sites. So site administrators may ask Webxygen at any time for an analysis of the users behavior

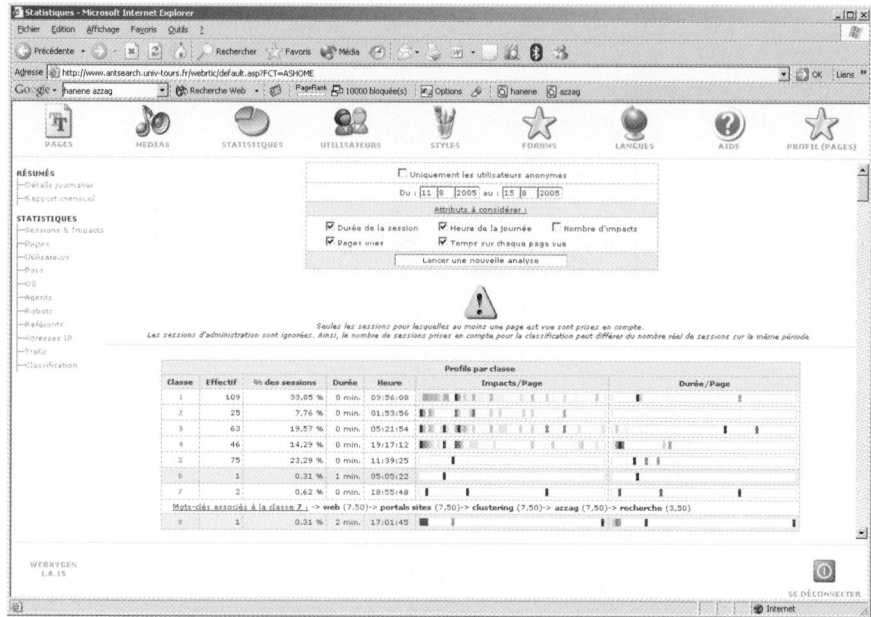

Fig. 7.10. On-line analysis of user behavior when visiting a web site: AntTree$_{NO-THRESHOLDS}$ is used to cluster together hundreds of user sessions, in real time, in order to quickly provide typical and representative user behaviors to the site administrator. Each line corresponds to a cluster (group of sessions). The bars on the right indicates the time spent on each page (intensity of blue color). When the cursor is placed on a bar, the interface displays the name of the corresponding page and the time spent.

in terms of visited pages. This requires a fast clustering algorithm such as AntTree (which as mentioned in section 7.4, is as efficient as AHC and much faster).

We have selected AntTree$_{NO-THRESHOLDS}$ for its speed, its determinism (probabilistic algorithms may not provided exactly the same result for the same data for different runs), and also because it creates slightly more clusters than necessary (and thus provides a finer analysis of those clusters). Figure 7.10 shows a typical result obtained with hundreds of sessions (and less than 1 minute of computation with VB6 implementation). Flat clusters are created using the method depicted in Figure 7.9. We have checked experimentally that the found clusters are relevant. One may also check this property by looking at the visualization of the clusters centers: each bar represents a given page and the color intensity of the bar is proportional to the time spent on this given page (averaged over all sessions in this cluster). Clusters are well separated and the time needed to compute these clusters is well adapted to the on-line constraint.

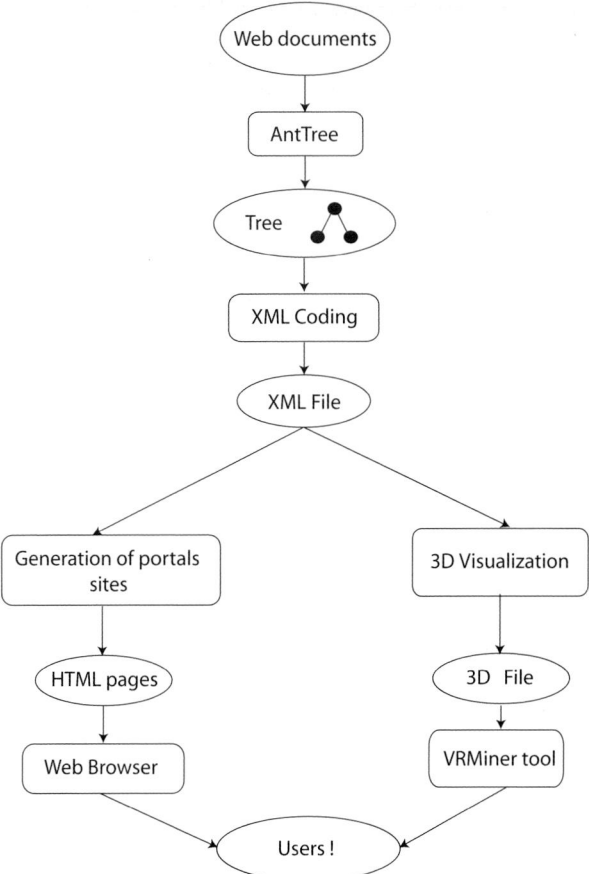

Fig. 7.11. The generation of a portal site from the tree of documents learned by AntTree.

7.5.3 Generation and interactive exploration of a portal site

Once the texts have been clustered in a tree, then it is straightforward to generate the corresponding portal site. As shown in Figure 7.11, we have developed two possibilities to explore the portal site. In both cases, we translate our tree of documents in an XML format. The first possibility for exploring this site (on the left of Figure 7.11) consists in encoding the tree in a database (using PHP language) in a few seconds and to dynamically generate the HTML pages in order to explore the portal site with a conventional browser. Figure 7.12 gives a typical example of the portal home page obtained for AntSearch (332 documents). In addition, we have integrated a search engine based on a word index. This index is automatically generated in the database and allows the user to select specific documents (see Figure 7.13).

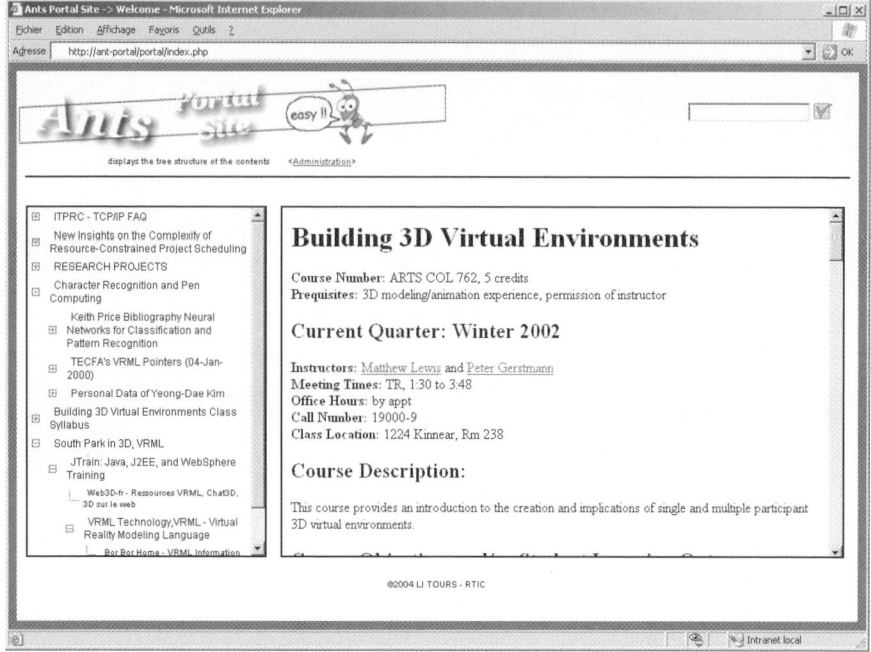

Fig. 7.12. Exploration of a portal with a conventional browser (here with the AntSearch database).

The second possibility we have studied is to integrate the results of AntTree in our virtual reality data mining tool VRMiner [39]. This tool allows a domain expert to explore multimedia databases (numeric, symbolic, sounds, images, videos and web sites) using a 3D stereoscopic display, a 3D sensor with six degrees of freedom for navigating through the data and a data glove with gesture recognition for requesting information from the system. VRMiner can use two displays: the first display is devoted to the 3D representation of the data, and the second display is used to visualize larger media associated to the data, like web sites for instance (see 7.14 and Figure 7.15). Our idea is 1) to visualize the portal site on the main computer, 2) to let the user navigate trough the data and possibly click on nodes (web pages), 3) to dynamically display the selected page in the second computer (see Figure 7.16). In this way, we hope to be able to browse large databases, and to let the user easily interact with the data and perceive both the details and the global context or shape of the tree.

One may find in the literature several algorithms for visualizing hierarchical structures which are mostly 2D. One may cite for instance the Tree maps method which recursively maps the tree to embedded rectangles ([40, 41]). Hyperbolic displays have also been studied in 2D and 3D (see for instance [42]). An other important example is the cone tree [43]: the root of the tree is the top of a cone. The subtrees of this root are all included in this cone. The size of a cone may depend

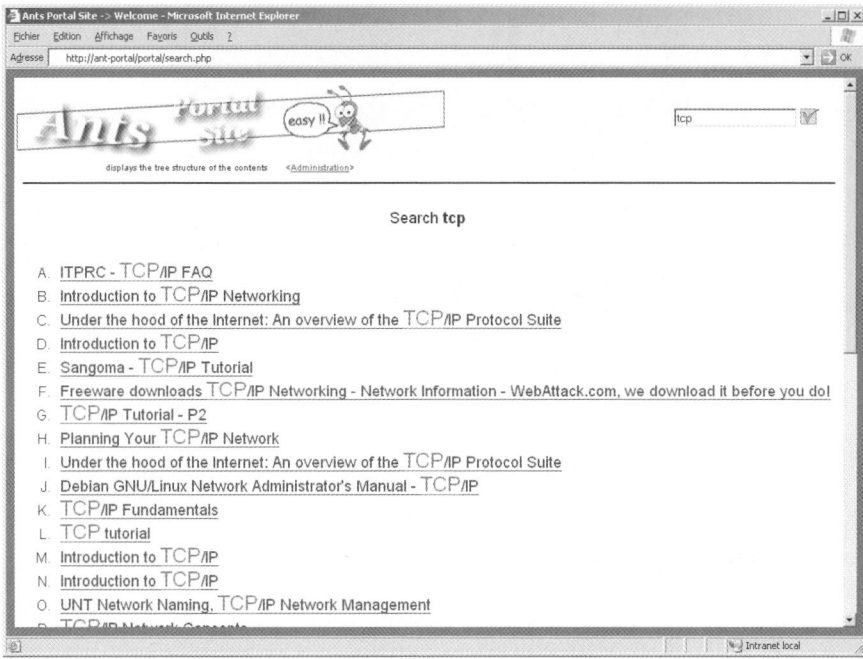

Fig. 7.13. The search engine which is automatically generated with the portal site.

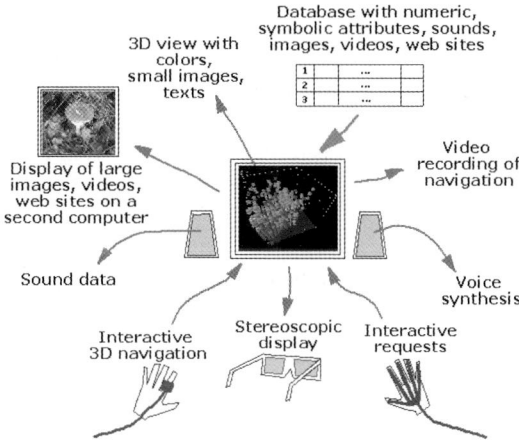

Fig. 7.14. Schematic view of VRMiner, an interactive tool for multimedia data exploration.

Fig. 7.15. Real view of VRMiner.

on the number of nodes which are present in each subtrees. We have used in this work a similar representation.

The problem to be solved now is the following one: we must define the position of each nodes in 3D. We have decided to let a constant distance from one level of the tree to the next level. For a given node a_i, the position of the subtrees ("subcones") is such that: 1) all daughters of a_i are located on a circle which center is vertically aligned with a_i location, 2) the radius of each "subcone" is proportional to the number of nodes in the corresponding subtree, 3) subtrees do not overlap. We have implemented this visualization and we show typical results in Figures 7.17 and 7.18.

In the future, and since VRMiner can easily handle textures and texts, we will map a thumbnail view of the pages on the cubes and display above each page a short text (which can be either keywords representing the category, or the title of the page, etc).

7.6 Incremental clustering of a large data set

7.6.1 Principles of AntTree$_{INC}$

Dealing with very large databases in data mining has become necessary due to the increasing volume of available information. This has lead recently to the study of specific clustering algorithms which do not need all the data at once. Rather, they may handle the data as a flow of information, where only a subpart of the whole data set is avaible at a given time. Such incremental clustering algorithms exist since 20 years now [44], and recent advances in this area are for instance Birch [45] or Cure [46]. We present in this section a version of AntTree called AntTree$_{INC}$ which is under study and which aims at solving such problems. It checks many of the critera defined in [47] for handling large databases. Another motivation for this part of our

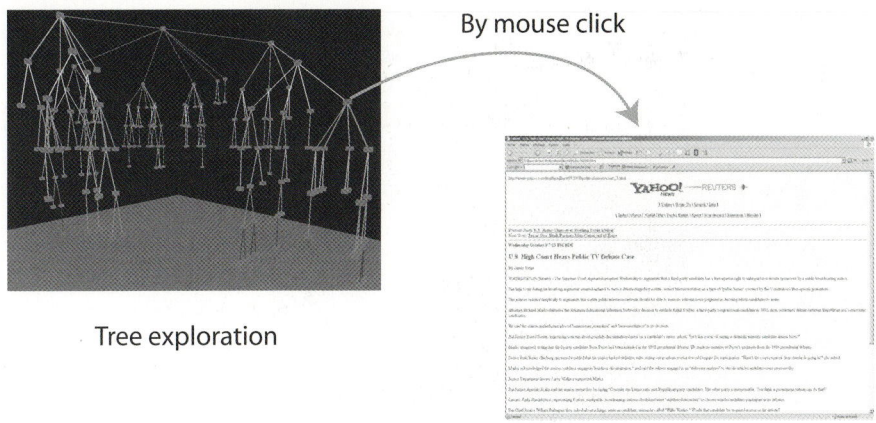

Fig. 7.16. Dynamic display of a web page during 3D navigation.

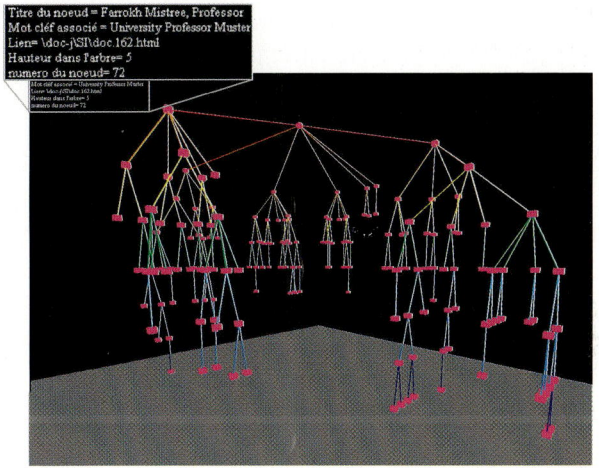

Fig. 7.17. 3D view of a tree generated from the WebAce1 database (185 nodes).

work is to show that biomimetic algorithms can handle large data sets and that they can be incremental.

AntTree$_{INC}$ is based on AntTree$_{NO-THRESHOLDS}$ because this algorithm is fast, it does not need any parameters (see previous sections) and it provides good results. The aim of AntTree$_{INC}$ is to cluster a large database (100 000 examples at this time) where these data are presented one by one, and where the system can store a maximum of 10 000 data, which is the actual limitation of AntTree$_{NO-THRESHOLDS}$ implementation. As in Birch or Cure, AntTree$_{INC}$ proceeds in two steps:

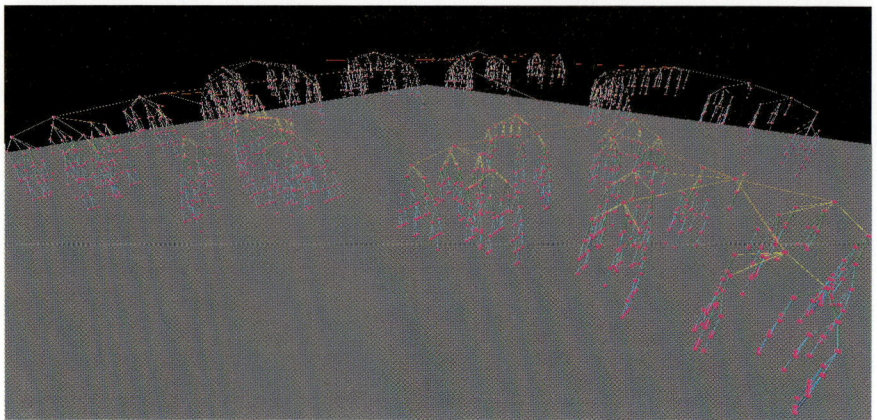

Fig. 7.18. 3D view of a tree generated from the WebAce2 database (2340 nodes).

- Step 1 (Pre-clustering phase): this consists in collecting the first data, to compute the similarity matrix, to sort these data in decreasing order and to run $AntTree_{NO-THRESHOLDS}$ in order to obtain a first tree. So these phase is not different from the descriptions presented in the previous sections,
- Step 2 (Incremental clustering phase): the new data arrive one by one. For a given new ant a_i, $AntTree_{INC}$ moves this ant in the tree until it becomes connected. Then a new data may be processed. The algorithm used for connecting the ant is the following one:

R_1 ($a_{pos} = a_0$ ($AntTree_{Inc}$ second phase: the support))
  1. If $Sim(a_i, a^+) < T_{Dissim}(a_0)$ Then connect a_i to a_0
  2. Else move a_i toward a^+

R_2 ($a_{pos} =$ ($AntTree_{Inc}$ second phase: an other ant))
  1. If no ant connected to a_{pos} Then connect a_i to a_{pos}
  2. Else
     a) If $a_{pos} = a^+$ Then connect a_i to a^+
     b) Else move a_i toward a^+

This algorithm is a simplified and faster version of $AntTree_{NO-THRESHOLDS}$. Intuitively, one must notice first that a_i can create a new cluster directly below the support or deeper in the tree. This allows the appearing of new clusters (see the experimental results). Second, a new ant follows the path of maximum similarity in the tree, and becomes connected to the closest ant it may find. This procedure is very fast due to the few tests which are necessary and the fact that the tree structure avoids to scan all the data (but simply a path in the tree, an important property which is also used in Birch).

7.6.2 Results with incremental and large data sets

We have generated artificial databases for evaluating $AntTree_{INC}$. We have used simple uniform laws to generate large databases with specific characteristics (see

Table 7.17): from 10000 to 100000 data, 5 real classes (with equal number of data), 10 numeric attributes. Test1 and Test2 differs from the order of the data: in Test1, the data are ordered according to their real classes, and the first 6000 data are used in the pre-clustering phase but not the 4000 remaining data (which belong to new clusters). In this way, we want to evaluate the ability of our incremental method to create clusters which were not present in the initial sample used for the pre-clustering phase. In Test2, all data are randomly ordered, and we use 6000 data in the pre-clustering phase. In Test3, we use 10000 data in the pre-clustering phase, and the remaining 90000 data in the second incremental phase. In Test2 and Test3 all five clusters are present in the initial sample.

Table 7.17. Databases used for the evaluation of $AntTree_{INC}$. Test1 and Test2 differ from the order of the data (see text for more explanation).

DataBases	C_R	M	N
Test1	5	10	10 000
Test2	5	10	10 000
Test3	5	10	100 000

Table 7.18. Comparative results between $AntTree_{INC}$ (incremental) and $AntTree_{NO-THRESHOLDS}$ (non incremental). "Pre." denotes the pre-clustering phase and "Cla." the second incremental phase.

Databases	$AntTree_{INC}$						$AntTree_{NO-TH}$		
	E_C		C_f		P_R		E_C	C_f	P_R
	Pre.	Cla.	Pre.	Cla.	Pre.	Cla.			
Test1	0,02	0,1	3	4	0,98	0,79	0,09	9	0,89
Test2	0,1	0,09	7	7	0,87	0,87	0,09	9	0,89
Test3	0,1	0,1	6	6	0,86	0,86	-	-	-

Table 7.19. Processing times (in seconds) needed by $AntTree_{INC}$.

Databases	Pre.	Cla.
Test1	7,09	0,859
Test2	6,875	3,61
Test3	20,00	185,00

We present the results in Table 7.18. One must notice first that $AntTree_{INC}$ and $AntTree_{NO-THRESHOLDS}$ obtain similar performances, while the first algorithm is incremental and the second one can deal with all the data at once. So this is a very positive point for our incremental version. Furthermore, AntTree$_{INC}$ seems to better approximate the real number of clusters. For Test1, AntTree$_{INC}$ discovers 3 classes in the first phase, but only four in the second phase. This algorithm is able to generate new and unseen clusters, but we are currently studying how to obtain 5 final clusters. We have represented in Table 7.19 the processing times needed by AntTree$_{INC}$ for both phases. It takes only 4 minutes to deal with 100000 data. These results are very encouraging.

7.7 Conclusions

We have presented in this chapter a new model for data clustering which is inspired from the self-assembly behavior observed in real ants. We have developed several algorithms for solving a hierarchical clustering problem, which was not solved yet by ants-based algorithms. We have shown that two algorithms, a stochastic and a deterministic one, can be competitive with a very standard and widely used method, namely the ascending hierarchical clustering. Our algorithms obtain similar performances in terms of quality, but with a much lower complexity and processing times. They do not require specific knowledge about the data, and one method is parameter-free. We have thus been able to apply our methods to three real world applications. We have shown in this context that the learned trees could also be visualized and interactively explored. We have presented an important extension toward the incremental processing of large databases. The first results that we have obtained are encouraging, both in terms of quality and processing times.

There are many perspectives to study after this first series of algorithms adn results. The self-assembly model introduces a new ants-based computational method which is very promising. We are currently studying how to generate graphs, because these structures are important for problem solving in computer science. As far as clustering is concerned, we would like to introduce some heuristics in order to evaluate their influence on the results, like removing clusters with a small number of instances. We also want to prepare a comparison between our incremental version and existing and efficient algorithms such as Birch. We would like to deal with larger databases (1 million data).

References

1. Kumar, R., Raghavan, P., Rajagopalan, S., Tomkins, A. (2001) On semi-automated web taxonomy construction. In: Proceedings of the Fourth International Workshop on the Web and Databases (WebDB), Santa Barbara
2. Sanderson, M., Croft, W.B. (1999) Deriving concept hierarchies from text. In: Research and Development in Information Retrieval 206–213
3. McCallum, A.K., Nigam, K., Rennie, J., Seymore, K. (2000) Automating the construction of internet portals with machine learning. Information Retrieval **3** 127–163

4. Goss, S., Deneubourg, J.L. (1991) Harvesting by a group of robots. In F.Varela, P.Bourgine, eds.: Proceedings of the First European Conference on Artificial Life, Paris, France, Elsevier Publishing 195–204
5. Lumer, E., Faieta, B. (1994) Diversity and adaptation in populations of clustering ants. In Cliff, D., Husbands, P., Meyer, J., W., S., eds.: Proceedings of the Third International Conference on Simulation of Adaptive Behavior, MIT Press, Cambridge, Massachusetts 501–508
6. Deneubourg, J.L., Goss, S., Franks, N., Sendova-Franks, A., Detrain, C., Chretien, L. (1990) The dynamics of collective sorting: robot-like ant and ant-like robots. In: Proceedings of the First International Conference on Simulation of Adaptive Behavior. 356–365
7. Kuntz, P., Layzell, P., Snyers, D. (1997) A colony of ant-like agents for partitioning in vlsi technology. In Husbands, P., Harvey, I., eds.: Proceedings of the Fourth European Conference on Artificial Life. 417–424
8. Kuntz, P., Snyers, D., Layzell, P.J. (1998) A stochastic heuristic for visualising graph clusters in a bi-dimensional space prior to partitioning. J. Heuristics **5** 327–351
9. Abraham, A., Ramos, V. (2003) Web usage mining using artificial ant colony clustering and linear genetic programming. In: The Congress on Evolutionary Computation, Canberra, Australia, IEEE-Press 1384–1391
10. Handl, J., Knowles, J., Dorigo, M. (2003) On the performance of ant-based clustering. 204–213
11. N. Labroche, C. Guinot, G.V. (2004) Fast unsupervised clustering with artificial ants. In: Proceedings of the Parallel Problem Solving from Nature (PPSN VIII), Birmingham, England 1143–1152
12. Anderson, C., Theraulaz, G., Deneubourg, J. (2002) Self-assemblages in insect societies. Insectes Sociaux **49** 99–110
13. Lioni, A., Sauwens, C., Theraulaz, G., Deneubourg, J.L. (2001) The dynamics of chain formation in oecophylla longinoda. Journal of Insect Behavior **14** 679–696
14. Theraulaz, G., Bonabeau, E., Sauwens, C., Deneubourg, J.L., Lioni, A., Libert, F., Passera, L., Solé, R.V. (2001) Model of droplet formation and dynamics in the argentine ant (linepithema humile mayr). Bulletin of Mathematical Biology
15. Colorni, A., Dorigo, M., Maniezzo, V. (1991) Distributed optimization by ant colonies. In F.Varela, P.Bourgine, eds.: Proceedings of the First European Conference on Artificial Life, Paris, France, Elsevier Publishing (1991) 134–142
16. Roux, O., Fonlupt, C. (2000) Ant programming: Or how to use ants for automatic programming. From Ant Colonies to Artificial Ants: 2nd International Workshop on Ant Colony Optimization
17. Bianchi, L., Gambardella, L.M., Dorigo, M. (2002) An ant colony optimization approach to the probabilistic traveling salesman problem. In: Proceedings of PPSN-VII, Seventh International Conference on Parallel Problem Solving from Nature. Lecture Notes in Computer Science, Springer Verlag, Berlin, Germany
18. Ando, S., Iba, H. (2002) Ant algorithm for construction of evolutionary tree. In Langdon, W.B., ed.: GECCO 2002: Proceedings of the Genetic and Evolutionary Computation Conference, New York, Morgan Kaufmann Publishers 131
19. Murata, S., Kurokawa, H., Kokaji, S. (1994). In: IEEE International Conference on Robotics and Automation. 441–448
20. Pamecha, A., Ebert-Uphoff, I., Chirikjian, G. (1997) Useful metrics for modular robot motion planning

21. Murata, S., Yoshida, E., Kamimura, A., Kurokawa, H., Tomita, K., Kokaji, S. (2002) M-tran: Self-reconfigurable modular robotic system. IEEE/ASME Transactions on Mechatronics 431–441
22. Jorgensen, M.W., Ostergaard, E.H., Lund, H.H. (2004) Modular atron: Modules for a self-reconfigurable robot. In: IEEE/RSJ InternationalConference on Intelligent Robots and Systems (IROS). 2068–2073
23. Mondada, F., Pettinaro, G.C., Guignard, A., Kwee, I.W., Floreano, D., Deneubourg, J.L., Nolfi, S., Gambardella, L.M., Dorigo, M. (2004) Swarm-bot: A new distributed robotic concept. Auton. Robots **17** 193–221
24. Azzag, H., Guinot, C., Oliver, A., Venturini, G. (2005) A hierarchical ant based clustering algorithm and its use in three real-world applications. In Wout Dullaert, Marc Sevaux, K.S., Springael, J., eds.: European Journal of Operational Research (EJOR). Special Issue on Applications of Metaheuristics.
25. Eiben, A.E., Hinterding, R., Michalewicz, Z. (1999) Parameter control in evolutionary algorithms. IEEE Trans. on Evolutionary Computation **3** 124–141
26. Blake, C., Merz, C. (1998) UCI repository of machine learning databases
27. Guinot, C., Malvy, D.J.M., Morizot, F., Tenenhaus, M., Latreille, J., Lopez, S., Tschachler, E., Dubertret, L. (2003) Classification of healthy human facial skin. Textbook of Cosmetic Dermatology Third edition
28. Fowlkes, E.B., Mallows, C.L. (1983) A method for comparing two hierarchical clusterings. J. American Statistical Associationn **78** 553–569
29. Jain, A., Dubes, R. (1988) Algorithms for Clustering Data. Prentice Hall Advanced Reference Series
30. Monmarché, N. (2000) Algorithme de fourmis artificielles : applications à la classification et à l'optimisation. Thèse de doctorat, Université de Tours
31. Labroche, N. (2003) Modélisation du système de reconnaissance chimique des fourmis pour le problème de la classification non-supervisée : application à la mesure d'audience sur Internet. Thèse de doctorat, Laboratoire d'Informatique, Université de Tours
32. Han, E.H., Boley, D., Gini, M., Gross, R., Hastings, K., Karypis, G., Kumar, V., Mobasher, B., Moore, J. (1998) Webace: a web agent for document categorization and exploration. In: AGENTS '98: Proceedings of the second international conference on Autonomous agents, New York, NY, USA, ACM Press (1998) 408–415
33. Filo, D., Yang, J. (1997) Yahoo!
34. Base de WebKb http://www-2.cs.cmu.edu/ webkb/.
35. Salton, G., Buckley, C. (1988) Term-weighting approaches in automatic text retrieval. Inf. Process. Manage. **24** 513–523
36. Zipf, G.K. (1949) Human behaviour and the principle of least effort. Addison-Wesley, Cambridge, Massachusetts
37. Salton, G., McGill, M.J. (1983) Introduction to Modern Information Retrieval. McGraw-Hill, Inc., New York, NY
38. Cooley, R. (2000) Web Usage Mining: Discovery and Application of Interesting Patterns from Web Data. Ph.d. thesis, University of Minnesota
39. Azzag, H., Picarougne, F., Guinot, C., Venturini, G. (2005) Vrminer: a tool for multimedia databases mining with virtual reality. In Darmont, J., Boussaid, O., eds.: Processing and Managing Complex Data for Decision Support. to appear.
40. Johnson, B., Shneiderman, B. (1991) Tree-maps: A space-filling approach to the visualization of hierarchical information structures. In: Proc. of Visualization'91, San Diego, CA 284–291

41. Shneiderman, B. (1992) Tree visualization with tree-maps: A 2-D space-filling approach. ACM Transactions on Graphics **11** 92–99
42. Carey, M., Heesch, D.and Rüger, S. (2003) Info navigator: A visualization tool for document searching and browsing. In: Proceedings of the 9th International Conference on Distributed Multimedia Systems (DMS'2003)
43. Robertson, G.G., Mackinlay, J.D., Card, S.K. (1991) Cone trees: animated 3d visualizations of hierarchical information. In: CHI '91: Proceedings of the SIGCHI conference on Human factors in computing systems, New York, NY, USA, ACM Press 189–194
44. Fisher, D.H. (1991) Knowledge Acquisition via Incremental Conceptual Clustering. Machine Learning **2** 139–172
45. Tian, Z., Raghu, R., Miron, L. (1996) Birch: An efficient data clustering method for very large databases. In Jagadish, H.V., Mumick, I.S., eds.: Proceedings of the 1996 ACM SIGMOD International Conference on Management of Data, Montreal, Quebec, Canada, June 4-6, 1996, ACM Press 103–114
46. Sudipto, G., Rajeev, R., Kyuseok, S. (1998) CURE: an efficient clustering algorithm for large databases. In Haas, L.M., Tiwary, A., eds.: Proceedings ACM SIGMOD International Conference on Management of Data, Seattle, Washington, USA, ACM Press 73–84
47. Domingos, P., Hulten, G. (2001) Catching up with the data: Research issues in mining data streams

8

Swarm Clustering Based on Flowers Pollination by Artificial Bees

Majid Kazemian, Yoosef Ramezani, Caro Lucas and Behzad Moshiri

Control and Intelligent Processing Center of Excellence, School of Electrical and Computer Engineering, University of Tehran, Tehran, Iran am.kazemian@ece.ut.ac.ir

Summary. This chapter presents a new swarm data clustering method based on flowers pollination by artificial bees which is named as FPAB[1]. FPAB does not require any parameter settings and any initial information such as the number of classes and the number of partitions on input data. Initially, in FPAB, bees move the pollens and pollinate them. Each pollen will grow in proportion to its garden flowers. Better growing will occur in better conditions. After some iteration, natural selection reduces the pollens and flowers and the gardens of the same type of flowers will be formed. The prototypes of each gardens are taken as the initial cluster centers for Fuzzy C Means algorithm which is used to reduce obvious misclassification errors. In the next stage, the prototypes of gardens are assumed as a single flower and FPAB is applied to them again. Results from three small data sets show that the partitions produced by FPAB are competitive with those obtained from FCM[2] or AntClass.

8.1 Introduction

The aim of clustering is to separate a set of data points into self-similar groups. Many standard clustering approaches like K-Means or ISODATA are limited because they require a priori knowledge of a number of classes or an initial partition of input data. In addition, the draw back of some methods based on the hill climbing search such as iterative version of Fuzzy C-Means is sensitivity to initialization of cluster centers. Many approaches have been proposed to improve these standard methods such as genetic-based approaches [4, 6, 13], hybridization of GA and K-Means [2]. Some of the recent approaches are based on swarm concept which can cover some of these deficiencies. Ant-based approaches [3, 8, 9], AntClass [12, 13], Fuzzy Ants [7] and clustering based on the chemical recognition system of ants [10] are the instances of swarm clustering approaches. In ant based approaches each ant is an agent with simple probabilistic behavior of picking up and dropping

[1] Flower Pollination by Artificial Bees
[2] Fuzzy C-Means

down objects. This behavior depends on similarity between the object and its belonging region and similarity measure threshold that effects on the number of clusters and misclassification errors, use of heterogeneous ants will remove complex parameter settings [10], but any parameter settings effects on number of clusters and misclassification errors. FPAB is an extension of AntClass cite12, which does not need any threshold because it introduces a growing factor and also uses natural selection to remove inappropriate object in clusters. In addition, FPAB move the instances of objects rather than the objects themselves. The remaining of this article is organized as follows: Next section describes the data clustering and its classification from the data partitioning aspect. Moreover, it illustrates the swarm clustering concept. Section 8.3 explains the FPAB algorithm in details; Section 8.4 explains the data sets used and the corresponding experimental results. Conclusions and future works are brought in Section 8.5.

8.2 Clustering

8.2.1 What is clustering?

Clustering is a division of data into groups of similar objects. Each group, called cluster, consists of objects that are similar among themselves and dissimilar to objects of other groups. The clustering problem has been addressed in many contexts and by researchers in many disciplines ranging from sociology and psychology, to commerce, biology, and computer science. This reflects its wide appeal and usefulness as one of the steps in exploratory data analysis, grouping, decision making, data mining, document retrieval, image segmentation, and pattern classification.

Cluster analysis is an unsupervised learning method that constitutes a basis of an intelligent data analysis process. It is used for the exploration of inter-relationships among a collection of patterns, by organizing them into homogeneous clusters. It is called unsupervised learning because unlike classification (known as supervised learning), no priori labeling of some patterns is available to use in categorizing others and inferring the cluster structure of the whole data.

There are several clustering methods, which differ not only in the principles of the algorithm used, but also in many of their most basic properties, such as the data handling, assumptions on the shape of the clusters, the form of the final partitioning and the parameters that have to be provided (e.g., the correct number of clusters). From the aspect of data partitioning, clustering algorithms can be classified as follow:

- Exclusive clustering in which data are grouped in an exclusive way. If certain data belong to a cluster, then they could not be fitted in another cluster. K-mean is a typical exclusive clustering algorithm.
- Overlapping clustering in which data can be grouped in two or more clusters depending on the membership value. Fuzzy C-mean is an exponent of overlapping cluster algorithms.

- Hierarchical clustering in which in the beginning of data processing, every data is a cluster and the nearest two clusters will be merged to one. The algorithm stops when all data are in one cluster or a certain threshold is reached.
- Probabilistic clustering, this clustering approach is completely based on probability. Two of these kinds are: Mixture of Gaussians and swarm-based algorithms.

8.2.2 Why swarm intelligence?

Data clustering is the hardest problem in the knowledge discovery systems. It seems that in natural systems, beings (e.g. insects, ants, bees, termites) are well succeeded in clustering of things collectively. This collective behavior caused the emergence of intelligence in the view of whole system. Swarm Intelligence can be applied to a variety of combinatorial optimization problems like Traveling Salesman Problem, Quadratic assignment problem, graph coloring, clustering, and sorting. Swarm Intelligence can be defined more precisely as: Any attempt to design algorithms or distributed problem-solving methods inspired by the collective behavior of the social insect colonies or other animal societies. The main properties of such systems are flexibility, robustness, decentralization, and self-organization. Moreover, the interaction among insects (hereafter called agents) can be direct or indirect and each agent is relatively simple and agile. Bonabeau et al. [1] have described the purpose of swarm based algorithms as follow:

"[. . .] at a time when the world is becoming so complex that no single human being can understand it, when information (and not the lack of it) is threatening our lives, when software systems become so intractable that they can no longer be controlled, swarm intelligence offers an alternative way of designing an intelligent system, in which autonomy, emergence, and distributed functioning replaces control, preprogramming, and centralization [1]."

8.2.3 Swarm clustering

Swarm clustering points out the algorithms inspired by the collective behavior of living things in natural systems used for clustering and sorting. For example, several species of insects use clusters of corpses to form cemeteries for cleaning up nests. Another instance of such natural process is brood sorting: the workers of some breed of ants gather larvae according to their size, that is, smaller larvae tend to be dropped in the center whereas bigger larvae tend to be in the periphery. The behavior is not fully understood, but a simple model, in which agents pick or deposit items depending on the local information of the cluster, can be accordingly established. Items with similar attributes build a homogeneous cluster and give a positive feedback to agents. This feedback attracts the agents to either drop or keep the item to amplify the cluster. Using this simple idea, some artificial models are proposed. In next section, we briefly describe some of these models.

8.2.4 Some artificial models

Data clustering is one of the problems in which real ants can suggest very interesting heuristics for computer scientists. Deneubourg et al. [3] proposed an agent-based model to explain how ants manage to cluster the corpses of their dead nest mates. Artificial ants (or agents) are moving randomly on a square grid of cells on which some items are scattered. Each cell can only contain a single item. Whenever an unloaded ant encounters an item, this item is picked up with a probability, which depends on an estimation of the density of items of the same type in the neighborhood. If this density is high, the probability of picking up the item will be low. When a loaded ant encounters a free cell on the grid, the probability that this item is dropped also depends on an estimation of the local density of items of the same type. However, when this density is high, the probability of dropping the load will be high. Simulations show that all objects of the same type are eventually clustered together.

Lumer and Faieta [9] extended the model of Deneubourg et al., using a dissimilarity-based evaluation of the local density, in order to make it suitable for data clustering. Unfortunately, the resulting number of clusters is often too high and convergence is slow. Therefore, a number of modifications were proposed, by Lumer and Faieta themselves as well as by others [5, 15]. Since two different clusters can be adjacent on the grid, heuristics are necessary to determine which items belong to the same cluster.

Monmarché proposed an algorithm in which several items are allowed to be on the same cell. Each cell with a nonzero number of items corresponds to a cluster. Each ant "a" is endowed with a certain capacity $c(a)$. Instead of carrying one item at a time, an ant "a" can carry a heap of $c(a)$ items. Probabilities for picking up, at most $c(a)$ items from a heap and for dropping the load on a heap are based on characteristics of the heap, such as the average dissimilarity among items of the heap. When an ant decides to pick up items, the $c(a)$ items whose dissimilarity to the centre of the heap under consideration is highest, are chosen. Two particularly interesting values for the capacity of an ant "a" are $c(a) = 1$ and $c(a) = \infty$.

Monmarché [14] proposed to apply this algorithm twice. The first time, the capacity of all ants is one, which results in a high number of tight clusters. Subsequently the algorithm is repeated with the clusters of the first pass as atomic objects and ants with infinite capacity. After each pass, k-means clustering is applied for handling small classification errors.

In a similar way, in [7] an ant-based clustering algorithm is combined with the fuzzy c-means algorithm. Furthermore, some work has been done on combining fuzzy rules with ant-based algorithms for optimization problems [11]. Moreover, in [16] fuzzy rules have been used to control the behaviors of artificial ants in a clustering algorithm.

8.3 FPAB

FPAB is a swarm clustering method based on flowers pollination by artificial bees. It is based on the fact in nature that each species of plants has better growth in special region and agglomeration of these species in such places is observable. In one area, natural conditions such as water and soil effect on growth of plants. In addition, natural selection process selects better fitted plants to survive. This fact could be seen as compatibility among same species of plants in nature. This compatibility and climate spark the same species to grow better and gradually constitute the garden of same type of flowers. The schematic of FPAB is shown in Figure 8.1.

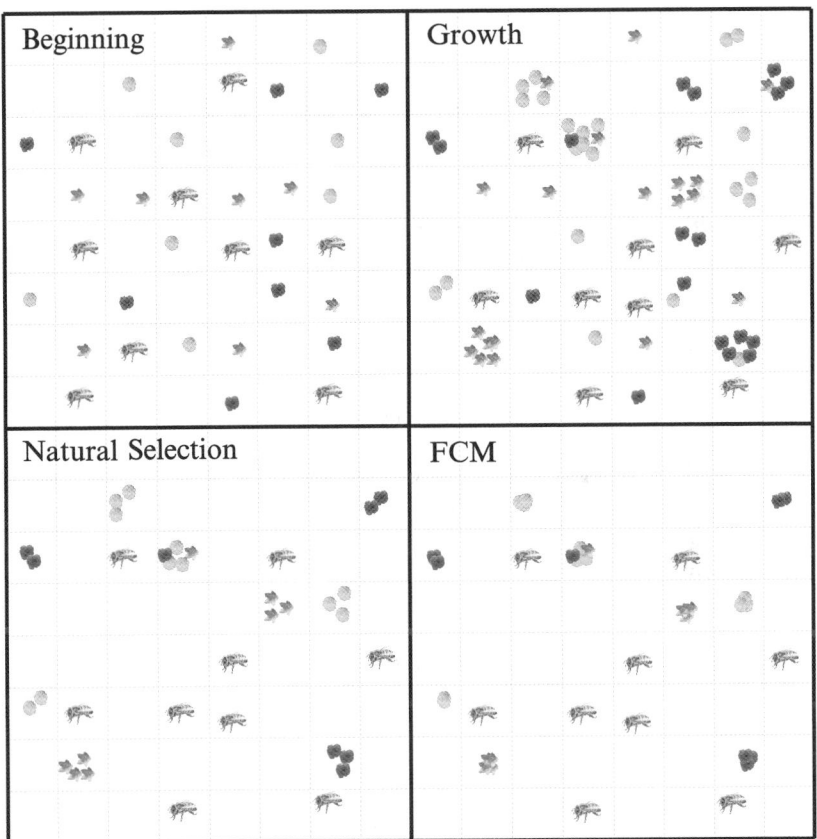

Fig. 8.1. FPAB Clustering schematic

In FPAB, Each individual artificial bee is behaviorally a simple agent with a one-slot memory (such as individual ants in AntClass). The memory is used to save the growth of carrying pollen in its source. The bees will pick up the pollen of flower with lowest growth and pollinate the pollen where it will grow better. Each pollen grows in proportion to its neighbor flowers (hereafter called garden) and after some iteration the natural selection will select the flower with best growth of one species to survive and will sear others. Using these simple stochastic behaviors and limited memory, the artificial bees can perform complicated task such as clustering.

We use FCM to remove some obvious clustering errors (such as single flowers which have zero growth) then we select one instance (not always a real instance) from each garden to continue clustering in higher levels of hierarchy by applying FPAB to it again because without this stratagem we have small clusters of very similar data. In FPAB each flower is equal to one object and the pollens are instances of that object. All of computation in this method is relative so it does not require any parameter settings. The details of FPAB are described in next section.

8.3.1 FPAB underlying algorithms

Initially, flowers scattered randomly on $2D$ discrete grid, which named as jungle. This jungle can be considered as a matrix of $m \times m$ cells. The matrix is toroidal which allows the bees to fly from one end to another easily and there is no initially more than one flower in each cell of jungle. The size of the jungle depends on the number of flowers. We have used a jungle of $m \times m$ such $m^2 = 4n$ that where n is the total number of flowers (objects) to be clustered (Figure 8.2). The general outline of this type of grid is used in [12].

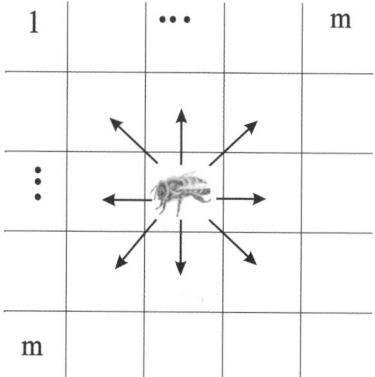

Fig. 8.2. Jungle schematic

Initially, the bees are randomly scattered throughout the jungle. We use $n/3$ bees, where n is the total number of flowers to be clustered. The bees constitute garden with juxtaposition the pollens which will grow in proportion to similarity to others. Each garden defined as a collection of 2 or more flowers (or pollens) and for simplicity of algorithm each garden spatially located in a single cell. Consider a garden G with n_g flowers. We define the following parameters:

- The compatibility between two flowers in the garden

$$C(pi, pj) = 1 - D(pi, pj) \tag{8.1}$$

Where D is normalized Euclidean distance between the attributes of flowers pi, pj
- The center of garden G

$$O_{center}(G) = \frac{1}{n_g} \sum_{O_i \in G} O_i \tag{8.2}$$

- The growth of the pollen in a garden

$$G(pi) = C(pi, O_{center}) \tag{8.3}$$

- The flower with minimum growth in a garden G

$$p_{min} = \arg \min G(pi) \tag{8.4}$$

where $pi \in G$

Initially, the bees are randomly scattered into the jungle and also randomly pick up or pollinate when they reach a new garden in the jungle. Here, the stopping criterion for the bees is the number of times through the repeat loop. The main FPAB algorithm is presented in Figure 8.3:

1. Randomly scatter bees and flowers into the jungle (at most one flower per cell)
2. Set the initial growth of all flowers to zero
3. Repeat 4,5,6 until stopping criteria
4. for each bee do
    a) fly the bee in jungle
    b) **if** the bee does not carry any pollen **then** it possibly picks up a pollen **else** the bee possibly pollinates
5. Next bee
6. Natural selection

Fig. 8.3. The main algorithm of FPAB

The algorithms of picking up, pollinating and natural selection are explained below.

Picking up pollen

When the bee does not carry any pollen, it looks for possible pollens to pickup looking at 8 neighbor cells around its current location. If a garden or a flower exists there, it could possibly pick up pollen. We consider three cases here: if there was only one flower in the cell, the bee picks up its pollen with probability P_{pickup}. This probability can be set to any number between [0, 1] and it effects merely on the speed of the algorithm. If there was a garden of two flowers, the bee picks up the pollen of one flower randomly from the garden with probability P_{dist}, this probability is set the same as P_{pickup}. If there was a garden of with more than two flowers, the bee must pick up the pollen of flower with lowest growth in that garden. The algorithm of picking up pollen is shown in Figure 8.4:

1. mark the 8 neighbor cells around the bee as "unexplored"
2. Repeat
   a) consider the next unexplored cell around the bee
   b) **if** the cell is not empty **then**
      i. if the cell contains a single flower, then pick up its pollen with probability P_{pickup}
      ii. if the cell contains a garden of 2 flowers then randomly pick up a pollen of ones with probability P_{dist}
      iii. else if the cell contains a garden of more than 2 flowers then pick up the pollen of flower with lowest growth
   c) label the cell as "explored"
3. until all 8 neighbor cells have been explored or one pollen has been picked up

Fig. 8.4. The picking up algorithm

Pollinating

When the bee carries a pollen, it examines 8 neighbor cells around its current location. Three cases may occur: if the cell is empty, the bee do nothing and consider the next neighbor cell. If the bee encounters a flower with lower growth than its carrying pollen growth, the bee discards the carrying pollen and picks up the pollen with lower growth. If the carrying pollen does not exist here and its growth will be better, the bee will pollinate the pollen. The algorithm of pollinating is shown in Figure 8.5:

1. mark the 8 neighbor cells around the bee as "unexplored"
2. Repeat
   a) consider the next unexplored cell around the bee
      i. if the cell is empty then continue
      ii. if the cell contains a flower with lower growth than carrying pollen then discard the carrying pollen and pick up the pollen of flower with lower growth
      iii. if the carrying pollen has better growth here than its growth in its source garden then pollinate in this garden
   b) label the cell as "explored"
3. until all 8 neighbor cells have been explored or one pollen has been picked up

Fig. 8.5. Pollinating algorithm

Natural selection

After pollinating, there maybe exists more than one instance of one flower in different gardens, which have different growth. Now we use natural selection process to sear instances with lower growth of the same type and allow others to survive. The flowers with the same characteristics are agglomerated in one garden i.e. same type flowers are clustered. We use here hard natural selection, which deletes all flowers of the same type with lower growth than maximum growth of that flower. The algorithm of natural selection is shown in Figure 8.6:

1. for all species of flowers find maximum growth of that flower
2. for each garden in jungle
    a) for each flower in garden
       If the flower is lower than its maximum growth in the same type then delete flower and update growth of the garden
    b) next flower
3. Next garden

Fig. 8.6. Natural selection algorithm

Merge algorithm

Artificial bees constitute several clusters of very similar flowers in limited iterations and objects will be clustered unsupervisedly without any parameter settings and any initial information about clusters. For removing obvious misclassification errors such as single flowers which does not belong to any garden, we use FCM and then we find the centroids of these gardens and assume them as single flowers and then iteratively apply FPAB to these new flowers to achieve lower number of gardens until stopping criterion (here after sufficient levels of clustering). The total algorithm of clustering is shown in Figure 8.7:

1. Randomly scatter flowers and artificial bees into the jungle at most one flower per cell
2. for i=1 to 100
    a) for each bee do
        i. fly the bee in jungle
        ii. if the bee does not carry any pollen then it possibly picks up pollen
        iii. else the bee possibly pollinates
    b) Next bee
    c) Natural selection
3. Next i
4. Apply Fuzzy C Means based on the centers of clusters obtained from the previous step
5. Assume the centers of clusters as new single flowers i.e. a single flower per garden
6. Repeat 2-5 to reach appropriate number of clusters

Fig. 8.7. The total clustering + merging algorithm

8.4 Experimental results

Three small data sets were used the results were compared with AntClass. These databases are classical databases in machine learning. The information about these three databases is shown in Table 8.1:

The reported results are averaged for 50 runs of the experiments and in each run bees and flowers randomly scattered in jungle. In this method, we have no parameter setting which results on less misclassification errors. In addition, we

Table 8.1. The information of databases

Database Name	number of examples	number of Attributes	number of classes	Description
Iris plant	150	4 Numeric	3	One class is linearly separable from the other 2 ones
Wine Recognition	178	13 continuous	3	The chemical analysis of wines grown in a region in Italy
Glass Identification	214	9 continuous	6	

use the following formula for measuring misclassification errors that measures the difference between the obtained partition and the real one.

$$E_c = \frac{2}{N(N-1)} \times \sum_{(i,j)\in\{1,2,...,N\}^2, i<j} \varepsilon_{i,j}$$

where $\varepsilon_{i,j} = \{0$
$if (C(Oi) = C(Oj) \wedge C'(Oi) = C'(Oj)) \vee$

$(C(Oi) \neq C(Oj) \wedge C'(Oi) \neq C'(Oj))$

1

else (8.5)

Which $C(O)$ is the expected cluster identifier for object "O" in the original data and $C'(O)$ is the cluster found by evaluated algorithm. The results of experiments are shown in Table 8.2:

Table 8.2. The Results of experiments

Database name	number of classes (AntClass)	number of classes (FPAB)	number of misclassification Error (AntClass)	number of misclassification Error (FPAB)
Iris plant	3.02	3.1	15.4%	8.2%
Wine Recognition	3.06	3.1	5.38%	4.3%
Glass Identification	7.7	6.9	4.48%	4.9%

8.5 Conclusion and future works

In this chapter, we surveyed some artificial models of data clustering and introduced a novel approach based on swarm concept. This technique was inspired by nature in a specific sense and it successfully acts as it is done in nature. We tried to mimic some of the behaviors of the bees but in general, this approach may not yield the exact pass, which bee colonies use for clustering plants. It is, however, competitive with AntClass and Fuzzy ants Methods in two aspects: firstly, this method does not require any parameter settings. Parameter setting in previous approaches makes them sensitive to datasets in other mean each parameter must be tune for each dataset. Secondly, the FPAB has less misclassification errors than previous methods. The experiments depicted that this method does not require many iterations to converge.

FPAB uses the hard natural selection, which means that it eliminates all instances that have grown less than maximum ones. This hard selection may cause some trials to waste. Using soft selection instead of hard selection is promising in that way, moreover it may eliminate the FCM phase.

Natural selection is the bottleneck of making the clustering algorithm fully distributed because it needs to find the maximum growth of one species via whole environment. Here each artificial bee has one slot memory for saving the features of the pollen of the flower, which is carrying. If each bee has more than one slot memory it can save the growth information of species that seen before so it can simulate the natural selection process in distributed manner.

In FPAB, each artificial bee carries a pollen of one species. If each bee could carry more than one pollen, the iterations via jungle will be reduced. Moreover, each bee can carry pollens of a small cluster of flowers at once and merging phase will be simpler than before.

References

1. E. Bonabeau, M. Dorigo, and G. Theraulaz (1999) Swarm Intelligence: From natural to artificial systems. Oxford University press, NY.
2. G.P. Babu , M. N. Murty (1993) A near-optimal initial seed value selection in K-Means algorithm using a genetic algorithm. Pattern Recognition Letters, v.14 n.10 pp. 763–769.
3. J.L. Deneubourg, S. Goss, N. Francs, A. Sendova-Franks, C. Detrain and L. Chretien (1991) The dynamics of collective sorting:Robot-Like Ant and Ant-Like Robot. In Proceedings First Conference on Simulation of adaptive Behavior: from animals to animats, edited by J.A. Meyer and S.W. Wilson, pp. 356–365. Cambridge, MA: MIT press.
4. L.O. Hall, I.B. Ozyurt, and J.C. Bezdek (1999) Clustering with a Genetically Optimized Approach. IEEE Trans-actions on Evolutionary Computation, Volume 3, No. 2, pp. 103–112.
5. J. Handl, B. Meyer (2002) Improved Ant-Based Clustering and Sorting in a Document Retrieval Interface. Proc. of the 7th Int. Conf. on Parallel Problem Solving from Nature. pp. 913–923.
6. D. R. Jones and M. A. Beltramo (1991) Solving partitioning problems with genetic algorithms. Proc. of the 4th ICGA, pp. 442–450.
7. M. Kanade, L. O. Hall (2003) Fuzzy Ants as a Clustering Concept. Proc. of the 22nd Int. Conf. of the North American Fuzzy Information Processing Society. pp. 227–232.

8. P. Kuntz, P. Layzell and D. Snyers (1997) A Colony of Ant-like Agents for Partitioning in VLSI Technology. Fourth European Conference on Artificial Life, MIT Press, pp. 417 – 424.
9. E. Lumer and B. Faieta (1994) Diversity and Adaptation in Populations of Clustering Ants. In Proceedings Third International Conference on Simulation of Adaptive Behavior: from animals to animats 3, Cambridge, Massachusetts MIT press, pp. 499-508.
10. N. Labroche, N. Monmarché and G. Venturini (2002) A new clustering algorithm based on the chemical recognition system of ants. Proceedings of the European Conference on Artificial Intelligence, Lyon, France, pp. 345–349.
11. P. Lucic (2002) Modeling Transportation Systems using Concepts of Swarm Intelligence and Soft Computing. PhD thesis, Virginia Tech.
12. N. Monmarché, M. Silmane and G. Venturini (1999) AntClass:discovery of clusters in numeric data by an hybridization of an ant colony with k-means algorithm. Internal Report no. 213, Laboratoire 'Informatique de l'Universite.
13. N. Monmarché, M. Slimane and G. Venturini (1999) On improving clustering in numerical databases with artificial ants. Advances in Artificial Life. 5th European Conference, ECAL'99. Proceedings Lecture Notes in Artificial Intelligence Vol. 1674, pp. 626–635.
14. N. Monmarché (2000) Algorithmes de Fourmis Artificielles: Applications à la Classification et à l'Optimisation. PhD thesis, Université France Rabelais.
15. V. Ramos, F. Muge and P. Pina (2002) Self-Organized Data and Image Retrieval as a Consequence of Inter-Dynamic Synergistic Relationships in Artificial Ant Colonies. Soft Computing Systems: Design, Management and Applications. 87, pp. 500–509.
16. S. Schockaert, M. De Cock, C. Cornelis and E. E. Kerre (2004) Efficient Clustering with Fuzzy Ants. in Applied Computational Intelligence, World Scientific Press.
17. D. Stirnimann and T. Lukacevic (2004) Adaptive Building Intelligence. Term Project, University of Applied Science Rapperswil.

9

Computer study of the evolution of 'news foragers' on the Internet

Zsolt Palotai, Sándor Mandusitz, and András Lőrincz*

Department of Information Systems
Eötvös Loránd University
Pázmány Péter sétány 1/C Budapest, Hungary H-1117
zspalotai@vnet.hu, santi@inf.elte.hu, andras.lorincz@elte.hu

Summary. We populated a huge scale-free portion of Internet environment with news foragers. They evolved by a simple internal selective algorithm: selection concerned the memory components, being finite in size and containing the list of most promising supplies. Foragers received reward for locating not yet found news and crawled by using value estimation. Foragers were allowed to multiply if they passed a given productivity threshold. A particular property of this community is that there is no direct interaction (here, communication) amongst foragers that allowed us to study compartmentalization, assumed to be important for scalability, in a very clear form. Experiments were conducted with our novel scalable Alife architecture. These experiments had two particular features. The first feature concerned the environment: a scale-free world was studied as the space of evolutionary algorithms. The choice of this environment was due to its generality in mother nature. The other feature of the experiments concerned the fitness. Fitness was not predetermined by us, but it was implicitly determined by the unknown, unpredictable environment that *sustained* the community and by the evolution of the competitive individuals. We found that the Alife community achieved fast compartmentalization.

9.1 Introduction

Dynamical hierarchies of nature are in the focus of research interest [4, 23]. We have developed a novel artificial life (Alife) environment populated by intelligent individuals (agents) to detect 'breaking news' type information on a prominent and vast WWW domain. We turned to Alife to achieve efficient division of labor under minimal communication load (i.e., no direct interaction) between individuals. All components, but direct interaction, seem necessary in our algorithm to achieve efficient cooperation. There are many different aspects of this work, including evolution (for reviews on relevant evolutionary theories, see, e.g., [7, 11] and references therein), the dynamics of self-modifying systems (see, e.g., [9, 13] and references therein), and swarm intelligence (for a review, see [14]).

*Corresponding author

A particular feature of this study is that evolution occurs in a scale-free world, the WWW [3, 15]. A graph is a scale-free network if the number of incoming (or outgoing or both) edges follows a power-law distribution ($P(k) \propto k^{-\gamma}$, where k is integer, $P(k)$ denotes the probability that a vertex has k incoming (or outgoing, or both) edges and $\gamma > 0$). The direct consequence of the scale-free property is that there are numerous URLs or sets of interlinked URLs, which have a large number of incoming links. Intelligent web crawlers can be easily trapped at the neighborhood of such junctions as it has been shown previously [16, 18]. The intriguing property of scale-free worlds is their abundance in nature [1]: Most of the processes, which are considered evolutionary, seem to develop and/or to co-occur in scale-free structures. This is also true for the Internet, the major source of information today. Thus, for the next generation of information systems, it seems mandatory to consider information gathering in scale-free environments.

The other feature of our study is that fitness is not determined by us and that fitness is implicit. Similar concepts have been studied in other evolutionary systems, where organisms compete for space and resources and cooperate through direct interaction (see, e.g., [21] and references therein.) Here also, fitness is determined by the external world *and* by the competing individuals together. Also, our agents crawl by estimating the long-term cumulated profit using reinforcement learning (for a review, see, e.g., [27]). In this estimation, a function approximation method and the so called temporal difference learning are utilized. Reinforcement learning has also been used in concurrent multi-robot learning, where robots had to learn to forage together via direct interaction [19]. The lack of explicit measure of fitness, the lack of our control over the environment, the value estimation executed by the individuals, and the *lack of direct interaction* distinguish our work from most studies using genetic, evolutionary, and reinforcement learning algorithms.

The paper is organized as follows. We review the related web crawler tools, including those [16, 18, 22] that our work is based upon, in Section 9.2. We describe our algorithms and the forager architecture in Section 9.3. This section contains the necessary algorithmic details imposed by the task, the search of the Web. Experimental results are provided in Section 9.4 followed by a discussion (Section 9.5) Conclusions are drawn in Section 9.6.

9.2 Related work

There are three main problems that have been studied in the context of crawlers. Angkawattanawit, Rungsawang [2], and Menczer [20] study topic specific crawlers. Risvik et al. [24] address research issues related to the exponential growth of the Web. Cho and Gracia-Molina [6], Menczer [20] and Edwards et. al [10] study the problem of different refresh rates of URLs (possibly as high as hourly or as low as yearly).

An introduction to and a broad overview of topic specific crawlers are provided in [2]. They propose to learn starting URLs, topic keywords and URL ordering through consecutive crawling attempts. They show that the learning of starting URLs and the use of consecutive crawling attempts can increase the efficiency of the crawlers. The used heuristic is similar to the weblog algorithm [22], which also finds good starting URLs and periodically restarts the crawling from the newly learned ones. The main limitation of this work is that it is incapable of addressing the freshness (i.e., modification) of already visited Web pages.

Menczer [20] describes some disadvantages of current Web search engines on the dynamic Web, e.g., the low ratio of fresh or relevant documents. He proposes to complement the search engines with intelligent crawlers, or web mining agents to overcome those disadvantages. He introduces the InfoSpider architecture that uses genetic algorithm and reinforcement learning, also describes the MySpider implementation of it, which starts from the 100 top pages of AltaVista. Our weblog algorithm uses local selection for finding good starting URLs for searches, thus not depending on any search engines. Dependence on a search engine can be a suffer limitation of most existing search agents, like MySpiders. Note however, that it is an easy matter to combine the present algorithm with URLs offered by search engines.

Risvik and Michelsen [24] overview different dimensions of web dynamics and show the arising problems in a search engine model. The main part of the paper focuses on the problems that crawlers need to overcome on the dynamic Web. As a possible solution the authors propose a heterogenous crawling architecture. The main limitation of their crawling architecture is that they must divide the web to be crawled into distinct portions manually before the crawling starts. A weblog like distributed algorithm – as suggested here – may be used in that architecture to overcome this limitation.

Cho and Garcia-Molina [6] define mathematically the freshness and age of documents of search engines. They propose the Poisson process as a model for page refreshment. The authors also propose various refresh policies and study their effectiveness both theoretically and on real data. They present the optimal refresh policies for their freshness and age metrics under the Poisson page refresh model. The authors show that these policies are superior to others on real data, too. Although they show that in their database more than 20 percent of the documents are changed each day, they disclosed these documents from their studies. Their crawler visited the documents once each day for 5 months, thus can not measure the exact change rate of those documents. While in our work we definitely concentrate on these frequently changing documents.

9.3 Forager architecture

There are two different kinds of agents: the foragers and the reinforcing agent (RA). The fleet of foragers crawl the web and sends the URLs of the selected documents to the reinforcing agent. The RA determines which forager should work for the RA and how long a forager should work. The RA sends reinforcements to the foragers based on the received URLs.

We employ a fleet of foragers to study the competition among individual foragers. A forager has simple and limited capabilities, like a stack for a limited number of starting URLs and a simple, content based URL ordering. The foragers compete with each other for finding the most relevant documents. In this way they efficiently and quickly collect new relevant documents without direct interaction.

At first we present the basic algorithms, followed by the algorithms for the reinforcing agent and the foragers.

9.3.1 Algorithms

Our constraints on finding the minimal set of algorithms were as follows: The algorithms should (i) allow the identification of unimportant parameters, (ii) support the specialization of the individuals (the foragers), (iii) allow the joining of evolutionary learning and individual learning, (iv) minimize communication as much as possible. We shall return to these points in the discussion (Section 9.5).

Weblog algorithm and starting URL selection

A forager periodically restarts from a URL randomly selected from the list of starting URLs. The sequence of visited URLs between two restarts forms a path. The starting URL list is formed from the 10 first URLs of the weblog. In the weblog there are 100 URLs with their associated weblog values in descending order. The weblog value of a URL estimates the expected sum of rewards during a path after visiting that URL. The weblog update algorithm modifies the weblog before a new path is started. The weblog value of a URL already in the weblog is modified toward the sum of rewards ($sumR$) in the remaining part of the path after that URL:

$$newValue = (1-\beta)\,oldValue + \beta\,sumR,$$

where β was set to 0.3. A new URL has the value of actual sum of rewards in the remaining part of the path. If a URL has a high weblog value it means that around that URL there are many relevant documents. Therefore it may worth it to start a search from that URL.

The weblog algorithm is a very simple version of evolutionary algorithms. Here, evolution may occur at two different levels: the list of URLs of the forager is evolving by the reordering of the weblog. Also, a forager may multiply, and its weblog, or part of it may spread through inheritance. This way, the weblog algorithm incorporates the basic features of evolutionary algorithms. This simple form shall be satisfactory for our purposes.

Reinforcement Learning based URL ordering

A forager can modify its URL ordering based on the received reinforcements of the sent URLs. The (immediate) profit is the difference of received rewards and penalties at any given step. Immediate profit is a myopic characterization of a step to a URL. Foragers have an adaptive continuous value estimator and follow the *policy* that maximizes the expected long term cumulated profit (LTP) instead of the immediate profit. Such estimators can be easily realized in neural systems [25, 27, 28]. Policy and profit estimation are interlinked concepts: profit estimation determines the policy, whereas policy influences choices and, in turn, the expected LTP. (For a review, see [27].) Here, choices are based on the greedy LTP policy: The forager visits the URL, which belongs to the *frontier* (the list of linked but not yet visited URLs, see later) and has the highest estimated LTP.

In the particular simulation each forager has a ($k = 50$) dimensional probabilistic term-frequency inverse document-frequency (PrTFIDF) text classifier [12], generated on a previously downloaded portion of the Geocities database. Fifty clusters were created by Boley's clustering algorithm [5] from the downloaded documents. The PrTFIDF classifiers were trained on these clusters plus an additional one, the $(k+1)^{th}$, representing general texts from the internet. The PrTFIDF outputs were non-linearly mapped to the interval [-1,+1] by a hyperbolic-tangent function.[2] The classifier was applied to reduce the texts to a small dimensional representation. The output vector of the classifier for the page of URL A is **state(A)** $= (state(A)_1, \ldots, state(A)_k)$. (The $(k+1)^{th}$ output was dismissed.) This output vector is stored for each URL.

A linear function approximator is used for LTP estimation. It encompasses k parameters, the *weight vector* **weight** $= (weight_1, \ldots, weight_k)$. The LTP of document of URL A is estimated as the scalar product of **state(A)** and **weight**:

$$value(A) = \sum_{i=1}^{k} weight_i \, state(A)_i.$$

During URL ordering the URL with highest LTP estimation is selected. (For more details, see, [22].)

The weight vector of each forager is tuned by temporal difference learning (TD-learning) [26, 28, 25]. Let us denote the current URL by URL_n, the next URL to be visited by URL_{n+1}, the output of the classifier for URL_j by **state(URL$_j$)** and the estimated LTP of a URL URL_j by $value(URL_j) = \sum_{i=1}^{k} wegiht_i \, state(URL_j)_i$. Assume that leaving URL_n to URL_{n+1} the immediate profit is r_{n+1}. Our estimation is perfect if

$$value(URL_n) = value(URL_{n+1}) + r_{n+1}.$$

Future profits are typically discounted in such estimations as $value(URL_n) = \gamma value(URL_{n+1}) + r_{n+1}$, where $0 < \gamma < 1$. The error of value estimation is

$$\delta(n, n+1) = r_{n+1} + \gamma value(URL_{n+1}) - value(URL_n).$$

We used throughout the simulations $\gamma = 0.9$. For each step $URL_n \to URL_{n+1}$ the weights of the value function were tuned to decrease the error of value estimation based on the received immediate profit r_{n+1}. The $\delta(n, n+1)$ estimation error was

---

[2]For each class 0.5 was subtracted from the PrTFIDF value and it was then multiplied by 40. The resulting number underwent non-linear transformation by using the tanh function producing Yes/No-like outputs for the overwhelming majority of the inputs.

used to correct the parameters. The ith component of the weight vector, $weight_i$, was corrected by

$$\Delta weight_i = \alpha \delta(n, n+1) \, state(URL_n)_i$$

with $\alpha = 0.1$ and $i = 1, \ldots, k$. These modified weights would improve value estimation in stationary and observable environments (see, e.g., [27] and references therein), but were also found efficient in large Web environments [16, 22].

Document relevancy

A document or page is possibly relevant for a forager if it is not older than 24 hours and the forager has not marked it previously. These documents are selected out by the crawler and are sent to the RA for further evaluation.

Multiplication of a forager

During multiplication the weblog is randomly divided into two equal sized parts (one for the original and one for the new forager). The parameters of the URL ordering algorithm (the weight vector of the value estimation) are copied.

9.3.2 Reinforcing agent

A reinforcing agent controls the 'life' of foragers. It can start, stop, multiply or delete foragers. RA receives the URLs of documents selected by the foragers, and responds with reinforcements for the received URLs. The reward is 100 in arbitrary units for a relevant document and the cost is 1 (that is, reward is -1) for sending a document to the RA. A document is relevant if it is not yet seen by the reinforcing agent and it is not older than 24 hours. The reinforcing agent maintains the score of each forager working for it. Initially each forager has 100 (a.u.) score. When a forager sends a URL to the RA, the forager's score is decreased by 0.05. After each relevant page sent by the forager, the forager's score is increased by 1.

Figure 9.1 shows the reinforcement of a step in an example path: at start (time $t - 2$), the agent is at URL 'A', where documents of neighboring URLs 'B', 'C' and 'D' are downloaded. URL 'D' is visited next. Documents of URLs 'E', 'F' and 'G' are downloaded. Document of URL 'G' has an obsolete date. Documents of URLs 'E' and 'F' are sent to the center. Document of URL 'F' is novel to the center, so it is rewarded. In turn, profit 98 is received by the forager. The forager maintains a list of neighbors of visited URLs, called *frontier*. It can only visit URLs of the *frontier*.

When the forager's score reaches 200 then the forager is multiplied. That is a new forager is created with the same algorithms as the original one has, but with slightly different parameters. When the forager's score goes below 0 and the number of foragers is larger than 2 then the forager is deleted. Note that a forager can be multiplied or deleted immediately after it has been stopped by the RA and before the next forager is activated.

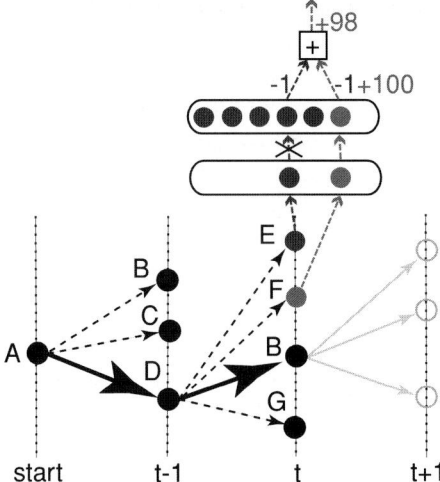

Fig. 9.1. Reward system. Central reinforcement system. Black circles: unselected documents. Blue circles: documents that the reinforcing agent has. Red circles: novel documents. Positive (negative) numbers: reward and profit (cost). -1: negative reward for sending a document. +100: positive reward for sending a novel document. Vertical dashed lines: consecutive time steps. Dots on the $(t+k)^{th}$ dashed line: documents available at time step $t+k-1$. Solid arrows connect URLs visited by the crawler. Dashed and solid arrows point to URLs that were downloaded by the crawler at the corresponding time step.

Foragers on the same computer are working in time slices one after each other. Each forager works for some amount of time determined by the RA. Then the RA stops that forager and starts the next one selected by the RA.

9.3.3 Foragers

A forager is initialized with parameters defining the URL ordering, and either with a weblog or with a seed of URLs. After its initialization a forager crawls in search paths, that is after a given number of steps the search restarts and the steps between two restarts form a path. During each path the forager takes 100 steps, i.e., selects the next URL to be visited with a URL ordering algorithm. At the beginning of a path a URL is selected randomly from the starting URL list. This list is formed from the 10 first URLs of the weblog. The weblog contains the possibly good starting URLs with their associated weblog values in descending order. The weblog algorithm modifies the weblog and so thus the starting URL list before a new path is started. When a forager is restarted by the RA, after the RA has stopped it, the forager continues from the internal state in which it was stopped.

The URL ordering algorithm selects a URL to be the next step from the frontier URL set. The selected URL is removed from the frontier and added to the visited URL set to avoid loops. After downloading the pages, only those URLs (linked from the visited URL) are added to the frontier which are not in the visited set.

In each step the forager downloads the page of the selected URL and all of the pages linked from the page of selected URL. It sends the URLs of the possibly relevant pages to the reinforcing agent. The forager receives reinforcements on any previously sent but not yet reinforced URLs and calls the URL ordering update algorithm with the received reinforcements.

9.4 Experimental results

Multiple two-week long experiments were conducted for four months (between September and November). Apart from short breaks, monitoring was continuous for each two week period. Most of the figures in this article represent two weeks from November. The parameters of this experiment are representative to the entire series: The number of downloaded, sent and reinforced documents was 1,090,074, 94,226 and 7,423, respectively. The used Internet bandwidth (1Mbps on average) was almost constant, decreasing slowly by 10% towards the end of the experiment. The number of foragers increased from 2 to 22. Experiments were run on a single computer. Foragers run sequentially in a prescribed order for approximately equal time intervals on one PC. The foraging period is the time interval during which all foragers run once. Unfinished paths are continued in the next run. Within each foraging-period, the allocated time of every forager was 180s. Some uncertainty (50 ± 30s) arose, because foragers were always allowed to complete the analysis of the last URL after their allocated time expired. The net duration of a 100 step path was 350s.

9.4.1 Environment

The domain of our experiments (one of the largest news sites), as most WWW domains, was scale-free according to our measurements. The distributions of both incoming and outgoing links show a power distribution (Fig. 9.2). The inset shows the links and documents investigated by a forager when it takes a step. The distributions, shown in the figure, correspond to links investigated by the foragers. The news forager visits URL 'A', downloads the not-yet visited part of the environment (documents of URLs, which URLs have not been visited yet by that forager and are linked from URL 'A'). Downloading is followed by a decision, URL 'B' is visited, downloading starts, and so on.

9.4.2 Time lag and multiplication

In the presented experiments we applied the cost/reward ratio described in Section 9.3.2. Larger cost/reward ratio increased the foraging areas and we noted a sudden increase in extinction probability.

Time-lag between publishing news and finding those decreases already after a few days (see Fig. 3(a)): the ratio of maxima to minima of the curves increases; and also, fewer news published on Friday were picked up on Saturday, a day late,

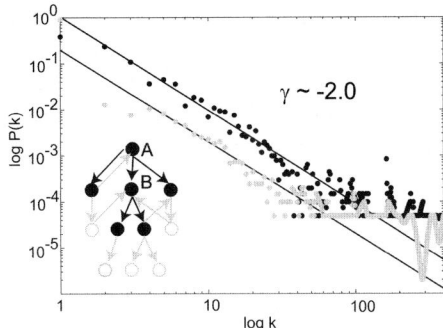

Fig. 9.2. Scale-free properties of the Internet domain. Log-log scale distribution of the number of (incoming and outgoing) links of all URLs found during the time course of investigation. Horizontal axis: number of edges ($\log k$). Vertical axis: relative frequency of number of edges at different URLs ($\log P(k)$). Black (gray) dots: incoming (outgoing) edges of URLs. Slope of the straight lines -2.0 ± 0.3. *Inset*: method of downloading. Black (gray) link: (not) in database. Solid (empty) circle: document (not) in database.

during the second weekend of the experiment than during the first. Further gains in downloading speed are indicated by the relative shift between lighter gray and darker gray peaks. The darker gray peaks keep their maxima at around midnight GMT, lighter gray peaks shift to earlier times by about 6 hours. The shift is due to changes in the number of sent documents. The minima of this number shifts to around 6:00 P.M. GMT. (Identical dates can be found for a 48 hour period centered around noon GMT.) Maxima of the relevant documents are at around 11:00 P.M. GMT (around 6:00 P.M. EST of the US). During the first week, the horizontal lines (average values of the corresponding curves during 2 workdays) are very close to each other. Both averages increase for the second week. The ratio of sent to reinforced documents increases more. At the beginning of the darker gray region, the relative shift of the two curves is large, at the end it is small, but becomes large again after that region, when multiplication slows down.

The multiplication of the foragers is shown in Fig. 3(b). Gray levels of this figure represent the value of the foragers, the range goes from 60 (darker) to 200 (lighter). In the time region studied, the values of the foragers have never fallen below 60. Upon bipartition new individuals are separated by horizontal white lines in the figure.

9.4.3 Compartmentalization

Division of work is illustrated by Fig. 9.4. Figure 4(a) show the early development of two then three and finally for four foragers. According to Fig. 4(b) large proportion of the sites are visited exclusively by not more than one forager. Only about 40% of the URLs are visited by more than one forager. Figure 4(a) demonstrates that new foragers occupy their territories quickly. Figure 4(b) shows that similar data were found for few (2-4) and for many (22) foragers (upper boundary is mainly between 0.6 and 0.7 throughout the figure). The figures depict the contributions of individual

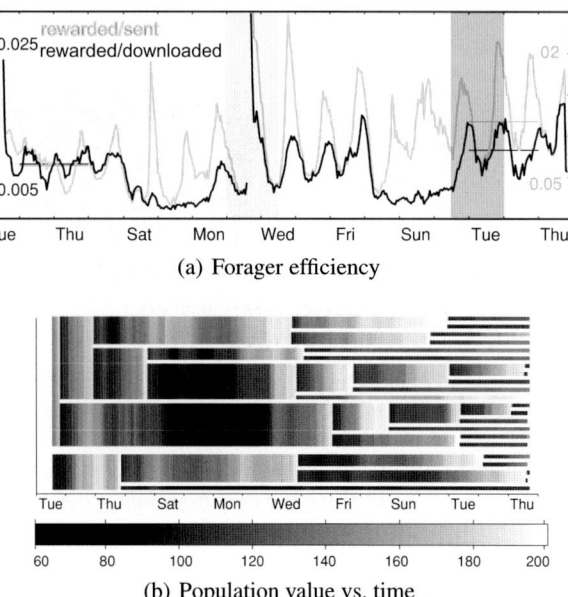

Fig. 9.3. Experimental results. (a): The rate of sent and rewarded documents showed daily oscillations. Lighter (darker) gray curve: the ratio of rewarded to sent (rewarded to downloaded) documents. Horizontal lines: average values of the two curves during two workdays. Light (darker) gray regions: number of foragers is 6 (number of foragers increases from 14 to 18). (b): Forager population and forager values vs. time. The scores of the foragers are given with different gray levels. Starting of white horizontal line means forager multiplication. (There was a break in our Internet connection in the middle of the second week.)

foragers: as new foragers start they quickly find good terrains while the older ones still keep their good territories. The environment changes very quickly, cca. 1200 new URLs were found every day. Time intervals of about 75 mins were investigated.

Figure 4(c) depicts the lack of overlap amongst forager trajectories. A relatively large number of sites were visited by different foragers. The question is if these foragers were collecting similar or different news. If they had the same 'goal' then they presumably made similar decisions and, in turn, their next step was the same. According to Figure 4(c) the ratio of such 2 step trajectories – i.e., the difference between the upper cover of the curves and value 1.0 – drops quickly at the beginning of the experiment, it remains very small and it decreases further as the number of foragers is growing. Given that the increase of the number of foragers gave rise to the decrease of individual foraging time, the actual numbers are only indicative. Nevertheless, the fast decrease at the beginning and the small ratio for few as well as for many foragers provides support that foragers followed different paths, that is, foragers developed different behaviors. Differences between hunting/foraging territories and/or differences between consumed food are called

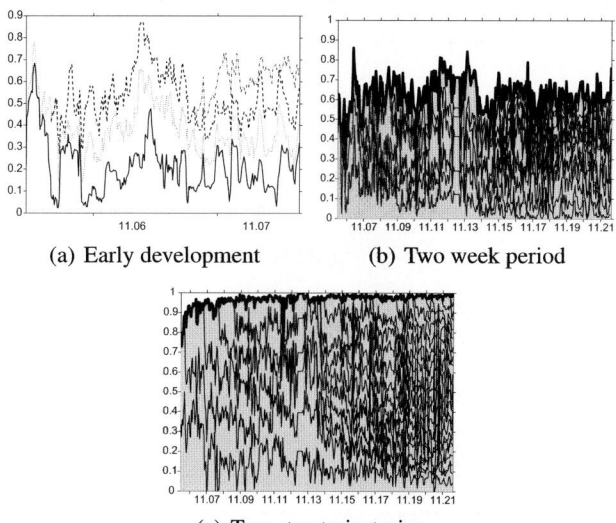

Fig. 9.4. Division of work. Horizontal axis in 'month.day' units. (a): Number of sites visited by only one forager relative to the number of all visited sites in a finite time period ($\approx 75 mins$). Contribution of a single forager is superimposed on cumulated contributions of older foragers. The difference between 1.0 and the upper boundary of the curves corresponds to the ratio of sites visited by more than one forager. Duration: about three days. (b): Same for 16 day period. (c): The ratio of different two step trajectories relative to all two step trajectories conditioned that the two step trajectories start from the same site, belong to different foragers and are in a finite time period ($\approx 75 mins$). Contribution of a single forager is superimposed on cumulated contributions of older foragers. The difference between 1.0 and the upper boundary of the curves corresponds to the ratio of the conditioned 2 step trajectories taken by more than one forager.

compartmentalization (sometimes called niche formation) [7, 9, 11, 13]. Figure 9.4 demonstrates that compartmentalization is fast and efficient in our algorithm.

The population of news foragers can be viewed as a rapidly self-assembling and adapting breaking news detector. The efficiency and speed of novelty detection is increasing while the structure of the environment is changing very quickly as it is shown in Fig. 9.5: The number of newly discovered URLs increases about linearly by time and new starting points keep emerging continuously. These drastic changes are followed by the breaking news detector, which continuously reassembles itself and maintains its monitoring efficiency. In fact, the efficiency of the detector increases with time; the quality of the assembly is improving.

9.5 Discussion

We have introduced an evolutionary Alife community comprising the following main features:

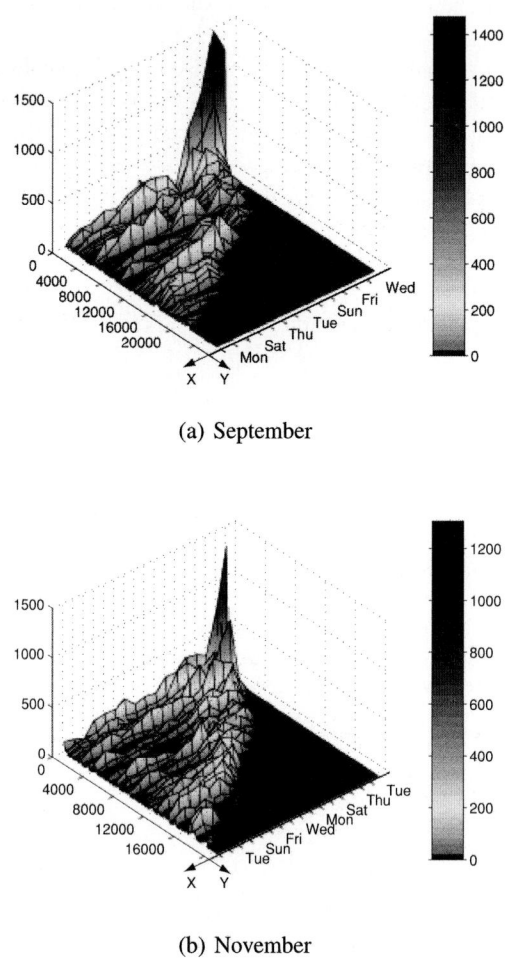

Fig. 9.5. Distribution of new starting points. Newly discovered URLs were indexed. X axis: time in days, Y axis: index of URLs. Vertical axis: number of *new* starting points summed over equal time domains and equal domains of URL indices. As time goes on, the forager community continuously discovers new regions. Runs are from September (a) and from November (b)

Feature 1. Individuals of the community could multiply. Multiplication was kept as simple as possible. It was a bipartition process that randomly shared the properties of the parent to the descendants.

Feature 2. Individuals had two adaptive memory components: (2.a): a discrete component, the list of good foraging places, subject to evaluation and selection and (2.b): a continuous component, the behavior forming value estimating module.

Feature 3. No fitness value is available, fitness is implicitly determined by the evolution of the community and by the environment, which was not under our control.

Feature 4. There is a central agent, that administers the immediate reward and cost.

Feature 2.a is necessary for scale free small worlds[3]. In such structures there is a large number of highly clustered regions. In these regions, the escape probability is very small. Intelligent crawlers get easily trapped in these regions (Fig. 9.6). This point has been studied thoroughly in our previous works [16, 18]. In fact, the well known PageRank algorithm had to introduce stochastic restarts to avoid the oversampling of these regions [17]. In general, feature 2.a is useful only if the forager 'knows' how to get to the foraging place. Here, this was a simple matter of memorizing the http addresses. Feature 2.a is crucial for efficient sharing of space.

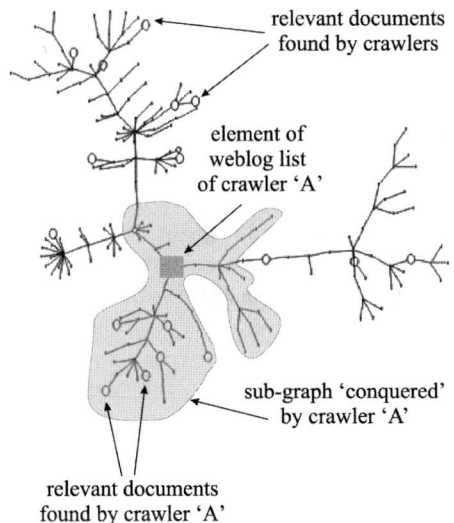

Fig. 9.6. Territory of crawler 'A' using one of the elements of its weblog list. Without other elements in the list the crawler is typically trapped in this domain [16, 18].

A typical path in the territory of a crawler is illustrated in Fig. 9.6. Without Feature 2.b, agents could divide the world into parts like the one on the illustration, but may not develop different behaviors. Feature 2.a and 2.b, together, enable different individuals to develop different function approximators and to behave differently even if they are launched from the same html page. This property has

---

[3]Note that the distinction between scale free worlds and scale free small worlds is delicate [8]. The Internet is considered scale free small world [3, 15].

been demonstrated in Fig. 4(c). That is individuals either live in different regions (i.e., have different weblogs), or behave differently (i.e., have different function approximators), or both. This is that we call *compartmentalization* in the context of our work: each individual has its environmental niche, where the word environment concerns the environment of the Alife community. However, these niches are apparent: those are determined jointly by the indirect competition, by the actual properties of the foragers and by the actual properties of the environment.

Our results indicate that agents with different behaviors have evolutionary advantages in the fast changing environment that we studied. Work sharing in the Alife community is efficient: foragers found new Internet pages continuously on the vast Internet news domain that we studied. Probably, in an environment, which is changing slowly and where food is hard to come by, the division of space and expertise in searching the owned environments could be the winning strategy. In our model, such strategy is possible. Regions of the Internet, poor in novel information, are currently under study to support this claim.

The population of news foragers can be viewed as a rapidly self-assembling and adapting news detector. The efficiency and speed of novelty detection is increasing. This occurs in spite of the fact that the structure of the environment is changing very quickly: the number of newly discovered URLs was about constant versus time. Such drastic changes are followed by the news detector, which continuously reassembles itself and improves its monitoring efficiency.

From the point of view of hierarchical organizations [4, 23], on the bottom side, the algorithm can be seen as a symbiotic collaboration between two components: the weblog is good in escaping traps, whereas reinforcement learning is good in searching neighborhoods. Such symbiotic pairs may multiply and adapt together in our algorithm. Adaptation produces a dynamic self-assembling system, which becomes closed by including the central reward administering agent into it. This closed architecture is a unit that could form a structured module at a higher level. Also, this dynamic hierarchy can be continued 'downwards': the structural modules, i.e, the selected list and the value estimator may be built from sub-components.

Finally we note that the distribution of work *without* direct interaction makes this model a possible starting point to study the emergence of interaction; the emergence of communication in our case. The intriguing point is that communication of all available information is impossible, communication of relevant information will scale badly by the number of participants if it is broadcasted, processing of communicated information consumes time, and all of these can be costly. On the other hand, properly selected and properly compressed information, which are verifiable at low risk and are robust against malicious attacks, and which can emerge and evolve under evolutionary pressure are of clear relevance in our distributed computational world.

9.6 Conclusions

An Alife algorithm has been introduced having individuals with two component memory. There is a discrete memory component providing a crude 'discretization' of the environment. This list develops together with the long-term cumulated value estimation memory component. Two component memory make individuals unique and efficient. Individuals behave differently, they move along different paths. Still, the efficiency of work-sharing is high and increases by time. This compartmentalization is appealing from the point of view of scalability of evolutionary systems.

The algorithm contributes to the understanding of memory components in evolutionary systems and hierarchies. Our results concern scale free small worlds, abundant in mother nature. Given that fitness is not provided by us, we see no straightforward way to show if these memory components are useful in grid worlds or not. It is possible that the speed and the efficiency of work sharing is mostly due to the highly clustered scale-free small world structure of the environment. We note that evolutionary processes seem to develop and/or to co-occur in scale-free structures. It is also true for the Internet, the major source of information today. Thus, for the next generation of information systems, it seems mandatory to consider information gathering in scale-free environments.

The algorithmic model is minimal. Many other algorithmic components, most importantly e.g., direct interaction could be included. Our point is that individuals of the community compete with each other in an *indirect* fashion and thus competition gives rise to work sharing, which looks like collaboration. Such apparent collaboration needs to be separated when the advantages of direct interaction are to be considered. It seems to us that our algorithm is a good starting point to study the advantages and the evolution of interaction by *adding* new features to the agents, which enable the development of such skills.

Acknowledgments

This material is based upon work supported by the European Office of Aerospace Research and Development, Air Force Office of Scientific Research, Air Force Research Laboratory, under Contract No. FA8655-03-1-3036. Any opinions, findings and conclusions or recommendations expressed in this material are those of the author(s) and do not necessarily reflect the views of the European Office of Aerospace Research and Development, Air Force Office of Scientific Research, Air Force Research Laboratory.

References

1. R. Albert and A.L. Barabási. Statistical mechanics of complex networks. *Reviews of Modern Physics*, 74:47–91, 2002.

2. N. Angkawattanawit and A. Rungsawang. Learnable topic-specific web crawler. In A. Abraham, J. Ruiz-del-Solar, and M. Köppen, editors, *Hybrid Intelligent Systems*, pages 573–582. IOS Press, 2002.
3. A.L. Barabási, R. Albert, and H. Jeong. Scale-free characteristics of random networks: The topology of the world wide web. *Physica A*, 281:69–77, 2000.
4. M. Bedau, J. McCaskill, N. Packard, S. Rasmussen, C. Adami, D. Green, T. Ikegami, K. Kaneko, and T. Ray. Open problems in artificial life. *Artificial Life*, 6:363–376, 2000.
5. D.L. Boley. Principal direction division partitioning. *Data Mining and Knowledge Discovery*, 2:325–244, 1998.
6. J. Cho and H. Garcia-Molina. Effective page refresh policies for web crawlers. *ACM Transactions on Database Systems*, 28(4):390–426, 2003.
7. C.W. Clark and M. Mangel. *Dynamic State Variable Models in Ecology: Methods and Applications*. Oxford University Press, Oxford UK, 2000.
8. P. Crucitti, V. Latora, M. Marchiori, and A. Rapisarda. Efficiency of scale-free networks: Error and attack tolerance. *Physica A*, 320:622–642, 2003.
9. V. Csányi. *Evolutionary Systems and Society: A General Theory of Life, Mind, and Culture*. Duke University Press, Durham, NC, 1989.
10. J. Edwards, K. McCurley, and J. Tomlin. An adaptive model for optimizing performance of an incremental web crawler. In *Proceedings of the tenth international conference on World Wide Web*, pages 106–113, 2001.
11. J.M. Fryxell and P. Lundberg. *Individual Behavior and Community Dynamics*. Chapman and Hall, London, 1998.
12. T. Joachims. A probabilistic analysis of the Rocchio algorithm with TFIDF for text categorization. In Douglas H. Fisher, editor, *Proceedings of ICML-97, 14th International Conference on Machine Learning*, pages 143–151, Nashville, US, 1997. Morgan Kaufmann Publishers, San Francisco, US.
13. G. Kampis. *Self-modifying Systems in Biology and Cognitive Science: A New Framework for Dynamics, Information and Complexity*. Pergamon, Oxford UK, 1991.
14. J. Kennedy, R.C. Eberhart, and Y. Shi. *Swarm Intelligence*. Morgan Kaufmann, San Francisco, USA, 2001.
15. J. Kleinberg and S. Lawrence. The structure of the web. *Science*, 294:1849–1850, 2001.
16. I. Kókai and A. Lőrincz. Fast adapting value estimation based hybrid architecture for searching the world-wide web. *Applied Soft Computing*, 2:11–23, 2002.
17. R. Lempel and S. Moran. The stochastic approach for link-structure analysis (salsa) and the tkc effect. *Computer Networks*, 33, 2000.
18. A. Lőrincz, I. Kókai, and A. Meretei. Intelligent high-performance crawlers used to reveal topic-specific structure of the WWW. *Int. J. Founds. Comp. Sci.*, 13:477–495, 2002.
19. M.J. Mataric. Reinforcement learning in the multi-robot domain. *Autonomous Robots*, 4(1):73–83, 1997.
20. F. Menczer. Complementing search engines with online web mining agents. *Decision Support Systems*, 35:195–212, 2003.
21. E. Pachepsky, T. Taylor, and S. Jones. Mutualism promotes diversity and stability in a simple artificial ecosystem. *Artificial Life*, 8(1):5–24, 2002.
22. Zs. Palotai, B. Gábor, and A. Lőrincz. Adaptive highlighting of links to assist surfing on the internet. *Int. J. of Information Technology and Decision Making*, 4:117–139, 2005.
23. S. Rasmussen, N.A. Baas, B. Mayer, M. Nilsson, and M.W. Olesen. Ansatz for dynamical hierarchies. *Artificial Life*, 7(4):329–354, 2001.
24. K. M. Risvik and R. Michelsen. Search engines and web dynamics. *Computer Networks*, 32:289–302, 2002.

25. W. Schultz. Multiple reward systems in the brain. *Nature Review of Neuroscience*, 1:199–207, 2000.
26. R. Sutton. Learning to predict by the method of temporal differences. *Machine Learning*, 3:9–44, 1988.
27. R. Sutton and A.G. Barto. *Reinforcement Learning: An Introduction*. MIT Press, Cambridge, 1998.
28. I. Szita and A. Lőrincz. Kalman filter control embedded into the reinforcement learning framework. *Neural Computation*, 2003. (in press).

10

Data Swarm Clustering

Christian Veenhuis and Mario Köppen

Fraunhofer IPK
Department Pattern Recognition
Pascalstr. 8-9, 10587 Berlin, Germany
{christian.veenhuis, mario.koeppen}@ipk.fhg.de

Summary. Data clustering is concerned with the division of a set of objects into groups of similar objects. In social insects there are many examples of clustering processes. Brood sorting observed in ant colonies can be considered as clustering according to the developmental state of the larvae. Also nest cleaning by forming piles of corpse or items is another example. These observed sorting and cluster capabilities of ant colonies have already been the inspiration of an ant-based clustering algorithm.

Another kind of clustering mechanism can be observed in flocks of birds. In some rainforests mixed-species flocks of birds can be observed. From time to time different species of birds are merging to become a multi-species swarm. The separation of this multi-species swarm into its single species can be considered as a kind of species clustering.

This chapter introduces a data clustering algorithm based on species clustering. It combines methods of Particle Swarm Optimization and Flock Algorithms. A given set of data is interpreted as a multi-species swarm which wants to separate into single-species swarms, i.e., clusters. The data to be clustered are assigned to datoids which form a swarm on a two-dimensional plane. A datoid can be imagined as a bird carrying a piece of data on its back. While swarming, this swarm divides into sub swarms moving over the plane and consisting of datoids carrying similar data. After swarming, these sub swarms of datoids can be grouped together as clusters.

10.1 Introduction

In nature a swarm is an aggregation of animals as, e.g., flocks of birds, herds of land animals or schools of fishes. The formation of swarms seems to be advantageous to protect against predators and increase the efficiency of foraging. To maintain the structure of the swarm, each swarm-mate behaves according to some rules as, e.g., keep close to your neighbors or avoid collisions. Even mixed-species flocks (i.e., multi-species swarms) of birds can be observed in nature [10].

Data clustering is concerned with the division of data into groups (i.e., clusters) of similar data. If a single species is considered as a type of data, then a multi-species swarm can be considered as a set of mixed data. The rules of the presented Data Swarm Clustering (DSC) algorithm divide this multi-species swarm of data into several single-species swarms consisting of similar data. This way a data clustering is performed. Each single-species swarm represents a cluster of similar data.

Also in social insects clustering processes can be observed [2]. For instance, brood sorting in ant colonies can be considered as clustering according to the developmental state of the larvae. Nest cleaning by forming piles of corpse or items is another example.

Deneubourg et al. introduced in [3] two models of larval sorting and corpse clustering. These models have been the inspiration of an ant-based clustering algorithm. Ant-based clustering [4] simulates the clustering and sorting behavior of ants. The data items to be clustered are assigned to items on a two-dimensional grid. These items on the grid are piled up by the simulated ants building this way the data clusters. The ants walk randomly over the grid and every time they find an isolated or wrong placed item they pick it up and put it down close to the first randomly found similar item somewhere else.

Monmarche et al. hybridized the Ant-based clustering approach with the k-means algorithm as introduced in [9]. They applied the Ant-based clustering algorithm for a fixed number of iterations to create an initial partitioning of the data. Afterwards, they refined this initial partition by using the k-means algorithm. In [6] a very similar approach to [9] is used. But instead of k-means the fuzzy c-means algorithms is used to refine the initial partition created by the ants.

Another type of clustering was introduced by Omran et al. [12]. They used Particle Swarm Optimization [7] to determine a given number of cluster centroids (i.e., center of clusters). For this, they used particles which contain all centroid vectors in a row. If $\mathbf{c}_{ij}$ denotes the jth cluster centroid of particle x_i, then a particle encoding N clusters is defined as $x_i = (\mathbf{c}_{i1}, \cdots, \mathbf{c}_{ij}, \cdots, \mathbf{c}_{iN})$. This way a particle has the dimensionality of number of centroids N times dimension of centroids. The swarm consists of many possible centroid vector sets. In [15] van der Merwe and Engelbrecht hybridized this approach with the k-means algorithm. A single particle of the swarm is initialized with the result of the k-means algorithm. The rest of the swarm is initialized randomly.

In some rainforests mixed-species flocks of birds can be observed [10]. Different species of birds are merging to become a multi-species swarm. The separation of this multi-species swarm into its single species can be considered as a kind of species clustering.

The method presented in this chapter is based on a simulation of species clustering. The data items to be clustered are assigned to objects called datoids placed on a two-dimensional plane. Similar to ant-based clustering the data items aren't clustered in their attribute space, but in the space of the datoids. This is an advantage, because data items belonging to the same class don't need to be close in the attribute space. They get close in the space of the datoids, if they belong together. Instead of randomly moving, the datoids have an affinity to move to their similar neighbors. The similar neighbors are determined based on a similarity distance function. In contrast to ant-based clustering, the data items (datoids) move on the two-dimensional plane by themself and aren't moved by additional entities like, e.g., ants.

This chapter is organized as follows. Section 10.2 gives a brief overview of clustering. Flock Algorithms and Particle Swarm Optimization are described in sections 10.3 and 10.4. The Data Swarm Clustering algorithm is described in section 10.5. The used experiments and results are presented in sections 10.6 and 10.7. Finally, in section 10.8 some conclusions are drawn.

10.2 Data Clustering

The task of data clustering is to divide a set of data X into sub-sets $C_i \subseteq X$ containing similar data. In Figure 10.1 a set of data containing three different kinds of objects is presented.

Clustering produces a set $C = \{C_i\}$ of all sub-sets $C_i \subseteq X$ as shown in Figure 10.2. A sub-set C_i is called a cluster and all clusters together result in X (i.e., $X = \bigcup_i C_i$). That means that each element of X is assigned to a cluster. Usually clusters are pair-wise disjoint (i.e., $\bigcap_i C_i = \emptyset$). That means that each element of X is assigned to exactly one cluster. The data contained within one cluster C_i are similar in some way and dissimilar to the data contained in other clusters.

Often the data are points $d_i = (d_{i1}, \cdots, d_{in})^T \in A$ in an n-dimensional attribute space $A = A_1 \times \cdots \times A_n$. Each $d_{iv} \in A_v$ ($1 \leq v \leq n$) represents a single variable or attribute.

For instance, if a set of points $\{(x_i, y_i)\}$ on a two-dimensional plane $\mathbb{R}^2$ shall be clustered, then $\mathbb{R}^2$ is the attribute space, the dimensions are the attributes and the coordinates $x_i, y_i \in \mathbb{R}$ are attribute values.

Often simply the distances between points in the attribute space are used as similarity measure to determine whether or not they are belonging to the same cluster (e.g., as in k-means).

Clustering is applied in a lot of fields as, e.g., character recognition, image segmentation, image processing, computer vision, data and web mining.

Clustering algorithms can be grouped among others into hierarchical clustering, partitioning relocation clustering, density-based partitioning and grid-based methods. Berkhin gives an extensive survey of clustering algorithms in [1].

10.3 Flock Algorithms

Flock Algorithm is a generic term for algorithms mimicking the natural aggregation and movement of bird flocks, herds of land animals or schools of fishes. Reynolds developed a distributed behavior model to compute the movement of flocks of birds within virtual environments [13] to get rid of modelling each single bird explicitly. A single entity of this model is called boid (bird-oid) and implements the following behavior:

- **Collision Avoidance**
  Each boid has to avoid collisions with nearby flock-mates.

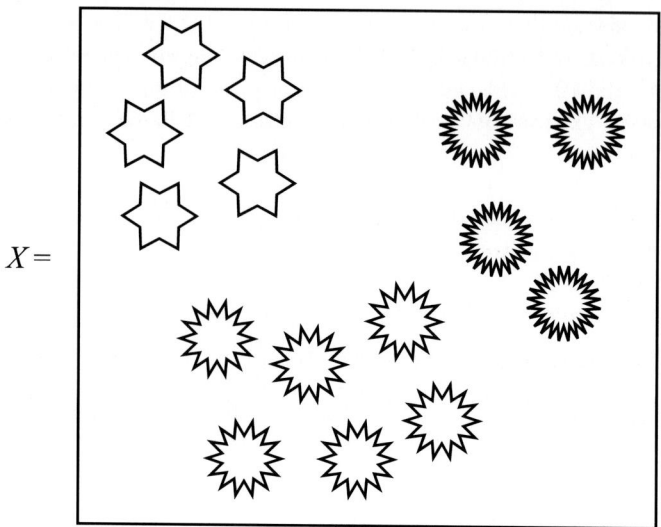

Fig. 10.1. A set of data containing three types of objects.

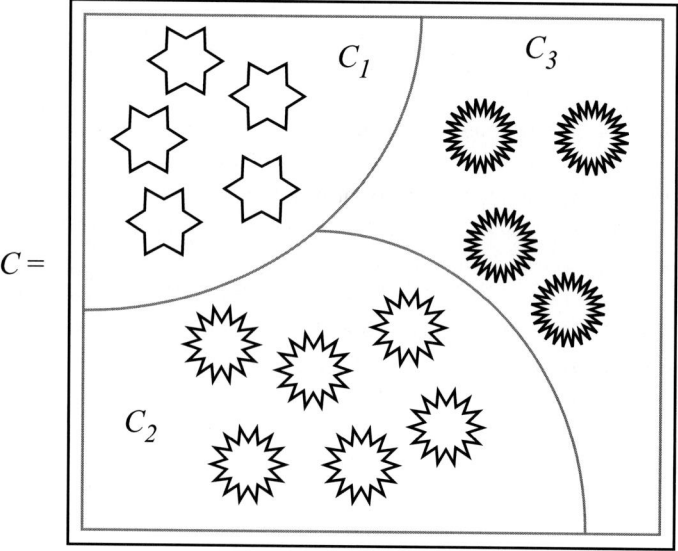

Fig. 10.2. A set of clusters $C = \{C_1, C_2, C_3\}$ containing three clusters ($|C| = 3$) of similar objects. Here, the similarity is based on the outer shape of the objects.

- **Velocity Matching**
  Each boid attempts to match the velocities of nearby flock-mates.
- **Flock Centering**
  Each boid tries to stay close to nearby flock-mates. For this, it steers to the center of the k nearest neighbors.
- **Obstacle Avoidance**
  Within virtual environments it is intended to avoid collision with obstacles (e.g., houses).
- **Following a Path**
  Usually the boids follow a specified path within the virtual environment.

Although this type of algorithm was thought for computer graphic purposes, it has been the inspiration for other algorithms as, e.g., the Particle Swarm Optimization.

10.4 Particle Swarm Optimization

Particle Swarm Optimization (PSO), as introduced by Kennedy and Eberhart [7] [8], is an optimization algorithm based on swarm theory. The main idea is to model the flocking of birds flying around a peak in a landscape.

In PSO the birds are substituted by particles and the peak in the landscape is the peak of a fitness function. The particles are flying through the search space forming flocks around peaks of fitness functions.

Let N_{dim} be the dimension of the problem (i.e., the dimension of the search space $\mathbb{R}^{N_{dim}}$), N_{part} the number of particles and $\mathcal{P}$ the set of particles $\mathcal{P} = \{P_1, \cdots, P_{N_{part}}\}$. Each particle $P_i = (x_i, v_i, l_i)$ has a current position in the search space ($x_i \in \mathbb{R}^{N_{dim}}$), a velocity ($v_i \in \mathbb{R}^{N_{dim}}$) and the locally best found position in history, i.e., the own experience ($l_i \in \mathbb{R}^{N_{dim}}$) of this particle.

In PSO, the set of particles $\mathcal{P}$ is initialized at time step $t = 0$ with randomly created particles $P_i^{(0)}$. The initial l_i are set to the corresponding initial x_i. Then, for each time step t, the next position $x_i^{(t+1)}$ and velocity $v_i^{(t+1)}$ of each particle $P_i^{(t)}$ is computed as shown in Eqns. (10.1) and (10.2).

$$v_i^{(t+1)} = w_I^{(t)} v_i^{(t)} + w_L r_1 (l_i^{(t)} - x_i^{(t)}) + w_N r_2 (n_i^{(t)} - x_i^{(t)}) \qquad (10.1)$$

$$x_i^{(t+1)} = x_i^{(t)} + v_i^{(t+1)} \qquad (10.2)$$

Here, r_1 and r_2 are random numbers in $[0,1]$. $n_i^{(t)} \in \mathbb{R}^{N_{dim}}$ represents the best found local position of the best neighbor particle at time t. Because there are several possibilities to define the neighborhood of a particle [8], the best neighboring particle can be, e.g., the best particle of a pre-defined neighborhood; the best particle of the nearest neighbors according to the distance in search space; the globally best particle etc. The inertia weight $w_I^{(t)}$ determines the influence of the particle's own velocity, i.e., it represents the confidence of the particle to its own position (typically

$w_I \in [0.1, 1.0]$). To yield a better convergence, this weight is decreased over time [14] [8]. The typical way to decrease the inertia weight $w_I^{(t)}$ is to subtract a fixed amount (e.g., 0.001) each iteration. w_L is the influence of the locally best position found so far. The influence of the best particle of the neighborhood is denoted with w_N.

After computing, the next velocity $v_i^{(t+1)}$ is added to the position of the particle to get the new position. To avoid chaotic behavior, this new velocity $v_i^{(t+1)}$ is clamped to a pre-defined interval $[-V_{max}, +V_{max}]$.

The fitness of a particle is determined by a fitness function F which maps a position vector to a real number (Eq. (10.3)).

$$F : \mathbb{R}^{N_{dim}} \to \mathbb{R} \qquad (10.3)$$

If the new position $x_i^{(t+1)}$ has a better fitness than the best solution found so far for particle P_i, it is stored in memory as shown in Eq. (10.4) (in case of minimization).

$$l_i^{(t+1)} = \begin{cases} x_i^{(t+1)}, & F(x_i^{(t+1)}) < F(l_i^{(t)}) \\ l_i^{(t)}, & otherwise \end{cases} \qquad (10.4)$$

The best solution of the run is found at particle P_b with the best local solution l_b. Best solution l_b is always element of the set of all best local solutions $\{l_i\}, \forall i \in \{1, \cdots, N_{part}\}$. Equation (10.5) determines the best fitness value $F(l_b)$ simply as the minimal fitness value of all local solutions.

$$F(l_b) = \min_{\forall i \in \{1, \cdots, N_{part}\}} \{F(l_i)\} \qquad (10.5)$$

10.5 Data Swarm Clustering

The Data Swarm Clustering (DSC) algorithm mimicks a sort of separation of different species forming one big multi-species swarm into sub-swarms consisting only of individuals of the same species (i.e., single-species swarms). The multi-species swarm can be considered as a set of mixed data X and a sub-swarm as cluster $C_i \subseteq X$.

To realize this, a PSO method with two-dimensional particles is used. The PSO particle is modified to hold one data object of X. Because a data object can be of every possible type or structure, an entity of DSC is called datoid (data-oid) instead of particle.

The separation (i.e., clustering) of a swarm of datoids can be described with three rules:

1. **Swarm Centering of Similar Datoids**
   Each datoid tries to stay close to similar nearby swarm-mates. For this, it steers to the center of the k nearest similar neighbors.
   $\Rightarrow$ This is the moving power to build the raw sub-swarms of similar datoids.

2. **Approach Avoidance to Dissimilar Datoids**
   Each datoid avoids to approach to the nearest dissimilar swarm-mate.
   ⇒ This helps to prevent dissimilar sub-swarms to merge.
3. **Velocity Matching of Similar Datoids**
   Each datoid attempts to match the velocity of the nearest similar neighbor.
   ⇒ This advantages the emergence of synchronized sub-swarms which build a better unity.

These rules are modified versions of the rules of Reynolds Flock Algorithm as described in section 10.3. The modifications include:

- The differentiation of similar and dissimilar neighbors.
- Collision avoidance is replaced by an approach avoidance to dissimilar datoids.
- Velocity is matched only to the nearest similar neighbor instead to all nearby flock-mates.

Data Swarm Clustering consists of three phases: (1) initialization, (2) multiple iterations and (3) retrieval of the formed clusters.

A swarm of datoids can be described quite similar as in PSO. Let N_{dat} be the number of datoids and $\mathcal{D}$ the set of datoids $\mathcal{D} = \{D(1), \cdots, D(N_{dat})\}$. Each datoid $D(i) = (x_i, v_i, o_i)$ consists of a current position on a two-dimensional plane ($x_i \in \mathbb{R}^2$), a velocity ($v_i \in \mathbb{R}^2$) and the data object bound to this datoid ($o_i \in X$). Each datoid is placed on a quadratic two-dimensional plane with a side length of X_{max}. That is, X_{max} restricts the two-dimensional plane in both dimensions. The maximal possible distance d_{max} between two datoids is simply the diagonal of the two-dimensional plane as computed in Eq. (10.6).

$$d_{max} = \sqrt{X_{max}^2 + X_{max}^2} \qquad (10.6)$$

If N_{iter} denotes the number of iterations, the main algorithm can be described as follows:

1. Initialize $\mathcal{D}$ with randomly created datoids (see section 10.5.1).
2. For $n = 1$ to N_{iter}: iterate swarm (see section 10.5.2).
3. Retrieve the clusters (see section 10.5.3).

10.5.1 Initialization

The initialization phase creates the set $\mathcal{D}$ with randomly created datoids. Because all data $d_i \in X$ have to be clustered, the number of datoids N_{dat} is set to the number of data in X (i.e., $N_{dat} = |X|$).

For each datoid $D(i) \in \mathcal{D}, \forall i \in \{1, \cdots, N_{dat}\}$ the following initialization is performed. First, the data object o_i bound to the datoid $D(i)$ is set to the corresponding data item $d_i \in X$.

Afterwards, the position of the datoid on the two-dimensional plane is set randomly. For this, each element of the position vector x_i is set to a random number between 0 and X_{max}.

At last, each element of the velocity vector v_i of datoid $D(i)$ is set to a random number between $-V_{max}$ and $+V_{max}$. V_{max} restricts the velocity as in PSO (see section 10.4).

The whole algorithm is shown in the following ($R_{a,b}$ means a random-number between a and b):

$N_{dat} \leftarrow |X|$
for $i = 1$ **to** N_{dat}
    $o_i \leftarrow d_i \in X$
    $x_i \leftarrow (R_{0,X_{max}}, R_{0,X_{max}})^T$
    $v_i \leftarrow (R_{-V_{max},+V_{max}}, R_{-V_{max},+V_{max}})^T$
end for i

10.5.2 Iteration

While iterating, a similarity function is needed to provide a similarity measure between two datoids. This similarity function as shown in Eq. (10.7) gets the data objects of two datoids and returns a value in $[0,1]$, whereby 1 means that both datoids are equal and 0 that both are maximal dissimilar.

$$S : X \times X \to [0,1] \quad (10.7)$$

This similarity function is problem-specific. For the datasets used in the experiments later on, the Euclidian distance as defined in Eq. (10.25) (section 10.7) is used.

The distance between two datoids on the two-dimensional plane is also determined by Euclidian distance as shown in Eq. (10.8).

$$d : \mathbb{R}^2 \times \mathbb{R}^2 \to \mathbb{R} \ , \ d(a,b) = \sqrt{\sum_{c=1}^{2} (a_c - b_c)^2} \quad (10.8)$$

An iteration step performs the following computation to each datoid $D(i) \in \mathcal{D}, \forall i \in \{1, \cdots, N_{dat}\}$.

First, the nearest similar neighbor $D(n_{i,similar})$ of $D(i)$ as depicted in Figure 10.3 is determined. For this, the similarity distance function $SD(i,j)$ as defined in Eq. (10.9) is used to compute the distance between two datoids.

$$SD(i,j) = S(o_i, o_j) d(x_i, x_j) + (1 - S(o_i, o_j)) d_{max} \quad (10.9)$$

The more similar two datoids are, the more the real distance is used as similarity distance. The more dissimilar they are, the more the maximal possible distance on the plane is used. This punish dissimilar neighbors by increasing their similarity distance. The index number of the nearest similar neighbor can be computed as in Eq. (10.10).

$$n_{i,similar} = n \quad \text{s.t.} \quad SD(i,n) = \min_{\substack{\forall j \in \{1,\cdots,N_{dat}\} \\ i \neq j}} \{SD(i,j)\} \qquad (10.10)$$

The nearest similar neighbor is needed for rule 3: *Velocity Matching of Similar Datoids*.

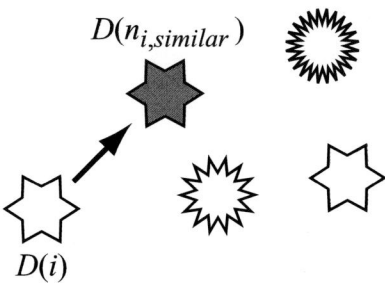

Fig. 10.3. The nearest similar neighbor $D(n_{i,similar})$. The gray-colored object is the nearest similar neighbor to $D(i)$. It is the one with the best similarity of the nearest neighbors.

Then, the nearest dissimilar neighbor $D(n_{i,dissimilar})$ of $D(i)$ as illustrated in Figure 10.4 is determined. For this, the dissimilarity distance function $DD(i,j)$ as defined in Eq. (10.11) is used to compute the distance between two datoids.

$$DD(i,j) = (1 - S(o_i, o_j))d(x_i, x_j) + S(o_i, o_j)d_{max} \qquad (10.11)$$

The more dissimilar two datoids are, the more the real distance is used as dissimilarity distance. The more similar they are, the more the maximal possible distance on the plane is used. This punish similar neighbors by increasing their dissimilarity distance. The index number of the nearest dissimilar neighbor can be computed as in Eq. (10.12).

$$n_{i,dissimilar} = n \quad \text{s.t.} \quad DD(i,n) = \min_{\substack{\forall j \in \{1,\cdots,N_{dat}\} \\ i \neq j}} \{DD(i,j)\} \qquad (10.12)$$

The nearest dissimilar neighbor is needed for rule 2: *Approach Avoidance to Dissimilar Datoids*.

For rule 1: *Swarm Centering of Similar Datoids* the center position of the k nearest similar neighbors $c_{i,similar}$ of $D(i)$ on the two-dimensional plane is needed. This is illustrated in Figure 10.5.

If $n = (n_1, \cdots, n_k)$ represents the sequence of indices of the k nearest similar neighbors (according to the similarity distance function SD), then Eq. (10.13) computes the needed center of those neighbors.

$$c_{i,similar} = \begin{pmatrix} \frac{1}{k}\sum_{j=1}^{k} x_{n_j 1} \\ \frac{1}{k}\sum_{j=1}^{k} x_{n_j 2} \end{pmatrix} \in \mathbb{R}^2 \qquad (10.13)$$

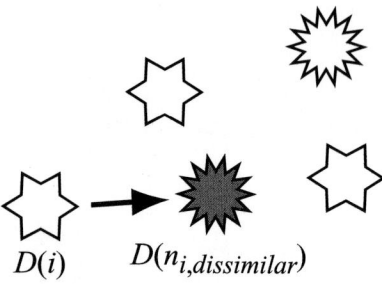

Fig. 10.4. The nearest dissimilar neighbor $D(n_{i,dissimilar})$. The gray-colored object is the nearest dissimilar neighbor to $D(i)$. It is the one with the best dissimilarity of the nearest neighbors.

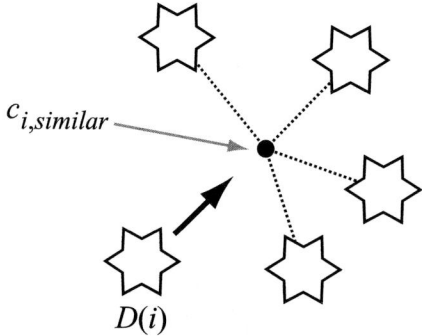

Fig. 10.5. The center position $c_{i,similar}$ of the k nearest similar neighbors to $D(i)$.

In Flock Algorithms a boid tries to match its velocity to the velocity of its neighbors. This mechanism is used in DSC to synchronize the sub-swarms to build a better unity. Instead of several neighbors, in DSC the velocity is matched only to the nearest similar neighbor as in Eq. (10.14). In Figure 10.6 this is presented by a line between the gray-colored datoid and its nearest similar neighbor.

The gray-colored datoid matches its velocity to its neighbor as defined in Eq. (10.14). The degree of matching is weighted with w_V.

$$\Delta_{velo} = w_V \left(v^{(t)}_{n_{i,similar}} - v^{(t)}_i \right) \qquad (10.14)$$

The swarm-mates in PSO and Flock Algorithms try to stay close to nearby swarm-mates. This behavior produces sub-swarms or one big compact swarm, depending on the size of neighborhood. In PSO there is often a pre-defined and pre-assigned neighborhood to which a particle tries to stay close. This reduces the computation time. But as shown in Figure 10.7 in DSC a datoid steers to the center of the k nearest similar neighbors as in Flock Algorithms.

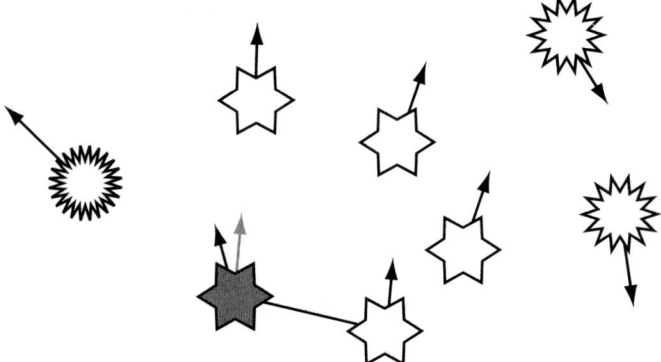

Fig. 10.6. Datoids match their velocity to their nearest similar neighbor. The new velocity is drawn as gray-colored arrow.

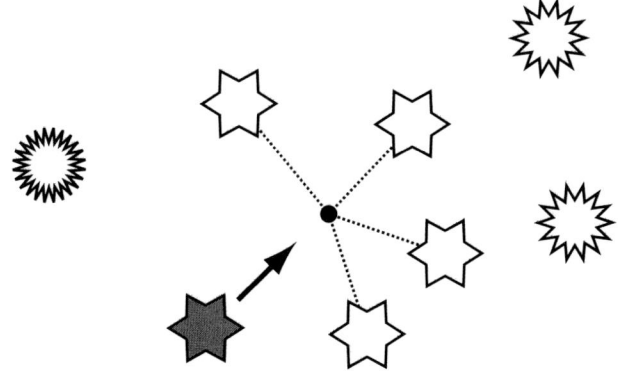

Fig. 10.7. Datoids steer to the center of the nearest similar neighbors.

This way the neighborhood is variable what is neccessary, because a pre-defined neighborhood would mean a pre-defined sub-swarm (i.e., cluster) of arbitrary datoids. This would work contrary to the clustering process.

$$\Delta_{neighbors} = S(o_i, o_{n_{i,similar}}) \, w_N \, R_{0,1} \, (c^{(t)}_{i,similar} - x^{(t)}_i) \tag{10.15}$$

The influence of the neighbors $\Delta_{neighbors}$ as shown in Eq. (10.15) is computed quite similar to PSO. The difference of the current position $x^{(t)}_i$ and the center of the similar neighbors $c^{(t)}_{i,similar}$ is weighted by w_N. Here, $R_{0,1}$ is a random number in $[0, 1]$. Additionally, the influence of the neighbors is weighted by the degree of similarity between them and the current datoid $D(i)$. As representant of the k nearest similar neighbors, the nearest similar neighbor $D(n_{i,similar})$ is used. The similarity between the data objects bound to the current datoid and the nearest similar neighbor $S(o_i, o_{n_{i,similar}})$ weights the influence of the neighbors.

To provide the separation of dissimilar datoids and to avoid the merging of dissimilar sub-swarms, the current datoid tries to move away from the nearest dissimilar neighbor, if the distance between them becomes too close (i.e., $d(x_i, x_{n_{i,dissimilar}}) < \tau_d$). This is illustrated in Figure 10.8.

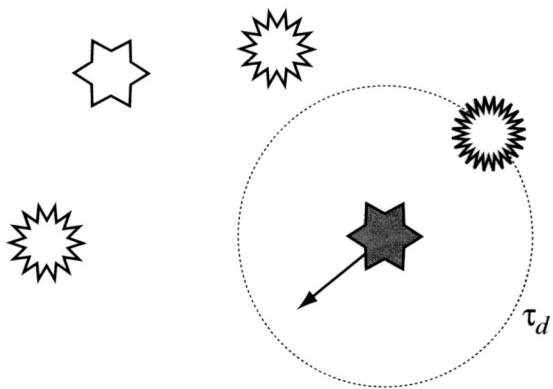

Fig. 10.8. Datoids try to move away from their nearest dissimilar neighbor.

For this, the difference between the current position $x_i^{(t)}$ and the nearest dissimilar neighbor $x_{n_{i,dissimilar}}^{(t)}$ is computed in a way that it points away from the nearest dissimilar neighbor (see Eq. (10.16)).

$$a = \begin{cases} x_i^{(t)} - x_{n_{i,dissimilar}}^{(t)} &, d(x_i, x_{n_{i,dissimilar}}) < \tau_d \\ 0 &, otherwise \end{cases}$$

$$\Delta_{avoidance} = (1 - S(o_i, o_{n_{i,similar}})) \, w_A \, a \qquad (10.16)$$

The avoidance vector a is weighted with w_A and set to a dependency to the influence of similar neighbors. That is, if a datoid has very similar neighbors, the avoidance to the dissimilar neighbor is reduced by $(1 - S(o_i, o_{n_{i,similar}}))$. On the other hand, if a datoid has few similar neighbors, the effect of avoidance is stronger.

After determining the velocity matching Δ_{velo}, the influence of the k nearest similar neigbors $\Delta_{neighbors}$ and the avoidance $\Delta_{avoidance}$ to dissimilar neighbors, the new velocity $v_i^{(t+1)}$ as in Eq. (10.17) can be computed.

$$v_i^{(t+1)} = w_I^{(t)} v_i^{(t)} + \Delta_{velo} + \Delta_{neighbors} + \Delta_{avoidance} \qquad (10.17)$$

The inertia weight $w_I^{(t)}$ determines the influence of the previous velocity. As in PSO this weight is decreased over time (see section 10.4). This decreasing procedure starts at iteration number s_I ($1 \leq s_I \leq N_{iter}$). Every iteration the inertia weight is decreased by an amount of a_I. Decreasing w_I results in more compact and dense sub-swarms which is an advantage in the latter cluster retrieval.

After computing the new velocity, the new position $x_i^{(t+1)}$ is computed as in Eq. (10.2).

Afterwards, the new velocity $v_i^{(t+1)}$, is clamped to $[-V_{max}, +V_{max}]$ by Eq. (10.18).

$$\forall c \in \{1,2\} : v_{ic}^{(t+1)} = \max[-V_{max}, \min(+V_{max}, v_{ic}^{(t+1)})] \qquad (10.18)$$

If the new position $x_i^{(t+1)}$ of the datoid lies outside the restricted plane, the datoid bounces against the border.

In the following the pseudo-code of the DSC iteration is shown:

---

for $i = 1$ **to** N_{dat}

    // Move datoid $D(i)$
    $n_{i,similar}$ ← index of nearest similar neighbor by Eq. (10.10)
    $n_{i,dissimilar}$ ← index of nearest dissimilar neighbor by Eq. (10.12)
    $c_{i,similar}$ ← center of k nearest similar neighbors by Eq. (10.13)

    Compute new velocity $v_i^{(t+1)}$ by Eq. (10.17)
    Compute new position $x_i^{(t+1)}$ by Eq. (10.2)

    // Ensure validity of datoid
    for $c = 1$ **to** 2
        // Clamp velocity
        $v_{ic}^{(t+1)} \leftarrow \max[-V_{max}, \min(+V_{max}, v_{ic}^{(t+1)})]$

        // Bounce against the borders
        if $x_{ic}^{(t+1)} < 0$
            $x_{ic}^{(t+1)} \leftarrow x_{ic}^{(t+1)} + 2(-x_{ic}^{(t+1)})$
            $v_{ic}^{(t+1)} \leftarrow -v_{ic}^{(t+1)}$
        else if $x_{ic}^{(t+1)} > X_{max}$
            $x_{ic}^{(t+1)} \leftarrow x_{ic}^{(t+1)} - 2(x_{ic}^{(t+1)} - X_{max})$
            $v_{ic}^{(t+1)} \leftarrow -v_{ic}^{(t+1)}$
        end if
    end for c
end for i

// Update parameters
if numIterations $\geq s_I \land w_I > 0.1$
    $w_I \leftarrow w_I - a_I$
end if
numIterations ← numIterations + 1

The variable numIterations represents the number of iterations performed so far and is initialized to 0 before calling the first iteration.

10.5.3 Cluster Retrieval

After a certain number of iterations, sub-swarms of similar datoids have formed. These sub-swarms represent the clusters. Therefore, the datoids of a given sub-swarm need to be grouped together as a cluster.

To realize this, a sort of an agglomerative clustering algorithm applied to the positions of the datoids on the two-dimensional plane is used.

All datoids $D(i), D(j) \in \mathcal{D}$ ($\forall i, j \in \{1, \cdots, N_{dat}\}, i \neq j$) whose Euclidean distance $d(x_i, x_j) \leq \tau_c$ is lower than a given threshold τ_c belong to the same cluster.

10.6 Experimental Setup

To evaluate the cluster capabilities of DSC it was tested on four datasets: two synthetical and two real life datasets. These datasets as well as the used parameterization of DSC are described in the following sections.

10.6.1 Synthetical Datasets

The synthetically generated datasets used are:

Corners
The dataset *Corners* contains 4 randomly created clusters in 200 records located at the 4 corners of a quadratic grid as presented in Figure 10.9. All clusters are separable by lines on the grid, i.e., in the attribute space.

The $N_{dat} = 200$ records are divided by four to create four corners of similar size. If $Int(x)$ denotes the integer part of x then the number n of records per class is computed as $n = Int(0.25 \cdot N_{dat})$.

Let X_{max} be the length of a side of the quadratic grid. Then, the side length of a single quadratic corner is computed as $s_{corner} = 0.4 \cdot X_{max}$.

The four corners can now be defined as relations:

Top Left is $TL = \{0, \cdots, s_{corner}\} \times \{0, \cdots, s_{corner}\}$
Top Right is $TR = \{X_{max} - s_{corner}, \cdots, X_{max} - 1\} \times \{0, \cdots, s_{corner}\}$
Bottom Left is $BL = \{0, \cdots, s_{corner}\} \times \{X_{max} - s_{corner}, \cdots, X_{max} - 1\}$
Bottom Right is $BR = \{X_{max} - s_{corner}, \cdots, X_{max} - 1\} \times \{X_{max} - s_{corner}, \cdots, X_{max} - 1\}$

The four clusters are created as follows:

Cluster 0 (top left): Randomly create n points $(x(i), y(i)) \in TL$.
Cluster 1 (top right): Randomly create n points $(x(i), y(i)) \in TR$.
Cluster 2 (bottom left): Randomly create n points $(x(i), y(i)) \in BL$.
Cluster 3 (bottom right): Randomly create $N_{dat} - 3n$ points $(x(i), y(i)) \in BR$.

Nested

As shown in Figure 10.10 the dataset *Nested* contains 2 randomly created clusters in 200 records, whereby one cluster is located at the center area of a quadratic grid and the other surrounds it. The clusters are not separable by lines on the grid, i.e., in the attribute space.

The $N_{dat} = 200$ records are divided into five sets for the four border areas as well as the center area. The number n_{border} of records per border area is computed as $n_{border} = Int(0.2 \cdot N_{dat})$. The number n_{center} of records for the center area is computed as $n_{center} = N_{dat} - 4 \cdot n_{border}$.

Again, let X_{max} be the side length of the quadratic grid. Then, the margin m of the center area is computed as $m = 0.4 \cdot X_{max}$.

The five sets for the four border areas and the center area can be defined as relations:

Border Top is $BT = \{0, \cdots, X_{max} - 1\} \times \{0, \cdots, 0.5 \cdot m\}$
Border Bottom is $BB = \{0, \cdots, X_{max} - 1\} \times \{X_{max} - 0.5 \cdot m, \cdots, X_{max} - 1\}$
Border Left is $BL = \{0, \cdots, 0.5 \cdot m\} \times \{0, \cdots, X_{max} - 1\}$
Border Right is $BR = \{X_{max} - 0.5 \cdot m, \cdots, X_{max} - 1\} \times \{0, \cdots, X_{max} - 1\}$
Center Area is $CA = \{m, \cdots, X_{max} - m\} \times \{m, \cdots, X_{max} - m\}$

The two clusters are created as follows:

Cluster 0 (borders): Randomly create n_{border} points $(x(i), y(i)) \in BT$, n_{border} points $(x(i), y(i)) \in BB$, n_{border} points $(x(i), y(i)) \in BL$ and n_{border} points $(x(i), y(i)) \in BR$.
Cluster 1 (center): Randomly create n_{center} points $(x(i), y(i)) \in CA$.

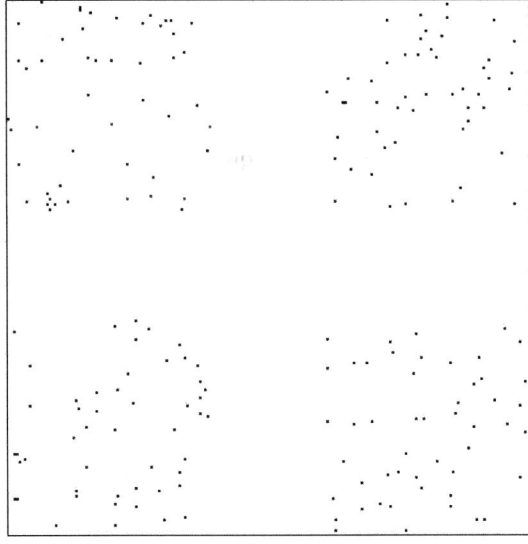

Fig. 10.9. Synthetical dataset *Corners*.

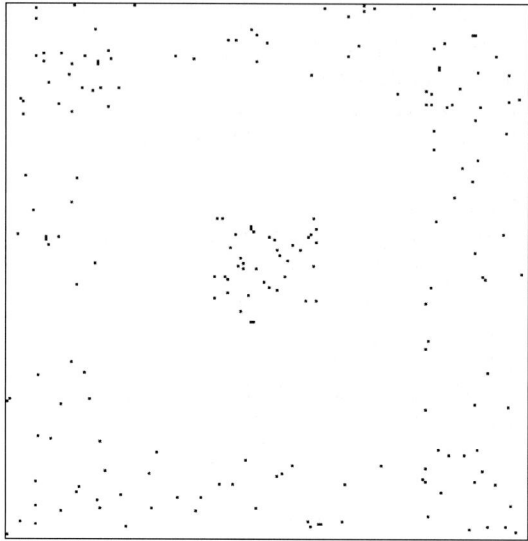

Fig. 10.10. Synthetical dataset *Nested*.

10.6.2 Real Life Datasets

The following real life datasets are used:

Iris: The dataset *Iris* contains 3 clusters in 150 records with 4 numerical attributes (sepal length, sepal width, petal length, petal width). Each of the 3 classes (Setosa, Versicolour, Virginica) contains 50 records.

WBC: The dataset Wisconsin Breast Cancer (*WBC*) contains 2 clusters of 2 classes (Benign, Malignant) in 683 records with 10 numerical attributes (Clump Thickness, Uniformity of Cell Size, Uniformity of Cell Shape, Marginal Adhesion, Single Epithelial Cell Size, Bare Nuclei, Bland Chromatin, Normal Nucleoli, Mitoses).

Both datasets are taken from the UCI Repository Of Machine Learning Databases [11] and the attribute values were normalized.

10.6.3 Parameters

In Table 10.1 the used parameter settings for DSC are shown. These parameters are determined by experimentation and have the following meaning:

N_{dat}	:= Number of datoids.
N_{iter}	:= Number of iterations.
X_{max}	:= Size of 2D plane.
V_{max}	:= Range of velocity.
k	:= Number of considered neighbors, i.e., size of neighborhood.
w_I	:= Start value of inertia weight.
s_I	:= Iteration number to start decreasing of inertia weight.
a_I	:= Amount of decreasing w_I.
w_V	:= Weight of velocity matching.
w_N	:= Weight of neighbors.
w_A	:= Weight of avoidance.
τ_d	:= Distance threshold to dissimilar datoids.
τ_c	:= Threshold for cluster retrieval.

	Corners	Nested	Iris	WBC
N_{dat}	200	200	150	683
N_{iter}	200	300	1500	100
X_{max}	400	400	400	400
V_{max}	10	10	10	10
k	20	20	15	68
w_I	1.0	1.0	1.0	1.0
s_I	0	0	500	0
a_I	0.001	0.001	0.001	0.01
w_V	0.5	0.5	0.5	0.5
w_N	0.5	2.0	0.5	0.5
w_A	0.5	0.5	0.5	0.5
τ_d	10	10	10	10
τ_c	5	10	5	10

Table 10.1. Used parameters for DSC.

The experiments showed that the parameters aren't too sensitive. In spite of little changes the clustering process works. Merely the numbers of correct clustered data items and of formed clusters change. Therefore, one can work towards a working parameter set for a given clustering problem.

10.7 Results

In section 10.6 the used datasets and their parameter settings are described. To evaluate the cluster capabilities, the DSC algorithm was applied to the datasets 50 times, i.e., 50 independent runs for each dataset. The results as shown in Table 10.2

are the averaged results over these 50 runs. The used measures are described in the following.

First, all correctly clustered data items are counted. For this it is neccessary to determine the cluster type of a cluster which means, to which of the real known classes of the dataset belongs the cluster. If C is the set of computed clusters $C_i \subseteq X$ and T the set of labels t of the real known classes of a dataset, then the class of cluster C_i is computed as shown in Eq. (10.19),

$$Class(C_i) = c \text{ s.t. } N_{ci} = \max_{t \in T}\{N_{ti}\}, \quad c \in T \tag{10.19}$$

where N_{ti} is the number of data items of class t within cluster C_i. The class of a cluster is the class of the biggest part of data items belonging to the same class. With this assumption the proportion of correctly clustered data items is just the number of data items which represent the class of the cluster summed over all clusters C_i as shown in Eq. (10.20).

$$Correct(C) = \frac{1}{|X|} \sum_{C_i \in C} \max_{t \in T}\{N_{ti}\} \tag{10.20}$$

$Correct(C)$ is to maximize. A second measure is just the number of found clusters $|C|$. This is important, because in DSC the number of clusters is not given by the user.

Another measure used is the entropy within a cluster as in Eq. (10.21),

$$Entropy(C_i) = -\frac{1}{\log(|X|)} \sum_{t \in T} \frac{N_{ti}}{N_i} \log(\frac{N_{ti}}{N_i}) \tag{10.21}$$

where N_i is the size of cluster C_i, i.e., $N_i = |C_i|$. The entropy measures the relative degree of randomness of cluster C_i. That is, it is 0 if the cluster contains data only from one class and 1 if the cluster is uniformly filled with data of all classes. The overall entropy $Entr(C)$ as given in Eq. (10.22) is the average over all clusters and is to minimize.

$$Entr(C) = \frac{1}{|C|} \sum_{C_i \in C} Entropy(C_i) \tag{10.22}$$

The last measure is the F-measure known from information retrieval as shown in Eq. (10.23). It uses the purity of the considered cluster C_i with $Prec(t, C_i) = \frac{N_{ti}}{N_i}$, i.e., how strong belongs cluster C_i completely to class t. Furthermore, it considers, how much of the data of class t are contained within cluster C_i with $Rec(t, C_i) = \frac{N_{ti}}{N_t}$ and N_t being the number of data in class t.

$$FMeasure(t, C_i) = \frac{2 \cdot Prec(t, C_i) \cdot Rec(t, C_i)}{Prec(t, C_i) + Rec(t, C_i)} \tag{10.23}$$

The best situation is to have each cluster consisting completely of data of the same class t ($Prec(t, C_i) = 1$) and for each class t having all data placed in just one cluster ($Rec(t, C_i) = 1$). This measure is limited to $[0, 1]$ and to be maximized. The overall F-measure value is determined as in Eq. (10.24).

$$FMeas(C) = \sum_{t \in T} \frac{N_t}{|X|} \max_{C_i \in C} \{FMeasure(t, C_i)\} \qquad (10.24)$$

All described measures are computed for each dataset and presented in Table 10.2. For each dataset, simply the Euclidian distance normalized by the maximal possible dissimilarity between two data items is used as similarity function as defined in Eq. (10.25).

$$S(d_i, d_j) = 1 - \frac{\sqrt{\sum_{c=1}^{n}(d_{i_c} - d_{j_c})^2}}{\sqrt{\sum_{v=1}^{n} \max(A_v)^2}} \qquad (10.25)$$

The synthetic dataset *Corners* is a very simple one having four separable classes. This works very well as expectable.

The synthetic dataset *Nested* is not so simple but can be solved very good by DSC. The reason is that the clustering doesn't occur in the attribute space, but on a plane where the data items are carried by datoids. The datoids interact according to their similarity and not only to their positions on the plane.

The real dataset *Iris* is not easy, because two attributes of two classes are strongly correlated. But the results of DSC are comparable with other clustering methods as can be seen in Table 10.3. There, DSC is compared to Ant-based clustering [4] and the well-known k-means algorithm. The comparative values are taken from [5]. On the one hand the determined number of clusters in average is a bit more stable in Ant-based clustering compared to DSC. In k-means the number of expected clusters is given *a priori*. Thus, this point is not comparable. On the other hand, the F-measure value of DSC is better than those of Ant-based clustering and k-means. That means, the computed clusters have a better purity.

The real dataset *WBC* is not solved well by DSC. An acceptable high number of similar data items are correctly clustered, but DSC produces too much clusters at all. This results in a bad F-measure value as revealed in Table 10.4, because the data items of a class are spread over several clusters. Ant-based clustering and k-means are better in clustering *WBC*.

	Corners	Nested	Iris	WBC
$Correct(C) \cdot 100$	100.0%	100.0%	94.786667%	94.585652%
standard deviation	0.0%	0.0%	3.496513%	5.754342%
$|C|$ (real number)	4 (4)	2 (2)	4 (3)	6 (2)
standard deviation	0	0	1.363818	1.462874
$Entr(C)$	0	0	0.026367	0.02042
standard deviation	0	0	0.006275	0.006602
$FMeas(C)$	1	1	0.830683	0.685732
standard deviation	0	0	0.073264	0.062420

Table 10.2. Averaged results over 50 independent runs for all datasets.

Iris	DSC	Ant-based Clustering	k-means
$\|C\|$	4	3.02	3 (given a priori)
standard deviation	1.36	0.14	0 (given a priori)
$FMeas(C)$	0.830683	0.816812	0.824521
standard deviation	0.073264	0.014846	0.084866

Table 10.3. Results for the dataset *Iris*. The comparative values are taken from [5].

WBC	DSC	Ant-based Clustering	k-means
$\|C\|$	6	2	2 (given a priori)
standard deviation	1.46	0	0 (given a priori)
$FMeas(C)$	0.685732	0.967604	0.965825
standard deviation	0.062420	0.001447	0

Table 10.4. Results for the dataset *WBC*. The comparative values are taken from [5].

The number of clusters and the F-measure values of the real datasets reveal one weak point of DSC. DSC sometimes produces several sub-swarms (i.e., clusters) which belong to the same class t while having a purity of 1 ($Prec(t,C_i) = Prec(t,C_j) = 1$, $C_i \neq C_j$). That is the data items of a class can be split up to several clusters and those clusters consist of data items of just this class.

A positive property of DSC is the transformation of the data items to the plane of the datoids. This is an advantage in problems like *Nested*.

The entropy of all datasets shows that each determined cluster is good dominated by data items of the same class. The clusters aren't mixed strongly. It seems that DSC produces clusters with a good purity.

10.8 Conclusion

The experimentations show that it is possible to cluster data by using swarm techniques. The power of swarm clustering is due to the local interaction between similar datoids. The data items to be clustered aren't clustered in their attribute space, but in the space of the datoids. Therefore, data items belonging to the same class don't need to be close in the attribute space. Datoids move in their space and have an affinity to their nearest similar neighbors. This allows the datoids to perform good on problems like the *Nested* dataset. The data items in the top region and bottom region of the *Nested* dataset aren't close in their space. But the datoids group together with their similar neighbors. The next near similar neighbors of the data items in the bottom region are the ones on both sides. And the next near similar neighbors of the side regions are the ones in the top region. Because of this behavior based on local interaction between similar datoids the data items of the four sides can be separated from the nested data items.

DSC uses a similarity function S to determine the similarity between two datoids. Thus, it can work with each data structure or attribute type, because DSC only gives the data objects carried by datoids to this similarity function. Therefore, a lot of properties of DSC depend on the used similarity function. One disadvantage of DSC is the great number of needed parameters. On the other hand, you don't need to specify the number of clusters *a priori*.

References

1. Berkhin P (2002) Survey of clustering data mining techniques. Technical report, Accrue Software, San Jose, California
2. Bonabeau E, Dorigo M, Theraulaz G (1999) Swarm Intelligence: From Natural to Artificial Systems. Oxford University Press, New York, NY
3. Deneubourg JL, Goss S, Franks N, Sendova-Franks A, Detrain C, Chretien L (1991) The Dynamics of Collective Sorting: Robot-like Ants and Ant-like Robots. In: Proc. First International Conference on Simulation of Adaptive Behaviour: From Animals to Animats, pp. 356-363, MIT Press, Cambridge, MA
4. Handl J, Knowles J, Dorigo M (2003) Ant-based clustering: a comparative study of its relative performance with respect to k-means, average link and 1D-som. Technical Report TR/IRIDIA/2003-24. IRIDIA, Universite Libre de Bruxelles, Belgium
5. Handl J, Knowles J, Dorigo M (2003) On the performance of ant-based clustering. In: Proc. 3nd International Conference on Hybrid Intelligent Systems, pp. 204-213, IOS Press, Amsterdam, The Netherlands
6. Kanade PM, Hall LO (2003) Fuzzy Ants as a Clustering Concept. In: Proc. 22nd International Conference of the North American Fuzzy Information Processing Society, pp. 227-232, Chicago, Piscataway, NJ: IEEE Service Center
7. Kennedy J, Eberhart RC (1995) Particle Swarm Optimization. In: Proc. IEEE International Conference on Neural Networks, pp. 1942-1948, Perth, Australia, IEEE Service Center, Piscataway, NJ
8. Kennedy J, Eberhart RC, Shi Y (2001) Swarm Intelligence. Morgan Kaufmann Publishers, San Francisco, ISBN: 1-55860-595-9
9. Monmarche N, Slimane M, Venturini G (1999) AntClass: discovery of clusters in numeric data by an hybridization of an ant colony with the kmeans algorithm. Internal Report No. 213, E3i, Laboratoire d'Informatique, Universite de Tours
10. Morse DH (1970) Ecological aspects of some mixed-species foraging flocks of birds. Ecological Monographs: Vol. 40, No. 1, pp. 119-168
11. Murphy PM, Aha DW (1994) UCI Repository of machine learning databases. [http://www.ics.uci.edu/~mlearn/MLRepository.html], Irvine, CA: University of California, Department of Information and Computer Science
12. Omran M, Salman A, Engelbrecht AP (2002) Image Classification using Particle Swarm Optimization. In: Proc. 4th Asia-Pacific Conference on Simulated Evolution and Learning, pp. 370-374, Singapore
13. Reynolds CW (1987) Flocks, herds and schools: a disctributed behavioral model. Computer Graphics 21, pp. 25-33
14. Shi YH, Eberhart RC (1998) A Modified Particle Swarm Optimizer. In: Proc. IEEE International Conference on Evolutionary Computation, pp. 69-73, IEEE Press, Piscataway, NJ
15. van der Merwe DW, Engelbrecht AP (2003) Data clustering using particle swarm optimization. In: Proceedings of the 2003 IEEE Congress on Evolutionary Computation, pp. 215-220, Piscataway, NJ: IEEE Service Center

11

Clustering Ensemble Using ANT and ART

Yan Yang[1], Mohamed Kamel[2], and Fan Jin[1]

[1] School of Information Science and Technology, Southwest Jiaotong University, Chengdu, Sichuan, 610031, China `yyang@home.swjtu.edu.cn`
[2] Pattern Analysis and Machine Intelligence Lab, Electrical and Computer Engineering, University of Waterloo, Waterloo, Ontario N2L 3G1, Canada `mkamel@uwaterloo.ca`

Summary. This chapter presents a clustering ensemble model using ant colony algorithm with validity index and ART neural network. Clusterings are visually formed on the plane by ants walking, picking up or dropping down projected data objects with different probability. The clustering validity index is used to evaluate the performance of algorithm, find the best number of clusters and reduce outliers. ART is employed to combine the clusterings produced by ant colonies with different moving speed. Experiments on artificial and real data sets show that the proposed model has better performance than that of single ant colony clustering algorithm with validity index, the ART algorithm, and the LF algorithm.

11.1 Introduction

Ant colony is a kind of social insects, which is capable of selforganization, pheromone communication, distribution, flexibility, and robustness. Researchers have designed a number of successful algorithms such as Ant Colony Optimization and Ant Colony Routing in diverse application fields such as combinatorial optimization, communications networks, and robotics [4]. The ant colony clustering algorithm is inspired by the behavior of ant colonies in clustering their corpses and sorting their larvae. Deneubourg et al. [9] proposed a basic model to explain the clustering behavior. In this model, artificial ants are allowed to randomly move, pick up and drop objects according to the number of similar surrounding objects so as to cluster them. Lumer and Faieta [21] expanded Deneubourg's model to the LF algorithm that is based on a local similarity density in order to make it suitable for data clustering. Ramos and Merelo [24] studied ant-clustering systems with different ant speeds for textual document clustering. Handl and Meyer [15] used inhomogeneous ant populations with "jumps" for document retrieval. Monmarche [22] described an AntClass algorithm in which several items are allowed to be on the same cell corresponding to a cluster. The AntClass algorithm uses stochastic principles of ant colony in conjunction with the deterministic principles of the K-means algorithm. In a similar way, in [30] CSIM algorithm combined CSI model

(Clustering based on Swarm Intelligence) and K-means algorithm. An ant-based clustering algorithm is aggregated with the fuzzy c-means algorithm in [20].

As the data to be clustered is usually unlabelled, measures that are commonly used for document classification such as the F-measure cannot be used here. Instead we need to use measures that reflect the goodness of the clustering. Cluster validity indices have been proposed in the literature to address this point [26]. Several clustering methods use validity index to find the best number of clusters [14,31,34]. Halkidi et al. [13] proposed multi representative clustering validity index that is suitable for non-spherical cluster shapes.

ART (Adaptive Resonance Theory) neural networks were developed by Grossberg [12] to address the problem of stability-plasticity dilemma. The ART network self-organizes in response to input patterns to form a stable recognition cluster. Models of unsupervised learning include ART1 [5] for binary input patterns, ART2 [6] and ART-2A [7] for analog input patterns, and fuzzy ART [8] for "fuzzy binary" inputs, i.e. analog numbers between 0 and 1. Many variations of the basic unsupervised networks have been adapted for clustering. Tomida et al. [27] applied fuzzy ART as a clustering method for analyzing the time series expression data during sporulation of Saccharomyces cerevisiae. He et al. [17] used fuzzy ART to extract document cluster knowledge from the Web Citation Database to support the retrieval of Web publications. Hussin and Kamel [18] proposed a neural network based document clustering method by using a hierarchically organized network built up from independent SOM (Self-Organizing Map) and ART neural networks.

Clustering ensembles have emerged as a powerful method for improving the quality and robustness of the clusterings. However, finding a combination of multiple clusterings is a more difficult and challenging task than combination of supervised classifications. Without the labeled pattern, there is no explicit correspondence between cluster labels in different partitions of an ensemble. Another intractable label correspondence problem results from different partitions containing different numbers of clusters. Recently a number of approaches have been applied to the combination of clusterings, namely the consensus function, which creates the combined clustering [29]. A co-association matrix was introduced for finding a combined clustering in [11]. Co-association values represent the strength of association between objects appearing in the same cluster. The combined clustering comes from the co-association matrix by applying a voting-type algorithm. Strehl and Ghosh [25] represented the clusters as hyperedges on a graph whose vertices correspond to the objects to be clustered, and developed three hypergraph algorithms: CSPA, HGPA, and MCLA for finding consensus clustering. Topchy et al. [28] proposed new consensus function based on mutual information approach. They employed combination of so-called weak clustering algorithm related to intra-class variance criteria. Recent approaches to combine cluster ensembles based on graph and information theoretic methods appear in [1] and [2]. An approach based on re-labeling each bootstrap partition using a single reference partition is presented in [10].

Neural network ensemble is a learning paradigm where several neural networks are jointly used to solve a problem. In [32], multistage ensemble neural network model was used to combine classifier ensemble results. Ensemble of SOM neural networks has been also used for image segmentation where the pixels in an image are clustered according to color and spatial features with different SOM neural networks, and the clustering results are combined as the image segmentation [19].

In this chapter, an ensemble model, i.e. combination of ant colony clustering algorithms with validity index (called ACC-VI) and ART network, is applied to clustering. Clusterings are visually formed on the plane by ants walking, picking up or dropping down projected data objects with different probability. The clustering validity index is used to evaluate the performance of the algorithm, find the best number of clusters and reduce outliers. ART network is employed to combine the clusterings. Experiments on artificial and real data sets show that the proposed model has better performance than that of the individual ACC-VI algorithm, the ART-2A algorithm, and the LF algorithm.

The rest of the chapter is organized as follows. Section 2 introduces the ACC-VI algorithm. Section 3 describes the ART algorithm. Section 4 presents the clustering ensemble model. Section 5 reports the results of the experiments conducted to evaluate the performance of the proposed model. Finally, Section 6 offers a conclusion of the chapter.

11.2 Ant Colony Clustering Algorithm with Validity Index (ACC-VI)

11.2.1 Ant Colony Clustering Algorithm

The ant colony clustering algorithm is based on the basic LF model and its added feature proposed by Lumer and Faieta [21] and the ant-based clustering algorithm by Yang and Kamel [33]. First, data objects are randomly projected onto a plane. Second, each ant chooses the object at random, and picks up or moves or drops down the object according to picking-up or dropping probability with respect to the similarity of current object within the local region by probability conversion function. Finally, clusters are collected from the plane.

Let us assume that an ant is located at site γ at time t, and finds an object o_i at that site. The local density of objects similar to type o_i at the site γ is given by

$$f(o_i) = \max\{0, \frac{1}{s^2} \sum_{o_j \varepsilon Neigh_{s \times s}(\gamma)} [1 - \frac{d(o_i, o_j)}{\alpha(1 + ((v-1)/v_{max}))}]\} \quad (11.1)$$

where $f(o_i)$ is a measure of the average similarity of object o_i with the other objects o_j present in its neighborhood. $Neigh_{s \times s}(\gamma)$ denotes the local region. It is usually a square of $s \times s$ sites surrounding site γ. $d(o_i, o_j)$ is the distance between two objects o_i and o_j in the space of attributes. The Cosine distance is computed as

$$d(o_i, o_j) = 1 - sim(o_i, o_j) \quad (11.2)$$

where $sim(o_i,o_j)$ reflects the similarity metric between two objects. It measures the cosine of the angle between two objects (their dot product divided by their magnitudes)

$$sim(o_i,o_j) = \frac{\sum_{k=1}^{q}(o_{ik} \cdot o_{jk})}{\sqrt{\sum_{k=1}^{q}(o_{ik})^2 \cdot \sum_{k=1}^{q}(o_{jk})^2}} \quad (11.3)$$

where q is the number of attributes. As the objects become more similar, the Cosine similarity $sim(o_i,o_j)$ approaches 1 and their Cosine distance approaches 0.

As shown in formula (1), α is a factor that defines the scale of similarity between objects. Too Large values of α will result in making the similarity between the objects larger and forces objects to lay in the same clusters. When α is too small, the similarity will decrease and may in the extreme result in too many separate clusters. On the other hand, parameter α adjusts the cluster number and the speed of convergence. The bigger α is, the smaller the cluster number and the faster the algorithm converges.

In formula (1), the parameter v denotes the speed of the ants. Fast mov-ing ants form clusters roughly on large scales, while slow ants group objects at smaller scales by placing objects with more accuracy. Three types of speed in different ant colonies are considered [33].

— *v is a constant.* All ants move with the same speed at any time;
— *v is random.* The speed of each ant is distributed randomly in $[1, v_{max}]$, where v_{max} is the maximum speed;
— *v is randomly decreasing.* The speed term starts with large value (form-ing clusters), and then the value of the speed gradually decreases in a random manner (helping ants to cluster more accurately).

The picking-up and dropping probabilities both are a function of $f(o_i)$ that converts the average similarity of a data object into the probability of picking-up or dropping for an ant. The converted approaches are based on: the smaller the similarity of a data object is (i.e. there aren't many objects that belong to the same cluster in its neighborhood), the higher the picking-up probability is and the lower the dropping probability is; on the other hand, the larger the similarity is, the lower the picking-up probability is (i.e. objects are unlikely to be removed from dense clusters) and the higher the dropping probability is. The sigmoid function is used as probability conversion function in our algorithm [33]. Only one parameter needs to be adjusted in the calculation.

The picking-up probability P_p for a randomly moving ant that is currently not carrying an object to pick up an object is given by

$$P_p = 1 - sigmoid(f(o_i)) \quad (11.4)$$

The dropping probability P_d for a randomly moving loaded ant to deposit an object is given by

$$P_d = sigmoid(f(o_i)) \quad (11.5)$$

where

$$\text{sigmoid}(x) = \frac{1}{1+e^{-\beta x}} \quad (11.6)$$

and β is a constant that can speed up the algorithm convergence if it is in-creased. Selecting larger values for β can help ants to drop faster the outliers at the later stages of algorithm [33].

11.2.2 Clustering Validity Index

Halkidi et al. [13] proposed multi representative clustering validity index that is based on cluster compactness and cluster separation. A clustering of data set into c clusters can be represented as $D = \{U_1, U_2, ..., U_c\}$, where $U_i = \{u_{i1}, u_{i2}, ...u_{ir_i}\}$ is the set of representative points of cluster i, r_i is the number of representative point of the ith cluster.

The standard deviation of the ith cluster is defined as [13]

$$stdev(U_i) = \sqrt{\frac{1}{n_i - 1} \sum_{k=1}^{n_i} d^2(x_k, m_i)} \quad (11.7)$$

where n_i is the number of data in the ith cluster, d is the distance between x_k and m_j. x_i is the data belonging to the ith cluster, and m_i is the mean of the ith cluster.

Intra-cluster density is defined as the average density within clusters, that is, the number of points that belong to the neighborhood of representative points of the clusters [13]. A bigger Intra-cluster density value indicates a more compacted cluster. It is defined by

$$Intra\_den(c) = \frac{1}{c} \sum_{i=1}^{c} \frac{1}{r_i} \sum_{j=1}^{r_i} \frac{density(u_{ij})}{stdev(U_i)}, c > 1 \quad (11.8)$$

The term $density(u_{ij})$ is defined by

$$density(u_{ij}) = \sum_{l=1}^{n_i} f(x_l, u_{ij}) \quad (11.9)$$

where x_l belongs to the ith cluster, u_{ij} is the jth representative point of ith cluster, n_i is the number of the ith cluster, and $f(x_l, u_{ij})$ is defined by

$$f(x_l, u_{ij}) = \begin{cases} 1 & , \quad d(x_l, u_{ij}) \leq stdev(U_i) \\ 0 & , \quad otherwise \end{cases} \quad (11.10)$$

Inter-cluster density is defined as the density between clusters [13]. For well-separated clusters, it will be significantly low. It is defined by

$$Inter\_den(c) = \sum_{\substack{i=1 \\ }}^{c} \sum_{\substack{j=1 \\ j \neq i}}^{c} \frac{d(close\_rep(i), close\_rep(j))}{stdev(U_i) + stdev(U_j)} density(z_{ij}), c > 1 \quad (11.11)$$

where $close\_rep(i)$ and $close\_rep(j)$ are the closest pair of representatives of the ith and jth clusters, z_{ij} is the middle point between the pair points $close\_rep(i)$ and $close\_rep(j)$. The term $density(z_{ij})$ is defined by

$$density(z_{ij}) = \frac{1}{n_i+n_j} \sum_{l=1}^{n_i+n_j} f(x_l, z_{ij}) \quad (11.12)$$

where x_l belongs to the ith and jth clusters, n_i is the number of the ith cluster, n_j is the number of the jth cluster, and $f(x_l, z_{ij})$ is defined by

$$f(x_l, z_{ij}) = \begin{cases} 1, & d(x_l, z_{ij}) \leq (stdev(U_i) + stdev(U_j))/2 \\ 0, & otherwise \end{cases} \quad (11.13)$$

Clusters' separation measures separation of clusters. It contains the distances between the closest clusters and the Inter-cluster density, and is defined as follows [13]

$$Sep(c) = \sum_{i=1}^{c} \sum_{\substack{j=1 \\ j \neq i}}^{c} \frac{d(close\_rep(i), close\_rep(j))}{1 + Inter\_den(c)}, c > 1 \quad (11.14)$$

Then the validity index $CDbw$, which is called "Composing Density Between and Within clusters" [13], is defined as

$$CDbw(c) = Intra\_den(c) \cdot Sep(c), c > 1. \quad (11.15)$$

11.2.3 ACC-VI Algorithm

A good clustering algorithm produces partitions of the data such that the Intra-cluster density is significantly high, the Inter-cluster density and Clusters' separation are significantly low, and the validity index $CDbw$ has a maximum, which corresponds to natural number of clusters. In [13], experiments showed that $CDbw$ can be used to find the optimal number of clusters at the maximum value. So we use $CDbw$ not only to evaluate the clustering algorithm, but also to find the best number of clusters.

In the ant colony clustering algorithm, the outliers with dissimilarity to all other neighborhood are dropped alone. The local clustering validity index is taken into account to reduce outliers in our algorithm. The process is described below. First, try to drop each outlier into each cluster, recalculate the new local $CDbw$, and compare to the old value for each cluster. Then, move the outlier to the cluster at the highest difference.

A pseudo code of ACC-VI algorithm is listed in Table 1. Essentially, the algorithm works as follows. Firstly, the data objects are projected onto a plane, that is, a pair of coordinates is given to each object randomly. Each ant is marked as unloaded and chooses an object at random initially. Secondly, the similarity $f(o_i)$ for each ant walking randomly is computed by formula (1). In the first case, each ant is unloaded, that is ants are not holding any objects. The picking-up probability P_p is calculated by formula (4). If P_p is greater than a random probability and an object is not picked up by the other ants simultaneously, the ant picks up this object, moves it to a new position, and marks itself as loaded. On the other hand, if P_p is less than a random probability, the ant does not pick up this object and re-selects another object randomly. In the second case, the ant is loaded, i.e. holding an object. The dropping probability P_d is calculated by formula (5). If P_d is greater than a random probability, the ant drops the object, marks itself as unloaded, and re-selects a new object randomly. Otherwise, the ant continues moving the object to a new position. The third step is to collect the clustering results on the plane. Whether crowded or isolated for an object can be determined by the number of its neighbor. If an object is isolated, that is the number of its neighbor is less than a given constant, the object is labeled as an outlier. On the other hand, if the object is in a crowd, that is the number of its neighbor is more than the given constant, it is given a labeling number denoting a cluster and is given same number recursively to those objects who are the neighbors of this object within a local region. At the fourth step, the validity index *CDbw* is calculated so as to find the optimal number of clusters. Finally, try to drop outlier at the cluster with the highest *CDbw* difference.

11.3 ART Algorithm

ART (Adaptive Resonance Theory) models are neural networks that develop stable recognition codes by self-organization in response to arbitrary sequences of input patterns. They are capable of solving well-known dilemma, stability-plasticity. How can a learning system be designed to remain plastic or adaptive in response to significant events and yet remain stable in response to irrelevant events? That means new clusters can be formed when the environment does not match any of the stored pattern, but the environment cannot change stored pattern.

A typical ART network consists of three layers: input layer (F0), com-parison layer (F1) and competitive layer (F2) with N, N and M neurons, respectively (see Fig. 11.1). The input layer F0 receives and stores the input patterns. Neurons in the input layer F0 and comparison layer F1 are one-to-one connected. F1 combines input signals from F0 and F2 layer to measure similarity between an input signal and the weight vector for the specific cluster unit. The competitive layer F2 stores the prototypes of input clusters. The cluster unit with the largest activation becomes the candidate to learn the input pattern (winner-take-all). There are two sets of connections, top-down and bottom-up, between each unit in F1 and each cluster unit in F2. Interactions between F1 and F2 are controlled by the orienting subsystem using a vigilance threshold ρ. The learning process of the network can be described as follows (refer to ART1 [3, 5]).

For a non-zero binary input pattern x (x_j {0, 1}, $j=1, 2, \ldots, N$), the network attempts to classify it into one of its existing clusters based on its similarity to the stored prototype of each cluster node. More precisely, for each node i in the F2 layer, the bottom-up activation T_i is calculated, which can be expressed as

$$T_i = \frac{|w_i \cap x|}{\mu + |w_i|} \qquad i = 1, \ldots, M \qquad (11.16)$$

Table 11.1. Algorithm 1

ACC-VI algorithm

Step 0. Initialize the number of ants: *ant_number*, maximum number of iteration:*Mn*, side length of local region: *s*, maximum speed of ants moving: v_{max} , and other parameters: α, β.

Step 1. Project the data objects on a plane, i.e. give a pair of coordinate (x, y) to each object randomly. Each ant that is currently unloaded chooses an object at random.

Step 2. For $i = 1, 2, \cdots, Mn$
for $j = 1, 2, \cdots,$ *ant_ number*
  2.1 Compute the similarity of an object within a local region by formula (1), where *v* is chosen as three kinds of speed : constant, random, and randomly decreasing for different colony;
  2.2 If the ant is unloaded, compute picking-up probability P_p by formula (4). If P_p is greater than a random probability, and this object is not picked up by the other ants simultaneously, then the ant picks up the object, labels itself as loaded, and moves the object to a new position; else the ant does not pick up this, object and reselect another object randomly;
  2.3 If the ant is loaded, compute dropping probability P_d by formula (5). If P_d is greater than a random probability, then the ant drops the object, labels itself as unloaded, and reselects a new object randomly; else the ant continues moving the object to a new position.

Step 3. For $i = 1, 2, ... , N$ // for all data objects
  3.1 If an object is isolated, or the number of its neighbor is less than a given constant, then label it as an outlier;
  3.2 Else give this object a cluster sequence number, and recursively label the same sequence number to those objects who is the neighbors of this object within local region, then obtain the number of clusters *c*.

Step 4. For $i = 1, 2, ... , c$// for *c* clusters
  4.1 Compute the mean of the cluster, and find four representative points by scanning the cluster on the plane from different direction of *x*-axis and *y*-axis;
  4.2 Compute the validity index *CDbw* by formula (15) as the foundation in finding the optimal number of clusters.

Step 5. For $i = 1, 2, \cdots, c$//for *c* clusters
  5.1 Try to drop outlier into cluster, recalculate the new *CDbw*, and compare to the old value for each cluster;
  5.2 Move the outlier to the cluster with the highest difference.

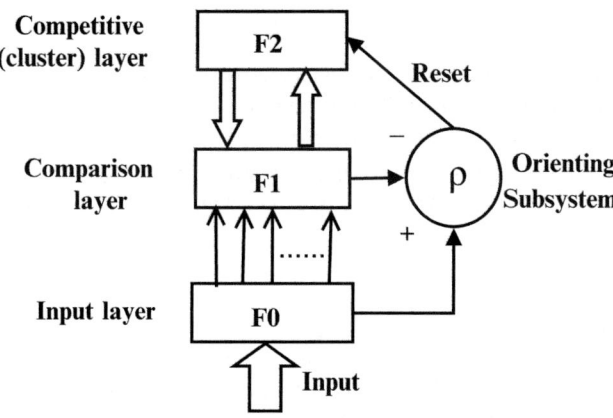

Fig. 11.1. The ART network architecture

where $|.|$ is the norm operator ($|x| = \sum_j^N x_j$), w_i is the binary weight vector of cluster i, in which case the bottom-up and top-down weights are identical for simplicity [5], and $\mu > 0$ is the choice parameter. Then the $F2$ node I that has the highest bottom-up activation, i.e. $T_I = \max\{T_i | i = 1, ..., M\}$, is selected winner-take-all. The weight vector of the winning node (w_I) will then be compared to the current input at the comparison layer. If they are similar enough, i.e. they satisfy the

$$\frac{|w_I \cap x|}{|x|} \geq \rho \qquad i = 1..., M \qquad (11.17)$$

matching condition, where ρ is a system parameter called vigilance ($0 < \rho \leq 1$), $F2$ node I will capture the current input and the network learns by modifying w_I:

$$w_I^{new} = \eta(w_I^{old} \cap x) + (1-\eta)w_I^{old} \qquad (11.18)$$

where η is the learning rate ($0 < \eta \leq 1$). All other weights in the network remain unchanged.

If, however, the stored prototype w_I does not match the input sufficiently, i.e. formula (11.17) is not met, the winning F2 node will be reset (by activating the reset signal in Fig. 11.1) for the period of presentation of the current input. Then another F2 node (or cluster) is selected with the highest T_i, whose prototype will be matched against the input, and so on. This "hypothesis-testing" cycle is repeated until the network either finds a stored cluster whose prototype matches the input well enough, or inserts the input prototype into F2 as a new reference cluster. Insertion of a new cluster is normally done by creating an all-ones new node in F2 as the winning node w_I and temporarily set the learning rate to 1.0, then learning takes place according to formula (18). It is important to note that once a cluster is found, the comparison layer F1 holds $|w_I \cap x|$ until the current input is removed.

The number of clusters can be controlled by setting ρ. The higher vigilance value ρ, the lager number of more specific clusters will be created. At the extreme, $\rho = 1$, the network will create a new cluster for every unique input.

ART is a family of different neural architectures. Except ART1 basic architecture, ART2 [6] is a class of architectures categorizing arbitrary sequences of analog input patterns. ART-2A [7] simplifies the learning process by using the dot product as similarity measure. A pseudo code of the ART-2A learning process is summarized in Table 11.2 [16].

Table 11.2. Algorithm 2

ART-2A learning process	
Step 0.	Initialize vigilance parameter $\rho(0 < \rho \leq 1)$; learning rate $\eta(0 < \eta \leq 1)$.
Step 1.	While stopping condition is false, do Steps 2-10.
Step 2.	For each training input do Steps 3-9.
Step 3.	Set activations of all F2 units to zero. Set activations of all F0 units to normalization input vector: $X = \Re x$, where $\Re x = \frac{x}{\|x\|} = \frac{x}{\sqrt{\sum_{i=1}^{N} x_i^2}}$
Step 4.	Send input signal from F0 to F1 layer.
Step 5.	For each F2 node that is not inhibited, calculate the bottom-up activation T_i If $T_i \neq -1$, then $T_i = X \cdot w_i, i = 1,...,M$.
Step 6.	While reset is true, do Steps 7-8.
Step 7.	Find I such that $T_I = max\{T_i | i = 1,...,M\}$ for all F2 nodes i. If $T_I = -1$, then all nodes are inhibited (this pattern cannot be clustered).
Step 8.	Test for reset: If $T_I < \rho$, then $T_I = -1$ (inhibit node I) and go to Step 6. If $T_I \geq \rho$, then proceed to Step 9.
Step 9.	Update the weights for node I: $w_I^{new} = \Re(\eta X + (1-\eta)w_I^{old})$.
Step 10.	Test for stopping condition.

11.4 Clustering Ensemble Model

11.4.1 Consensus Functions

Suppose we are given a set of N data points $X = \{x_1,....,x_N\}$ and a set of H partitions $\Pi = \{\pi_1, \pi_2,....,\pi_H\}$ of objects in X. Different partitions of X return a set of labels for each point $x_i, i = 1,....,N$ [29]

$$x_i \rightarrow \{\pi_1(x_i), \pi_2(x_i),....,\pi_H(x_i)\} \qquad (11.19)$$

where H indicates different clusterings and $\pi_j(x_i)$ denotes a label assigned to x_i by the j-th algorithm. A consensus function maps a set of partitions $\Pi = \{\pi_1, \pi_2,....,\pi_H\}$ to a target partition λ. Generally, there are four types of consensus functions:

- *Co-association matrix.* The consensus function operates on the coassociation matrix. A voting-type algorithm could be applied to the coassociation matrix to obtain the final clustering.
- *Hypergraph approach.* The clusters in the ensemble partitions could be represented as hyperedges on a graph with N vertices. Each hyperedge denotes a set of data points belonging to the same cluster. The problem of consensus clustering is then become to finding the minimum-cut of a hypergraph. Three hypergraph algorithms for ensemble clustering:CSPA, HGPA, and MCLA are presented in [25].
- *Mutual information algorithm.* The consensus function could be formulated as the mutual information between the empirical probability distribution of labels in the consensus partition and the labels in the ensemble.
- *Re − labeling approach.* All the partitions in the ensemble can be relabeled according to their best agreement with some chosen reference partition [29].

These existing consensus functions are complex and rely on uncertain statistical properties in finding consensus solutions. Neural network as an ensemble combiner is another method that motivates our study of ART ensemble aggregation. The next section introduces that model.

11.4.2 ART Ensemble Aggregation Model

Aggregation of ensemble of multiple clusterings can be viewed as a cluster-ing task itself. Fig. 2 shows an architecture diagram of ART ensemble model. In the first phase, three clustering components generate clustering result using ant colony algorithms with different moving speed such as constant, random, and randomly decreasing respectively. Each clustering in the combination is represented as a set of labels assigned by the clustering algorithm. The combined clustering is obtained as a result of ART clustering algorithm with validity index whose inputs are the cluster labels of the contributing clusterings.

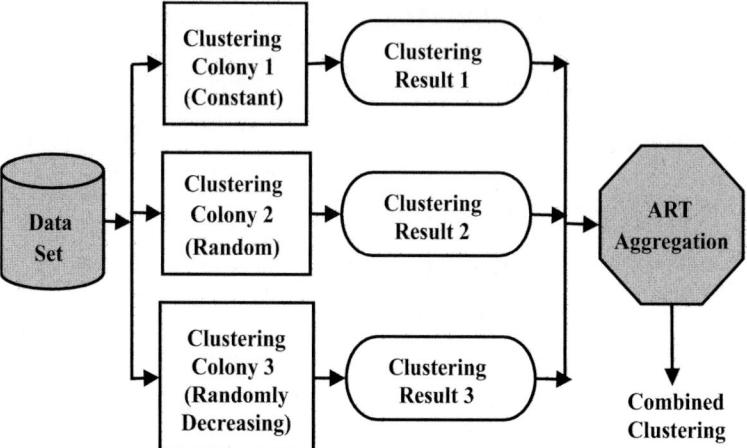

Fig. 11.2. System architecture of ART ensemble model

Let $X = \{x_1,....,x_N\}$ denote a set of objects, and a set of 3 partitions $\Pi = \{\pi_1, \pi_2, \pi_3\}$ of objects in X is obtained by ant colony algorithm with different settings. For each label vector $\pi_i \in N^n$ with $c^{(i)}$ clusters, the binary membership matrix $A^{(i)} \in N^{n \times c^{(i)}}$ is constructed, in which each cluster is represented as a row. All entries of a column in the binary membership matrix $A^{(i)}$ are 1, if the column corresponds to an object with known label. Columns for objects with unknown label are all zero. For example in Table 3 [25], there are 8 objects $x_i (i = 1,2,....,8)$ corresponding to 3 label vectors of clusterings. The first and second clusterings are logically identical. The third one involves a dispute about objects 3 and 5. These clusterings are represented as the binary membership matrixes A shown in Table 4, where $c^{(1,2,3)} = 3$.

The binary membership matrix A is as input of ART neural network. After ART clustering, final target clustering λ can be obtained. The clustering validity index is also used to find the best number of clusters and reduce outliers. For dispute points such as objects 3 and 5, the combined clustering result may match most cases in clustering, i.e. object 3 belongs to cluster 1 and object 5 belongs to cluster 2 like in clusterings 1 and 2.

More precisely, we use $\|x\|$ instead of 1 in matrix A that aims to enhance the accuracy of clustering ensemble. The idea is based on, combining several clustering results and nature attributes of data set. $\|x\|$ is defined by

$$\|x\| = \sqrt{\sum_{j=1}^{q} x_j^2} \tag{11.20}$$

where q is the number of attributes.

The algorithm for clustering ensemble using ART is summarized in Table 5.

Table 11.3. Label vectors

	π_1	π_2	π_3
x_1	1	2	1
x_2	1	2	1
x_3	1	2	2
x_4	2	3	2
x_5	2	3	3
x_6	3	1	3
x_7	3	1	3
x_8	3	1	3

Table 11.4. 3 binary membership matrixes A

	π_1	π_2	π_3	π_4	π_5	π_6	π_7	π_8
$A^{(1)}$	1	1	1	0	0	0	0	0
	0	0	0	1	1	0	0	0
	0	0	0	0	0	1	1	1
$A^{(2)}$	0	0	0	0	0	1	1	1
	1	1	1	0	0	0	0	0
	0	0	0	1	1	0	0	0
$A^{(3)}$	1	1	0	0	0	0	0	0
	0	0	1	1	0	0	0	0
	0	0	0	0	1	1	1	1

Table 11.5. Algorithm 3

Clustering ensemble algorithm	
Step 0.	Apply ant colony algorithm with different settings to generate diversity clusterings: $\Pi = \{\pi_1, \pi_2, ..., \pi_H\}$.
Step 1.	Compute the binary membership matrix A by label vectors π_i, $i = 1, 2, ..., H$, and use $\|x\|$ instead of 1 as input of ART network.
Step 2.	Use ART-2A model to ensemble clustering.
Step 3.	Calculate the validity index $CDbw$ by formula (15) so as to find the optimal number of clusters and reduce outliers.

11.5 Experimental Analysis

We have designed experiments to study the performance of the clustering ensemble model by comparing it with the ACC-VI algorithm, the ART-2A algorithm and the LF algorithm on various artificial and real data sets. We evaluated the clustering performance using cluster validity index *CDbw*.

11.5.1 Artificial Data Set (2D3C)

We artificially generated the data set (2D3C), containing three 2D-Gaussian distributed clusters of different sizes (50,100,75), different densities (variance) and shapes (one with elliptical Gaussian distributions in elongation level and rotated orientation) shown in Fig. 3.

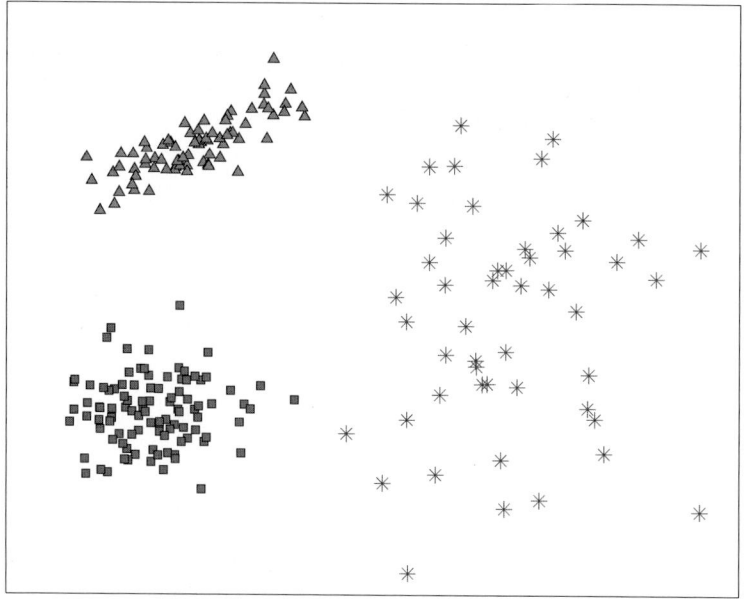

Fig. 11.3. Artificial data set (2D3C)

Table 6 presents the *CDbw* values on artificial data set (2D3C) for the proposed ensemble algorithm, the ACC-VI algorithm, the ART-2A algorithm, and the LF algorithm, respectively. It is noted that *CDbw* takes its maximum value 10.42 for the partitioning of three classes defined by the ACC-VI algorithm, 16.76 for the partitioning of three classes defined by the proposed ensemble algorithm, and 15.53 for the partitioning of three classes defined by the ART-2A algorithm, respectively. While the clustering results of the LF Algorithm into 5 clusters is presented by highlight 12.95 in the fourth column. It is obvious that 3 is considered as the correct number of clusters. This is also the number of actual clusters of (2D3C). The biggest value 16.76 on *CDbw* shows that the ensemble clustering algorithm is optimal.

11.5.2 Real Data Set (Iris)

The real data set used is the Iris data set, which has been widely used in pattern classification, downloaded from the UCI machine learning repository [23]. The data

Table 11.6. Optimal number of clusters found by *CDbw* for different clustering algorithm

No clusters	ACC-VI	Ensemble	ART-2A	LF Algorithm
2	1.26	5.52	8.22	1.10
3	**10.42**	**16.76**	**15.53**	9.02
4	4.17	12.50	11.98	8.70
5	9.85	13.03	12.78	**12.95**

set contains 3 classes of 50 instances each in a 4 dimensional space, where each class refers to a type of iris plant. One class is linearly separable from the other 2, the latter are not linearly separable from each other.

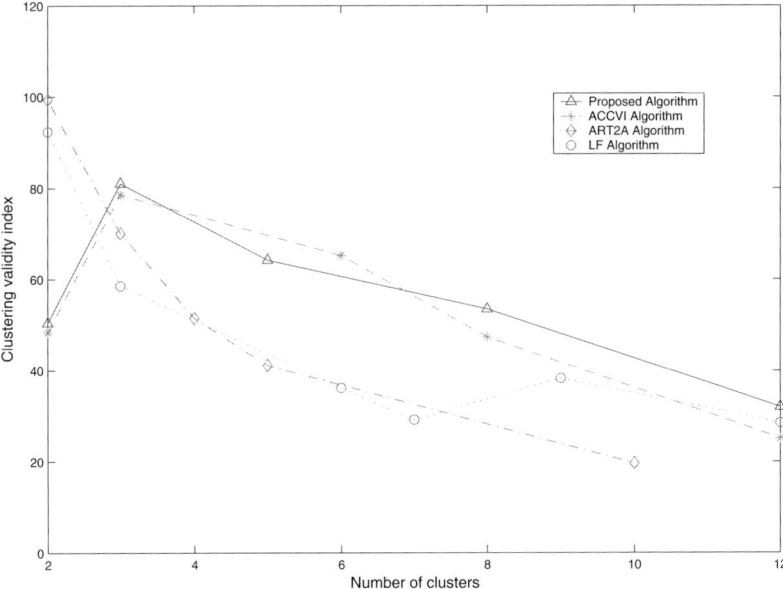

Fig. 11.4. *CDbw* as a function of number of clusters on Iris data set for the proposed algorithm, the ACC-VI algorithm, the ART-2A algorithm, and the LF algorithm, respectively

Fig. 11.4, *CDbw* indicates that the Iris data are divided into three clusters by the proposed ensemble algorithm and the ACC-VI algorithm. It is more consistent with the inherent three clusters of data, compared to two clusters by the ART-2A algorithm and the LF algorithm. The ensemble model is a little better than the ACC-VI algorithm with *CDbw* at its peak.

11.5.3 Reuter-21578 Document Collection

The Reuters-21578 document collection is a standard text-clustering corpus composed of 21578 news articles in 1987 [36]. We sampled 5 different document collections each of size 1000 that have only TOPICS labels. Each document is processed by removing a set of common words using a "stop-word" list, and the suffixes are removed using a Porter stemmer. Then the document is represented as a vector space model using TF-IDF-weighting [35].

Fig. 5-9 illustrates *CDbw* as a function of the number of clusters for the samples using the proposed algorithm, the ACC-VI Algorithms, the ART-2A Algorithm, and the LF Algorithm, respectively. The maximum value of *CDbw* indicates the optimal number of clusters for each algorithm. Table 7 summarized the highest *CDbw* in Fig. 5-9, where the highlighted results presented the optimal number of clusters. For example, the best number of clusters equals to 12 for the proposed ensemble algorithm and the ACC-VI algorithm, 21 for the ART-2A algorithm, and 8 for the LF algorithm, respectively. At the same time, *CDbw* can also be considered to evaluate the performance of different algorithms. From the results shown in Table 4, we can see that the proposed algorithm has produced the maximum *CDbw* value compared to the 3 other algorithms. Note that not all algorithms produced results for all the number of clusters considered.

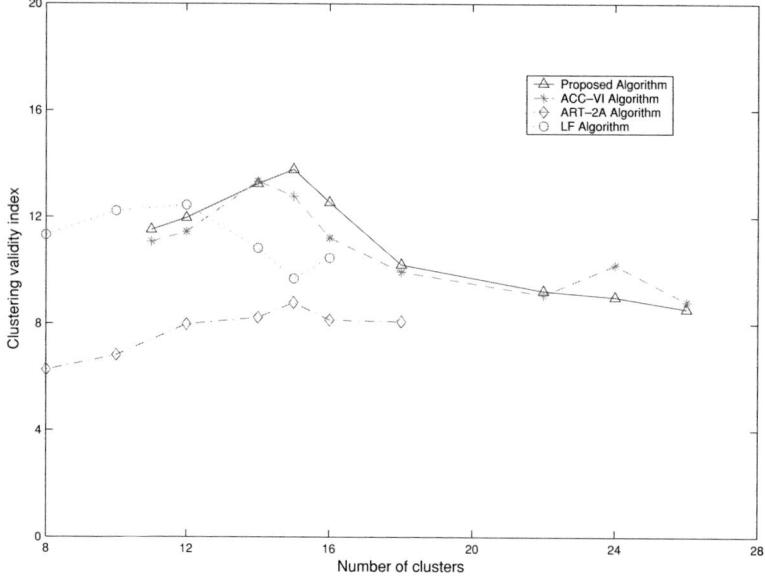

Fig. 11.5. *CDbw* as a function of number of clusters on the first sample collection of 1000 documents each for the proposed algorithm, the ACC-VI algorithm, the ART-2A algorithm, and the LF algorithm, respectively

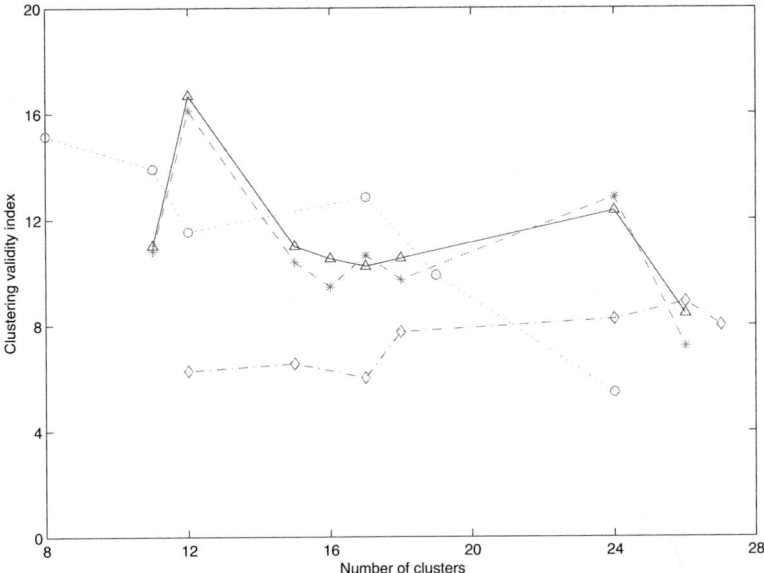

Fig. 11.6. *CDbw* as a function of number of clusters on the second sample collection of 1000 documents each for the proposed algorithm, the ACC-VI algorithm, the ART-2A algorithm, and the LF algorithm, respectively

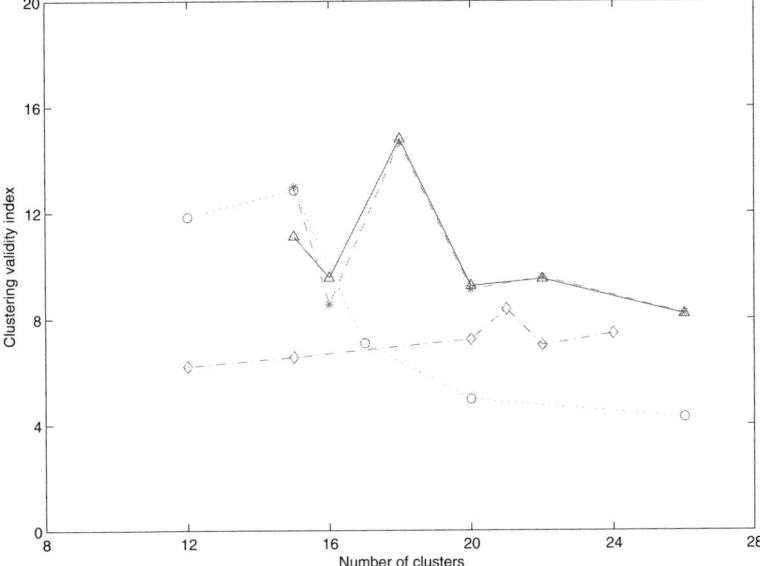

Fig. 11.7. *CDbw* as a function of number of clusters on the third sample collection of 1000 documents each for the proposed algorithm, the ACC-VI algorithm, the ART-2A algorithm, and the LF algorithm, respectively

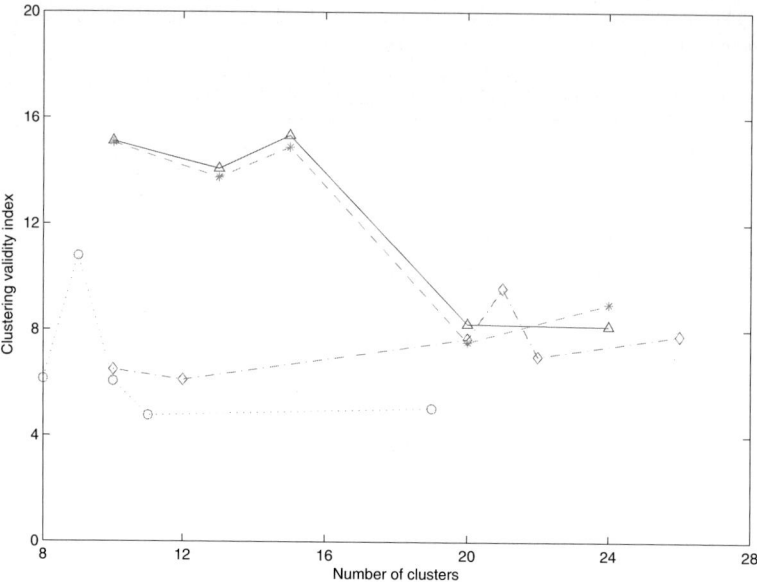

Fig. 11.8. *CDbw* as a function of number of clusters on the fourth sample collection of 1000 documents each for the proposed algorithm, the ACC-VI algorithm, the ART-2A algorithm, and the LF algorithm, respectively

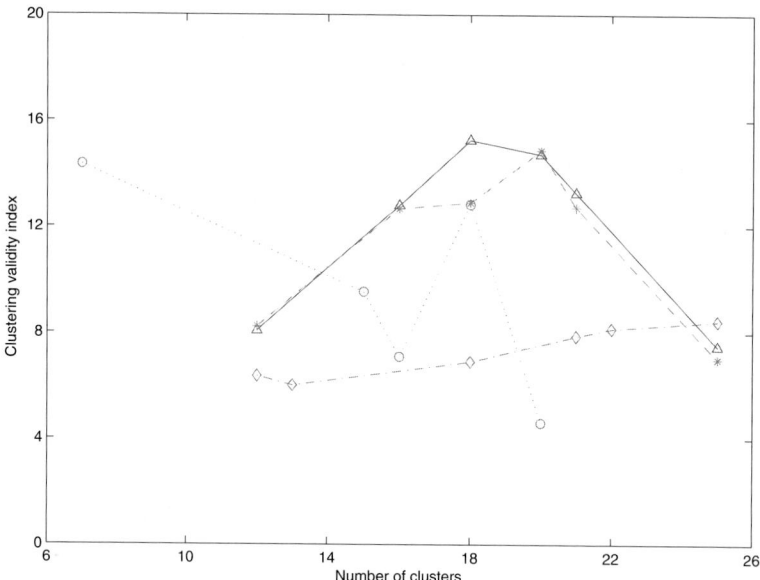

Fig. 11.9. *CDbw* as a function of number of clusters on the fifth sample collection of 1000 documents each for the proposed algorithm, the ACC-VI algorithm, the ART-2A algorithm, and the LF algorithm, respectively

Table 11.7. Optimal number of clusters found by *CDbw* on 5 sample collection of 1000 documents each for different clustering algorithm

No clusters	ACC-VI	Ensemble	ART-2A	LF Algorithm
7				14.35
8				**15.15**
9				10.79
10	15.05			
12	**16.11**	**16.68**		12.45
14	13.34			
15		15.33	8.79	12.85
15		13.78		
18		15.26		
18	14.66	14.78		
20	14.88			
21			8.36	
21			**9.57**	
25			8.44	
26			8.88	

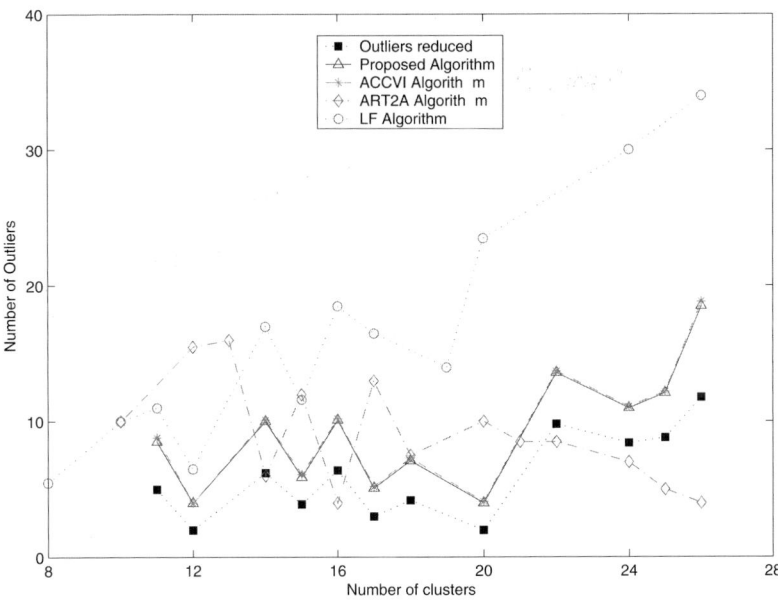

Fig. 11.10. The average number of outliers on 5 document collection of 1000 documents each

Fig. 10 gives the average number of outliers on the same data sets. It is noted that the proposed algorithm has lower outliers after using the outlier reduction strategy.

11.6 Conclusions

In this chapter we proposed a clustering ensemble model using ant colony algorithm with validity index and ART network. This model uses the parallel and independent ant colonies combined by ART network as well as clustering validity index to improve the performance of the clustering. As shown by the results of the experiment, the proposed ensemble model improved the quality of the clustering.

Acknowledgements

This work was partially funded by the Key Basic Application Founding of Sichuan Province (04JY029-001-4) and the Science Development Founding of Southwest Jiaotong University (2004A15).

References

1. Ayad H, Kamel M (2003) Finding natural clusters using multi-clusterer com-biner based on shared nearest neighbors. In: Multiple Classifier Systems: Fourth International Workshop, MCS 2003, UK, Proceedings, pp166-175
2. Ayad H, Basir O, Kamel M (2004) A probabilistic model using information theoretic measures for cluster ensembles. In: Multiple Classifier Systems: Fifth International Workshop, MCS 2004, Cagliari, Italy, Proceedings, pp144-153
3. Bartfai G (1996) An ART-based Modular Architecture for Learning Hierar-chical Clusterings. J Neurocomputing, 13:31-45
4. Bonabeau E, Dorigo M, Theraulaz G (1999) Swarm Intelligence - From Natural to Artificial System. Oxford University Press, New York
5. Carpenter G A, Grossberg S (1987a) A massively parallel architecture for a self-organizing neural pattern recognition machine. J Computer Vision, Graphics, and Image Processing, 37:54-115
6. Carpenter G A, Grossberg S (1987b) ART 2: Self-organization of stable cate-gory recognition codes for analog input patterns. J Applied Optics, 26(23):4919-4930
7. Carpenter G A, Grossberg S, Rosen D B (1991a) ART2-A: An Adaptive Resonance Algorithm for Rapid Category Learning and Recognition. J Neu-ral Networks, 4:493-504
8. Carpenter G A, Grossberg S, Rosen D B (1991b) Fuzzy ART: fast stable learning and categorization of analog patterns by an adaptive resonance sys-tem. J Neural Networks, 4:759-771
9. Deneubourg J ., Goss S, Franks N, Sendova-Franks A, Detrain C, Chretien L (1991) The Dynamics of Collective Sorting: Robot-like Ant and Ant-like Ro-bot. In: Meyer J A, Wilson S W (eds) Proc. First Conference on Simulation of Adaptive Behavior: From Animals to Animats. Cambridge, MA: MIT Press, pp356-365

10. Dudoit S, Fridlyand J (2003) Bagging to improve the accuracy of a clustering procedure. J Bioinformatics, 19(9):1090-1099
11. Fred A L N (2002) Finding Consistent Clusters in Data Partitions. In: Roli F, Kittler J (Eds) Proc. 3rd Int. Workshop on Multiple Classifier Systems, LNCS 2364, pp309-318
12. Grossberg S (1976) Adaptive pattern classification and universal recoding, I: Parallel development and coding of neural feature detectors II: Feedback, expectation,olfaction,and illusions.J Biological Cybernetics,23:121-134 187-202
13. Halkidi M, Vazirgiannis M (2002) Clustering validity assessment using multi representatives. In: Proc. of SETN Conference
14. Halkidi M, Vazirgiannis M, Batistakis Y (2000) Quality scheme assessment in the clustering process. In: Proc. 4th Eur. Conf. Principles and Practice of Knowledge Discovery in Databases (PKDD), pp165-276
15. Handl J, Meyer B (2002) Improved ant-based clustering and sorting in a document retrieval interface. In: Proceedings of the Seventh International Conference on Parallel Problem Solving from Nature. LNCS2439, Berlin, Germany: Springer-Verlag, pp913-923
16. He J, Tan A, Tan C (2004) Modified ART 2A Growing Network Capable of Generating a Fixed Number of Nodes. J IEEE Trans. on Neural Networks, 15(3):728-737
17. He Y, Hui S C, Fong A C M (2002) Mining a web citation database for docu-ment clustering. J Applied Artificial Intelligence, 16:283-302
18. Hussin M F, Kamel M (2003) Document clustering using hierarchical SOMART neural network. In: Proc of the Int'l Joint Conf on Neural Network, Portland, Oregon, USA, pp2238-2241
19. Jiang Y, Zhou Z (2004) SOM Ensemble-Based Image Segmentation. J Neural Processing Letters, 20:171-178
20. Kanade P M, Hall L O (2003) Fuzzy Ants as a Clustering Concept. In: Proc. of the 22nd Int. Conf. of the North American Fuzzy Information Processing Society, pp227-232
21. Lumer E, Faieta B (1994) Diversity and Adaptation in Populations of Clustering Ants. In: Proc. Third International Conference on Simulation of Adaptive Behavior: From Animals to Animats 3.Cambridge, MA: MIT Press,pp499-508
22. Monmarch¨| N, Slimane M, Venturini G (1999) Antclass: Discovery of Clus-ters in Numeric Data by a Hybridization of an Ant Colony with the Kmeans Algorithm. Technical Report 213, Laboratoire d'Informatique, E3i, Univer-sity of Tours
23. Murpy P M, Aha D W (1994) UCI repository of machine learning databases. Irvine, CA: University of California. [Online] Available: http://www.ics.uci.edu/mlearn/MLRepository.html
24. Ramos V, Merelo J J (2002) Self-organized Stigmergic Document Maps: Environment as a Mechanism for Context Learning. In: Alba E, Herrera F, Merelo J J (eds) AEB?2002 - 1st Spanish Conference on Evolutionary and Bio-Inspired Algorithms, Centro Univ. de M¨|rida, M¨|rida, Spain, pp284-293
25. Strehl A, Ghosh J (2002) Cluster ensembles - a knowledge reuse framework for combining multiple partitions. J Machine Learning Research, 3:583-617
26. Theodoridis S, Koutroubas K (1999) Pattern Recognition. Academic Press
27. Tomida S, Hanai T, Honda H, Kobayashi T (2002) Analysis of expression profile using fuzzy adaptive resonance theory. J Bioinformatics, 18(8):1073-1083
28. Topchy A, Jain A K, Punch W (2003) Combining Multiple Weak Clusterings. In: Proc. IEEE Intl. Conf. on Data Mining, Melbourne, FL, pp331-338

29. Topchy A, Jain A K, Punch W (2004) A Mixture Model of Clustering En-sembles. In: Proc. SIAM Intl. Conf. on Data Mining, pp379-390
30. Wu B, Zheng Y, Liu S, Shi Z (2002) CSIM: a Document Clustering Algo-rithm Based on Swarm Intelligence. In: IEEE World Congress on Computational Intelligence, pp477-482
31. Wu S, Chow T (2003) Self-organizing-map based clustering using a local clustering validity index. J Neural Processing Letters, 17:253-271
32. Yang S, Browne A, Picton P D (2002) Multistage Neural Network Ensembles. In: Roli F, Kittler J (Eds) Proc. 3rd Int. Workshop on Multiple Classifier Systems, LNCS 2364, pp91-97
33. Yang Y, Kamel M (2003) Clustering ensemble using swarm intelligence. In: IEEE Swarm Intelligence Symposium, Indianapolis, USA, pp65-71
34. Yang Y, Kamel M (2005) A Model of Document Clustering using Ant Col-ony Algorithm and Validity Index. In: Int. Joint Conf. on Neural Network (IJCNN'05), Montreal, Canada, pp1732-1737
35. Yang Y, Kamel M, Jin F (2005) Topic discovery from document using ant-based clustering combination. In: Web Technologies Research and Develop-ment - APWeb 2005, 7th Asia-Pacific Web Conference, Shanghai, China, LNCS3399, UK, Springer, pp100-108
36. [Online] Available: http://kdd.ics.uci.edu/databases/reuters21578/reuters21578.html

Index

α-adaptation, 109
$\mathcal{MAX}\text{-}\mathcal{MIN}$ ant system, 23

agents, 2
amount of clustering, 9
Ant Colonies Optimization, 4
ant colony, 1
ant colony classifier system, 16
Ant Colony Optimization, 56
ant colony optimization
    for rule induction, 75
        fuzzy, 75
ant system, 22
ant-based clustering, 103
Ant-based feature selection, 56, 60, 63
AntClass, 153
AntClust, 153
AntMiner, 25
AntMiner+, 21
artificial bees, 191
Ascending Hierarchical Clustering, 153
attribute space, 223

best position, 7
biomimetic methods, 170
bird flocks, 1
breast cancer, 13
breast cancer diagnosis, 35

C4.5, 34
cascading classifiers, 12
Classification, 11
classification, 21
classification rules
    fuzzy, 75

cluster, 223, 226, 234
Cluster analysis, 13
cluster retrieval, 110, 233
cluster validity criteria, 112
Clustering, 11
clustering, 4, 102, 126, 135
cognition component, 5
collective behavior, 1
collective dynamical behaviors, 2
collision, 1, 3
Collision Avoidance, 1
color image quantization, 130, 138
compartmentalization, 203
comprehensibility, 25
confusion matrix, 114
construction graph, 27
credit scoring, 33, 35

Data Clustering, 221
data clustering, 223
data mining, 10, 24
data pattern processing, 10
Data Swarm Clustering, 221, 226
datoid, 222, 226
degree of connectivity, 8
Dependency modeling, 11
distance
    dissimilarity, 229
    similarity, 228
Dynamic parallel group, 2

early stopping, 32
end-member, 132, 133, 141
evaporation phase, 10

Index

feature extraction, 116
fish schools, 1
Flock Algorithm, 221, 223
Forager architecture, 205
FPAB, 191, 195
Fuzzy
   equivalence classes, 50
      lower approximations, 50
      upper approximations, 50
fuzzy classification rules, *see* classification rules
fuzzy rule induction, *see* rule induction
Fuzzy-rough feature selection, 50, 52
Fuzzy-rough set, 50

gbest, 7
global search, 5

heuristic value, 23
Highly parallel group, 2
Homogeneity, 1
hybrid technique, 11

image processing, 125
image segmentation, 11
Incremental clustering, 182
independent component analysis, 117
information discovery, 10
intrusion detection, 102
Iris, 236

K-means algorithm, 14
k-nearest neighbor, 35
kdd-cup99 dataset, 112
Kmeans, 153
knowledge discovery, 11
knowledge extraction, 10

lbest, 7
local density of similarity, 104
local regional entropy, 107
local search, 5
Locality, 1
logistic regression, 35

microarray gene expression, 11
mixed-species flock, 222
multi-species swarm, 221, 226

natural selection, 191, 198
neighbor
   dissimilar, 229
   similar, 228
neighborhood, 225, 231
neighborhood topologies, 7
nest, 9
news foragers, 203
NP-hard problems, 4

parameter settings, 110
particle, 225, 226
particle swarm, 1
Particle Swarm Optimization, 4, 221, 225
particle swarm optimization, 134
pattern recognition, 125
pbest, 5
pheromone, 22, 108
   evaporation, 22, 30
   reinforcement, 30
   updating, 30
pheromone concentration, 9
pheromone trail, 9
pollination, 191, 197
position, 227
principle component analysis, 116

real-world benchmark datasets, 111
Recommender systems, 12
Regression, 11
Reinforcement Learning, 206
reinforcement phase, 10
reinforcing agent, 208
Rough set theory, 46
rule discovery, 21
rule induction
   fuzzy, 75

search space, 225
self organizing map, 16
self-organization, 107
short-term memory, 109
shortest paths, 9
similarity function, 228, 239
single-species swarm, 222, 226
small-world network, 8
social component, 5
social insects, 4

Species Clustering, 221
species clustering, 222
spectral unmixing, 132, 141
stigmergy, 22
sub-swarm, 226, 234
Summation, 11
support vector machine, 35
swarm, 2
Swarm Clustering, 221
swarm clustering, 191, 193, 195
swarm intelligence, 1, 22
swarm topologies, 9
Swarming agents, 14
Systems monitoring, 66

Takagi-Sugeno Fuzzy Systems, 13
temporal complexity, 168

textual databases, 172
time series segmentation algorithm, 15
Torus, 2
tournament selection, 110
trajectories, 7

unsupervised classification, 11

velocity, 4, 227
Velocity Matching, 1

Web classification, 63
Web usage mining, 177
Web usage patterns, 16
Weblog algorithm, 206
Wisconsin Breast Cancer, 236

Printing: Krips bv, Meppel
Binding: Stürtz, Würzburg